ENERGY CONSERVATION IN EAST ASIA

Towards Greater Energy Security

World Scientific Series on Energy and Resource Economics

(ISSN: 1793-4184)

Published

Vol. 1 Quantitative and Empirical Analysis of Energy Markets
by Apostolos Serletis

Vol. 2 The Political Economy of World Energy: An Introductory Textbook
by Ferdinand E Banks

Vol. 3 Bridges Over Water: Understanding Transboundary Water Conflict, Negotiation and Cooperation
by Ariel Dinar, Shlomi Dinar, Stephen McCaffrey & Daene McKinney

Vol. 4 Energy, Resources, and the Long-Term Future
by John Scales Avery

Vol. 5 Natural Gas Networks Performance after Partial Deregulation: Five Quantitative Studies
by Paul MacAvoy, Vadim Marmer, Nickolay Moshkin & Dmitry Shapiro

Vol. 6 Energy and International War: From Babylon to Baghdad and Beyond
by Clifford E. Singer

Vol. 7 Resource and Environmental Economics: Modern Issues and Applications
by Clement A. Tisdell

Vol. 8 Energy Conservation in East Asia: Towards Greater Energy Security
edited by Elspeth Thomson

Forthcoming

Historical Energy Statistics: Global, Regional and National Trends Since Industrialization
by T. S. Gopi Rethinaraj & Clifford E. Singer

The Implications of China's Rising Energy Use
by Peter Sheehan

World Scientific Series on Energy and Resource Economics – Vol. 8

ENERGY CONSERVATION IN EAST ASIA

Towards Greater Energy Security

Editors

Elspeth Thomson
Energy Studies Institute, National University of Singapore, Singapore

Youngho Chang
Nanyang Technological University, Singapore
National University of Singapore, Singapore

Jae-Seung Lee
Korea University, Korea

World Scientific

NEW JERSEY • LONDON • SINGAPORE • BEIJING • SHANGHAI • HONG KONG • TAIPEI • CHENNAI

Published by

World Scientific Publishing Co. Pte. Ltd.
5 Toh Tuck Link, Singapore 596224
USA office: 27 Warren Street, Suite 401-402, Hackensack, NJ 07601
UK office: 57 Shelton Street, Covent Garden, London WC2H 9HE

British Library Cataloguing-in-Publication Data
A catalogue record for this book is available from the British Library.

World Scientific Series on Energy and Resource Economics — Vol. 8
ENERGY CONSERVATION IN EAST ASIA
Towards Greater Energy Security

ISBN-13 978-981-277-177-3
ISBN-10 981-277-177-8

Typeset by Stallion Press
Email: enquiries@stallionpress.com

Printed in Singapore by World Scientific Printers.

Acknowledgements

The Network of East Asian Think Thanks (NEAT) was established in 2004 as a Track Two process to support and complement the ASEAN + 3 (APT) quest for greater East Asian cooperation.

The East Asian Institute (EAI) at the National University of Singapore, on behalf of NEAT Singapore, organised the NEAT Working Group (WG) Meetings on Energy Security Cooperation in East Asia. The first meeting (Phase 1) was held on 6 May 2005 and the second meeting (Phase 2) was held on 30 June 2006. Both meetings were held in Singapore.

There were two key themes at these meetings: energy conservation and the maritime dimension of energy security. This volume is based on the presentations devoted to energy conservation. In this area, the goals of the NEAT WG were to promote energy conservation and energy consumption efficiency, to share and learn from the experiences of the energy-efficient APT countries and to support an East Asian Community in the long run through regional energy cooperation.

We would like thank all the contributors to this volume for their excellent work and provision of detailed country data. We are also very grateful to the participants at the two meetings who gave us useful suggestions for improving our work, and former Director, Professor Wang Gungwu, former Research Director, Professor John Wong and the current Director, Professor Zheng Yongnian of EAI for their patience in awaiting this volume.

Elspeth Thomson, Youngho Chang and Jae-Seung Lee
September 2010

Introduction

Elspeth Thomson

Over the next two decades at least, energy consumption growth rates within the Association of Southeast Asian Nations (ASEAN) + 3 (China, Japan and Korea) are expected to be among the highest in the world. High energy consumption will be necessary to fuel anticipated industrialisation, urbanisation and trade. The region is certainly not without its own energy resources and there is considerable energy trade within ASEAN. Vast quantities of energy from ASEAN are exported to Northeast Asia. The main deficiency in the region as a whole at this time is adequate crude oil reserves.

There is an indubitable link between energy security and energy conservation and at the time of writing, all of the ASEAN + 3 governments were beginning to feel some urgency over the need to implement more and/or stiffer energy conservation measures. Strenuous efforts to secure sufficient energy supplies, diversify away from fossil fuels, protect energy transport routes, maintain strategic fossil fuel reserves, develop new fuels, harness renewable energies, etc., are all almost pointless if the efficiency with which the precious energy is consumed is very low and there are major losses through wastage. Saved energy should be regarded as a form of energy production itself. It is generally less expensive and less destructive environmentally than the procurement of new energy supplies either domestically or from abroad.

Besides enhancing national energy security, improving the energy consumption efficiency of all types of energy has many benefits for governments, such as decreased energy expenditure, increased economic competitiveness, greater sustainable development, higher incomes, improved trade balances, reduced need for new power plants, reduced environmental impact of energy use, etc.

Almost all of the ASEAN + 3 countries were affected to some degree by the oil shocks of the 1970s.[1] For Japan in particular, the oil shocks were major turning points which jolted the government into meticulously examining how energy had been used and taking immediate steps to eliminate wastage and improve consumption efficiency. Many of the other countries in the region, however, were either suffering from serious political problems at the time and it was not possible to immediately mount and execute such sweeping reforms, and/or the industrial sectors were yet small.

Almost all of the region was also affected by the Asian Financial Crisis which struck in July 1997. Several governments were making headway with energy conservation measures. However, following the crisis they had to largely abandon these and focus on helping businesses recover. At the time of writing, besides a dire warning from the International Energy Agency that crude oil and natural gas supplies will be tight in the coming years, the motivation to save energy is closely tied to mitigating the effects of climate

[1] The exception was China. The Chinese Government had been almost smug about its energy independence when the first two international oil crises struck in 1973 and 1979. The country used to be a major exporter of crude. At their peak in 1985, exports amounted to over 35 million tons. From 1965 to 1979, China's production of crude grew at an average annual rate of nearly 19 per cent. However, since 1994, the growth rate has fallen to less than 2 per cent while the growth rate of demand has averaged at just over 7 per cent. China became a net importer of petroleum products in 1993 and of crude oil in 1996. Since then, the government has aggressively competed globally not only for oil and oil product imports, but also in buying foreign oil fields and constructing of oil ports, pipelines and refineries.

change.[2] In the past, it was only the relatively more mature economies which were concerned about local and global environmental damage directly attributable to the use of energy. However, due to the perception that anthropomorphic activities have led to global warming, which has contributed to cataclysmic weather patterns that have taken high economic and human tolls in recent years and the prediction of worse to come, all governments in the region are seriously trying to do what they can to reduce the emission of greenhouse gases in particular.

The Cebu Declaration on East Asian Energy Security, passed on 15 January 2007, recognises the region's rapidly growing energy needs and the limited global reserve of fossil energy.[3] One of the specific goals is to "reduce dependence on conventional fuels through intensified energy efficiency and conservation programmes…", and to achieve it, "take concrete action toward improving efficiency and conservation, while enhancing international cooperation through intensified energy efficiency and conservation programmes" and "set individual goals and formulate action plans voluntarily for improving energy efficiency."

Very soon after this, the East Asia Summit (EAS) Energy Cooperation Task Force was formed on 1 March 2007, and then on 23 August 2007 the 1st EAS Energy Ministers' Meeting was held in Singapore. The EAS ECTF identified three energy cooperation work streams — energy efficiency and conservation, energy market integration, and bio-fuels for transport and other purposes — "as a starting point to focus their efforts and to work towards the goal of affordable, secured and sustainable energy at all economic levels."

The specific recommendations for EE&C were:[4]

a. The voluntary formulation of individual, quantitative, and where possible, sector-specific energy efficiency goals and action plans with medium term time frames;

[2] IEA, *Oil Market Report July 2007*.

[3] See ASEAN website: http://www.aseansec.org/19319.htm [Aug. 2007].

[4] Joint Ministerial Statement — First EAS Energy Ministers Meeting, held in Singapore, 23 August 2007 at http://app.sprinter.gov.sg/data/pr/20070823968.htm [Aug. 2007].

b. Presentation of a preliminary report on their development to the Second EAS Energy Ministers Meeting in 2008; and presentation of the goals and action plans at EAS EMM3 in 2009;
c. Utilising international and regional initiatives when developing energy efficiency indicators and compiling best practices in formulating energy efficiency goals;
d. Carrying out stock-takes of existing measures on energy efficiency and conservation with regular updates in cooperation with the Asian Energy Conservation Collaboration Centre;
e. Considering the full range of policies and measures covering key energy consuming sectors in formulating energy efficiency goals and action plans;
f. Promoting and encouraging market-based pricing;
g. Promoting effective monitoring through regular dialogues and communications on the progress of each country's energy efficiency goals in close cooperation with the International Energy Agency, the Asia Pacific Economic Cooperation and other related international organisations;
h. Enhancing regional dialogue, information sharing and cooperation to improve energy efficiency, with a focus on capacity building;
i. Encouraging international financial institutions to provide support for energy efficiency investments and to develop effective tools for facilitating investments;
j. Developing an EAS Energy Outlook; and
k. Promoting links with relevant international and regional institutions.

Specialised publications pertaining to certain aspects of energy conservation in various parts of Asia have been published in recent years.[5]

[5] Peter Rumsey and Ted Flanigan, *Compendium: Asian Energy Efficiency Success Stories*, published for the International Institute for Energy Conservation by Global Energy Efficiency Initiative, Washington, DC, 1995; UN, *Energy Efficiency: Compendium of Energy Conservation Legislation in Countries of the Asia and Pacific Region*, New York: UN, 1999; *Energy Efficiency Indicators: A Study of Energy Efficiency Indicators in APEC Economies*, published by the Asia Pacific Energy Research Center and Institute of Energy Economics, 2001; Ming Yang and Peter Rumsey, "Energy Conservation in Typical Asian Countries", *Energy Sources* 19, no. 5 (1997): 507–21.

The ASEAN website also provides detailed information about past and ongoing initiatives, as do the websites maintained by the Japanese Energy Conservation Centre, International Institute for Energy Conservation, Asia Alternative Energy Program (World Bank), etc.[6]

This book provides a concise overview of energy conservation activities in ASEAN + 3. Cambodia, Laos and Myanmar were excluded because they are still consuming relatively small quantities of commercial energy and there is not nearly enough energy to meet demand. A major difficulty in carrying out comparative work is the different ways in which the various governments measure energy production, consumption, imports and exports. Thus, the first three chapters, which consider the region as a whole, are based on data from international energy organisations which have converted the information into standard units. It is useful to see at a glance the relative positions of the countries in terms of domestic energy resources, import dependence, etc.

The first chapter outlines the scope and availability of known energy resources in the region, presents a set of forecasts to the year 2020 and discusses some basic requirements for regional cooperation in the provision of energy and for improving energy consumption efficiency. The first half of the second chapter examines the energy intensity and income elasticity of energy consumption for each country, then briefly describes the scope of the energy conservation programmes undertaken thus far and some of the difficulties encountered. The second half investigates how energy can be saved in the sector most relevant to energy security in Asia at this time, namely, transport fuel. In the third chapter, the concept of 'energy security' is explored and how an Asian energy partnership might be used to best advantage. This chapter also provides a brief history of energy conservation in Asia and details the potential for cooperation as well as the obstacles that ASEAN + 3 faces in energy policy harmonisation. Comparisons are made with the European Union and International Energy Agency.

[6] See for example, http://www.aseanenergy.org/; http://www.eccj.or.jp/index_e.html; http://www.iiec.org/; http://www.worldbank.org/astae/, and more websites.

The remainder of the book describes in detail the energy conservation policies of Brunei, China, Indonesia, Japan, Korea, Malaysia, the Philippines, Singapore, Thailand and Vietnam. In reading these, it is immediately apparent that the approaches taken by these governments to save energy share some common features but also differ considerably. This is not surprising given that the economic histories of ASEAN + 3 are greatly varied. At one extreme are the economies of Japan, Singapore and Korea which are relatively mature. They consume by far the most energy in the region but also have the least amount of domestic energy resources relative to demand. All of the others have large rural areas which are extremely poor.

The relative energy resource endowment is one determining factor in energy conservation policy formulation. For example, Brunei and Indonesia, with very significant fossil fuel resources, are now coming to grips with the inevitable depletion of these resources, and the need not only to conserve them but to develop new industrial foundations for their economies. Malaysia's energy conservation efforts were also initially geared to slowing down depletion of domestic energy supplies rather than trying to raise the output value of each unit of energy consumed. This is in contrast to Singapore, Japan and Korea, for example, which have either never had any energy resources or had minimal ones relative to demand. Japan, especially, has made unceasing efforts since the 1970s to minimise consumption, raise consumption efficiency and eliminate all possible wastage.

Another factor is the structure of the economy and relative growth rates of the various sectors. The larger the industrial sector, the higher the national consumption of energy, especially if there are many heavy industries such as steel, automobiles, cement, etc., as in China, Japan and Korea. Some of the countries in the region have large agricultural sectors which use relatively little fuel. However, many of these same countries are rapidly urbanising and have booming manufacturing sectors. Energy consumption growth is expected to soar.

In some economies, it is air conditioning or transport which are the largest energy consuming sectors. Some industries can go 'green' relatively easily, i.e., successfully introduce energy efficiency and conservation policies, and indeed gain from promoting themselves as 'green' while others, are inherently less flexible.

The differing climates of the countries also determine energy conservation priorities. In the central and northern parts of China, Japan and Korea the winters are long and cold. Energy is required for heating. By contrast, in much of ASEAN, fans or air conditioners must be kept going all year round in homes, schools and businesses to keep the population cool.

The countries' pricing systems also play a pivotal role in the energy conservation policies implemented thus far and their relative success. Any energy price subsidy makes conservation almost impossible because in most cases it is difficult to prevent the cheap energy from getting into the hands of people who can afford to pay real prices. Also, subsidies, discourage energy consumption efficiency, and may even spawn wasteful consumption practises. The extent of government involvement generally, varies considerably across the region. The Japanese government, for example, greatly influences the energy consumption behaviour of the business and civilian sectors alike. However, in other countries, either the mandate of the government is insufficiently strong to call for and strictly enforce energy conservation measures, or the governments are still much occupied trying firstly to meet basic human needs.

Fear of energy supply disruption due to terrorism or some other calamity is another motivating factor for energy conservation. The degree of this fear is a function of relative vulnerability. For example, whereas China can produce almost all of its electricity needs from domestic energy resources, most countries must import coal, oil or natural gas to produce large proportions of their required power.

Japan has achieved remarkable success in conserving energy and raising energy consumption efficiency. It has expressed keen desire to assist the region in these areas. However, will the policies that worked so well in Japan be feasible elsewhere? There are many hurdles for the ASEAN + 3 governments to overcome. Gathering and maintaining accurate energy consumption statistics are imperative. Governments must know the precise price elasticities of energy consumption. Legal infrastructure must be in place which prevents all theft, smuggling, illegal use, etc. Besides the painful elimination of all price subsidies, another very real difficulty is cost. How can governments and companies pay for new energy-saving equipment? Are they in a position to offer low-interest loans? How should all of the old, energy-wasting equipment be best disposed of? Armies of people in all sectors tasked with monitoring energy consumption use and enforcing targets are required. Training them may not be a problem, but who will pay their salaries? Organisations can be instructed to establish their own energy management teams, but what incentives and/or penalties should be levied to ensure government standards are consistently met? Should governments attempt to tackle energy waste in all sectors, or concentrate initially on only one industry perhaps? How can governments maintain progress? Should they wield a heavy hand or just launch programmes and hope that market forces continue the momentum?

For lasting energy consumption efficiency, the ASEAN + 3 governments face two fundamental tasks. Firstly, they must set a good example. All government offices must be seen to be operating at the highest efficiencies possible. Otherwise the public will become cynical and resentful. Secondly, there must be instilled within the public a desire to minimise energy wastage. In many of the wealthier parts of the world, where basic human needs have been abundantly met for some decades, there is public outcry against over-exploitation and waste of natural resources, including energy. Governments do not need to raise awareness or cajole people into adopting energy conserving habits. For the sake of their own health and that of the generations to come, they are adapting their household and work behaviour so as to minimise the use of resources, including energy. In some cases they are even putting pressure on their local governments to adopt energy-saving equipment, examine ways to minimise the use of commercial lighting as well as transport fuel, and to reduce greenhouse gas emissions as much as possible.

The situation in the poorer parts of the world, including much of ASEAN + 3, is completely different. Understandably, many people living in such areas yearn not only to substantially raise their living standards and make use of all of the commercial energy-powered conveniences that wealthier people use in their homes and places of work, they also want to take pride in their cities and see them rapidly modernised and filled with excitement.

The governments of ASEAN + 3 cannot rely solely on heavy penalties, rhetoric or moral suasion to eliminate wasteful energy consumption practises. Citizens must learn to abhor the wastage of all natural resources. This behaviour is not easily inspired through campaigns. It must be taught from a young age in the schools. Leaderships must, on the one hand, try to encourage minimisation of energy consumption, but on the other, not make it impossible for industries and economic growth, generally, to flourish.

In reading the country chapters of this book, it is apparent that almost all of the ASEAN + 3 governments have attempted to minimise energy consumption in two key areas, namely, air conditioning and refrigeration, and that they are attempting to establish various forms of demand-side management such as labelling of equipment and machines, auditing of buildings and designing new urban areas which help minimise energy use. A good start has been made, but much remains to be done.

The good news is that there are technologies available now which are entirely suitable for use in these countries which can drastically increase energy consumption efficiency. It is not necessary for them to waste time and effort developing new technologies. Many parts of ASEAN + 3 can potentially 'leapfrog' into high energy consumption efficiency and avoid the wasteful and highly polluting paths of development that most developed countries and the more mature areas of Asia have already suffered.

Contents

Chapter

1

East Asian Energy Supply, Demand and Cooperation Outlook

Youngho Chang and Elspeth Thomson

INTRODUCTION

The 21st century is being hailed as 'the Asian century'. Europe and North America have long since 'taken off' and matured economically, commercially and culturally and many believe it is now Asia's turn. Though the economies of Japan, Korea and Singapore are now mature, the rest of Asia can be categorised as being in either the 'pre-takeoff' or 'takeoff' stages.

The focus of this study is energy supply and demand in the ASEAN + 3 region, i.e., the ten ASEAN member countries plus China, Japan and Korea, over the next 15 years to 2020. Its purpose is to lay an information foundation for the subsequent chapters of this book. It is concerned primarily with the amounts and types of energy consumed in the region.

As this region is, and will increasingly have a large impact on global energy flows, it is useful to provide a concise overview of the quantities and types of energy available in the immediate area for consumption. However, any study such as this which attempts to compare so many countries is fraught with data problems. Needless to say, there are enormous historical, cultural, political and economic differences among these countries and the analyst is faced with a bewildering array of non-compatible data which is of uneven quality, units of measurement and currentness. Even for one country alone, there can be large discrepancies among the various reporting sources.[1]

Besides data problems is the reality that some countries do not want to reveal the full extent of their energy resources lest they be put under pressure to share these with neighbouring countries at subsidised rates. Other problems are the various sources' definition of 'Asia'. Some sources include India, others include Mongolia, etc. Moreover, the inclusion of China obviously skews the data. Unless looked at separately, the spectacular growth which has occurred in the rest of the region looks paltry by comparison.[2]

TOTAL ENERGY CONSUMPTION (TEC)

Before looking at what the future may hold for this region in terms of energy demand and supply, it is useful to look briefly at what has occurred in recent years. Table 1 clearly shows how ASEAN + 3's energy consumption growth rates have been much higher than those of the world as a whole since 1980. Primary energy consumption as a whole grew at an average annual rate of 6.0 per cent in ASEAN + 3

---

[1] The main sources for this paper were the websites maintained by the International Energy Agency (IEA) at http://www.iea.org/Textbase/stats/index.asp; Energy Information Administration (EIA) at http://www.eia.doe.gov/emeu/international/contents.html; ASEAN Centre for Energy at http://www.aseanenergy.org; Asia Pacific Energy Research Centre (APERC) at http://www.ieej.or.jp/aperc/; British Petroleum (BP) at http://www.bp.com; Institute of Energy Economics at http://www. http://eneken.ieej.or.jp/en/; and the Economic Research Institute for Northeast Asia at http://www.erina.or.jp/.

[2] Specific data for Hong Kong, Taiwan and Macau were not included in this study though ideally they should have been.

Table 1: Comparison Between Asian and World Average Annual Energy Demand Growth, 1980–2003 (%)

	ASEAN + 3	World
Primary Energy	6.0	1.7
Oil	4.4	1.6
Gas	17.2	1.9
Coal	6.7	0.9
Electricity	10.6	3.0
Per Capita Primary Energy	4.1	0.4

Source: Calculated from EIA website.

between 1980 and 2003 compared to 1.7 per cent worldwide, while the respective per capita figures were 4.1 and 0.4 per cent.[3]

Looking at TEC growth in the individual countries (see Table 2 and Figure 1), we see that Laos' and Vietnam's growth, averaging 17 per cent and 10 per cent, respectively, were far above the rest. The main reason for such high growth in both cases is the fact that energy consumption levels prior to the decade under examination here were extremely low. (Some may argue that it would be more useful to look at only the past five years, but we decided that a longer trend would absorb the effects of some exogenous factors which have influenced economic growth in recent years.)

As for TEC figures and the totals for each type of energy, China, Japan and Korea are of course at the top the list while Brunei, Cambodia and Laos are at the bottom due to their population size and/or economic structure (see Table 3).

As mentioned in the introduction, the inclusion of China skews the data considerably. It is the world's second largest energy consumer in the world after the US and the third largest energy producer after the US and Russia. Energy consumption growth within the country, however, is highly uneven. There are rural regions, as in most of the other countries in this study, which have no electricity at all.

[3] The difference between ASEAN's and the world's growth is especially large for gas consumption (17.2 per cent average growth versus 1.9 per cent worldwide) due to the introduction of the Philippines' Malampaya gas production in late 2001.

Table 2: Average Energy Consumption, GDP and Population Growth, 1994–2004

	Average Oil Consumption Growth	Average Coal Consumption Growth	Average Natural Gas Consumption Growth	Average Electricity Consumption Growth	Average Total Energy Consumption Growth	Average Per Capita Energy Growth*
Brunei*	3.29	0.00	10.10	5.73	8.26	4.14
Cambodia*	1.74	0.00	0.00	−2.95	—	—
China	7.70	3.16	9.48	9.22	5.57	2.96
Indonesia	3.54	13.72	3.55	8.42	4.59	3.35
Japan	−0.53	4.27	2.15	1.95	1.53	0.1
Korea	2.52	4.65	16.68	6.33	4.39	4.18
Laos*	4.88	0.00	0.00	57.81	16.83	13.51
Malaysia	4.55	20.52	8.73	7.54	7.43	4.95
Philippines	1.73	8.50	92.87	7.45	3.55	2.26
Myanmar*	8.46	21.34	8.79	7.47	6.03	5.46
Singapore	5.31	0.00	2.26	6.43	5.49	2.04
Thailand	5.13	5.20	11.35	6.53	6.65	1.38
Vietnam*	9.62	9.82	48.96	14.26	10.18	8.54
World Average	1.89	1.46	2.27	2.94	2.01	0.47

Sources: Calculated from EIA and BP websites.

Note: * Denotes 1994–2003 data.

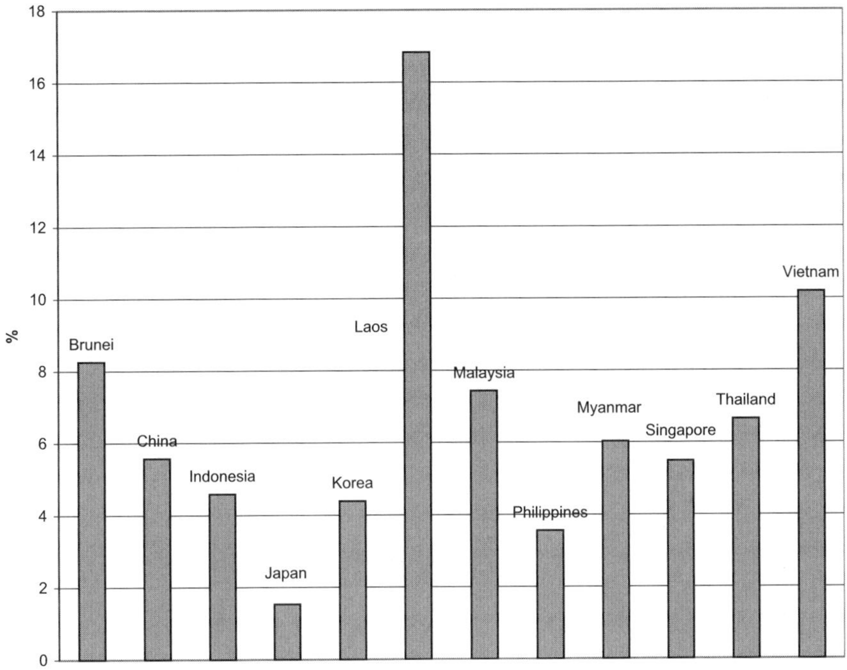

Figure 1: ASEAN Plus 3: Average Total Energy Consumption Growth, 1994–2004 (%)

Source: See Table 2.

The rank order of per capita energy consumption is quite different (see Table 3 and Figure 2 which shows the relative growth rates of per capita consumption) from the total consumption figures. China is in the lower half simply because of its enormous population. For Singapore and Korea, care must be taken in interpreting these figures because both have large oil refining industries and are major air traffic hubs. For countries such as Laos, Cambodia and Myanmar, it must be remembered that these economies still rely heavily on non-commercial forms of energy (not included here).

As for the breakdown of energy consumption, Table 4 amply reveals the very different structures of the economies under study. In some cases, the industrial sector accounts for the largest share while in others the residential sector dominates. This can be partly explained

Table 3: ASEAN + 3:Total Consumption Statistics, 2004 Ranked Order

Rank	Petroleum (thousand barrels per day)	Coal (million tons oil equivalent)	Natural Gas (billion cubic metres)	Electricity* (billion kilowatt-hours)	Total Energy (quadrillion btu)	Energy Per capita* (million btu)
1	China 6684	China 956.9	Japan 72.2	China 1682.37	China 1386.2	Singapore 400.1
2	Japan 5288	Japan 120.8	China 39.0	Japan 1000.25	Japan 514.6	Brunei 262.3
3	Korea 2280	Korea 53.1	Indonesia 33.7	Korea 275.31	Korea 217.2	Korea 186.8
4	Indonesia 2555	Indonesia 22.2	Malaysia 33.2	Thailand 100.28	Indonesia 109.6	Japan 173.4
5	Thailand 909	Thailand 10.2	Thailand 28.7	Indonesia 95.21	Thailand 81.5	Malaysia 106.2
6	Singapore 748	Vietnam 6.3	Korea 31.6	Malaysia 62.48	Malaysia 60.3	Thailand 54.2
7	Malaysia 504	Malaysia 5.7	Vietnam 2.5	Philippines 43.56	Philippines 25.0	China 34.9
8	Philippines 336	Philippines 5.0	Myanmar 2.4	Vietnam 36.66	Singapore 45.1	Indonesia 21.0

(*Continued*)

Table 3: (*Continued*)

Rank	Petroleum (thousand barrels per day)	Coal (million tons oil equivalent)	Natural Gas (billion cubic metres)	Electricity* (billion kilowatt-hours)	Total Energy (quadrillion btu)	Energy Per capita* (million btu)
9	Vietnam	Myanmar	Philippines	Singapore	Vietnam	Philippines
	191.1	0.2	2.0	30.79	24.8	16.0
10	Myanmar	Laos	Brunei	Myanmar	Myanmar	Vietnam
	34.8	0.00	1.8	6.13	4.2	12.0
11	Brunei	Brunei	Singapore	Laos	Brunei	Laos
	12.4	0.00	7.8	3.06	2.7	7.3
12	Cambodia	Cambodia	Laos	Brunei	Laos	Laos
	3.8	0.00	0.00	2.32	1.1	7.3
13	Laos	Singapore	Cambodia	Cambodia	Cambodia	Cambodia
	3.2	0.00	0.00	0.11	0.2	0.6

Sources: *British Petroleum Review of World Energy*, June 2005 and EAI website.
Note: Data for Brunei, Cambodia, Laos, Myanmar and Vietnam is 2003.
* 2003 Data.

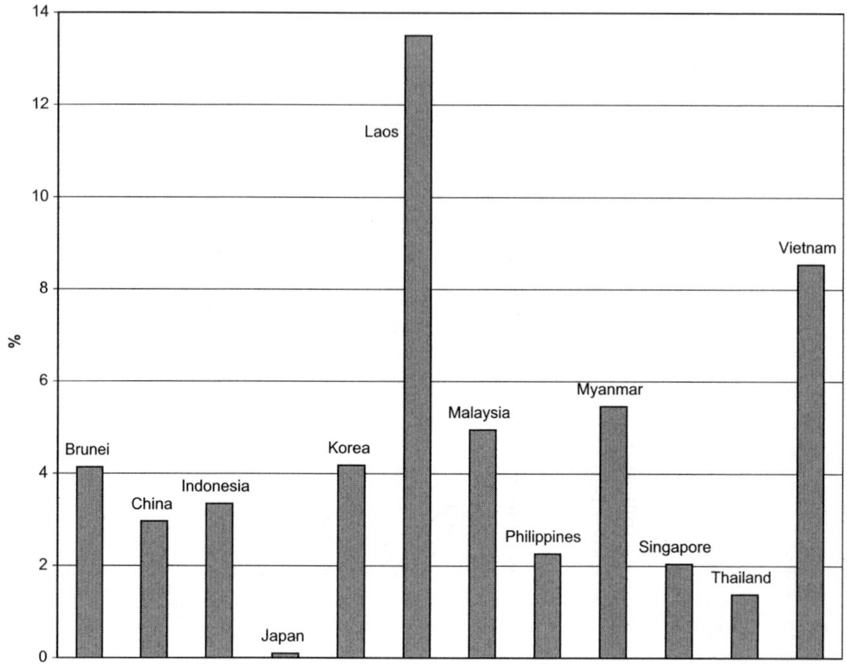

Figure 2: ASEAN Plus 3:Average Per Capita Energy Consumption Growth, 1994–2004 (%)

Source: See Table 2.

by the fact that some countries require much more fuel than others during the winter months, and that in some countries the residential sector still uses a lot of non-commercial fuels for cooking and heating.

Overall, the energy consumption breakdown by sector for the ASEAN + 3 in 2002 was (per cent):[4]

Industry 28
Transportation 26
Residential/commercial 41
Other 5

---

[4] IEA, *Energy Balances for Non-OECD Countries* and for *Energy Balances for OECD Countries 2001/2002*. Paris: OECD/IEA, 2003.

Table 4: Breakdown of Total Energy Consumption, 2002 (%)

	Industry	Transport	Agriculture	Commercial and Public Services	Residential	Non-specified	Non-energy
Brunei	10.6	55.6	—	19.6	10.0	2.8	1.4
Cambodia							
China	40.2	9.9	4.2	3.7	37.6	1.4	3.0
Indonesia	27.3	20.7	1.7	3.7	45.9	—	0.7
Japan	37.6	26.3	6.8	59.6	49.2	3.3	2.9
Korea	45.2	24.1	2.8	12.4	12.8	1.4	1.3
Laos	—						
Malaysia	41.6	39.3	0.3	7.4	9.6	—	1.8
Philippines	29.6	35.1	1.8	8.1	21.6	2.9	0.9
Myanmar	7.8	8.4	0.1	0.7	82.7	0.2	0.1
Singapore	32.5	39.8	—	7.6	4.6	12.1	3.3
Thailand	39.0	33.5	5.4	4.8	15.6	0.1	1.6
Vietnam	14.8	11.4	1.1	3.6	68.4	—	0.7

Sources: Calculated from IEA, *Energy Balances for Non-OECD Countries 2001/2002*, Paris: OECD/IEA, 2003 and IEA, *Energy Balances for OECD Countries 2001/2002*, Paris: OECD/IEA, 2003.

By 2020, the International Institute of Energy Economics (IEEJ) predicts it will be (per cent):[5]

Industry 40
Transportation 28
Residential/commercial 31
Other 2.1

Over the next decade and beyond, the IEEJ expects industry to account for a much larger share due to the lesser developed countries' need to develop more heavy industry for construction and economic expansion.

Figure 3 provides the energy consumption mix for most of the countries in 2003. This very much reflects the energy resource base, or lack thereof, of each country. At one extreme is Brunei which uses only 10 per cent of the energy (mostly gas) it produces, and exports the rest. At the other, Singapore has to import 100 per cent of its energy, South Korea about 97 per cent, the Philippines about 60 per cent, Thailand about 57 per cent, etc. Each of these countries is importing mainly oil and gas. Of all the countries, Indonesia has the most diversified indigenous energy mix.

Overall, the energy consumption breakdown by type for ASEAN + 3 in 2003 was (per cent):[6]

Coal 14.5
Petroleum 45.6
Natural gas 13.7
Nuclear 2.7
Hydropower 1.7
Geothermal, wind, solar 2.3
Combustible material 19.5

---

[5] IEEJ, *Asia/World Energy Outlook,* 385th Forum on Research Works, 10 Mar. 2004, p. 7. Note that their definition of Asia includes India.
[6] Calculated from IEA website.

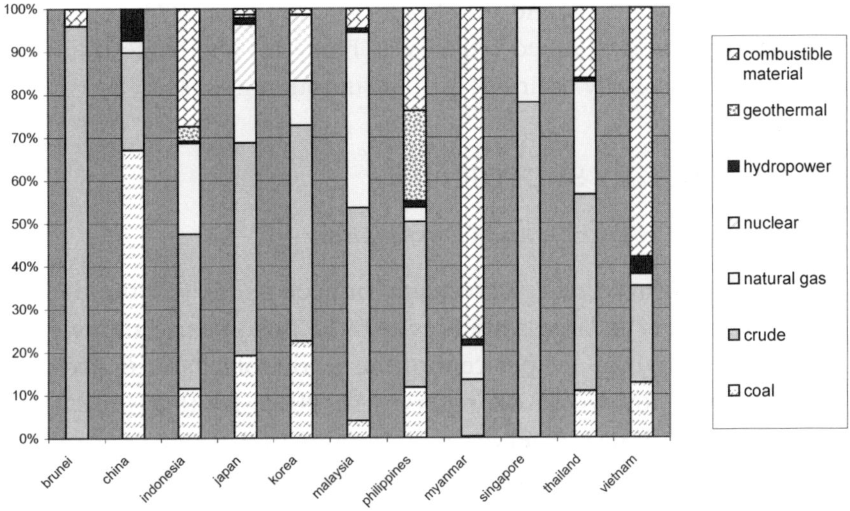

Figure 3: ASEAN Plus 3: Energy Consumption Mix, 2002 (% shares)

Sources: IEA, *Energy Balances of Non-OECD Countries*, Paris: OECD/IEA, 2003 and IEA, *Energy Balances of OECD Countries*, Paris: OECD/IEA, 2003.

This compares to the world average current breakdown of (per cent):[7]

Coal 23.9
Petroleum 38.8
Natural gas 23.0
Nuclear 6.5
Hydro 6.5
Geothermal 0.8
Combustible material 0.5

The world as a whole uses much more coal, as well as nuclear and hydro power, but less petroleum, geothermal power and combustible materials. The reason for these differences are the facts that the developed world uses virtually no biomass and has had the political stability and wealth to harness more hydropower potential and build more nuclear power plants. Also, whereas much of Asia was yet very poor

[7] Calculated from IEA website.

when the first and second oil shocks occurred in 1973 and 1979, respectively, the developed world was badly hit and since then has tried hard to diversify their energy consumption mixes.

ENERGY USE BY SECTOR

Oil Reserves, Production and Consumption

Table 5 gives proved oil reserves data for each country in the region. China has by far the largest reserves at 18.25 billion barrels. However, these amount to only 1.4 per cent of the world total. Table 6 gives the IEA estimates for world oil reserves in 2025. By that time, they estimate that China's reserves will amount to 1.8 per cent of the world total. OPEC will have 56.8 per cent and non-OPEC will have 43.2 per cent compared to 76.9 and 15.6 per cent, respectively now. Also, according to the IEA, total imports of oil into Asia in 2001 were 13.2 million metric barrels but by 2025 will be 30.8 (see Table 7).

Figure 4 gives average oil consumption growth from 1994 to 2004 for each of the thirteen countries. Consumption growth has been fastest in Vietnam, Myanmar and China at over 7 per cent per year. Japan's on the other hand, has been negative. Japan has to import almost all of its requirements (it is the second largest importer in the world), though it has worked strenuously to reduce the proportion of oil in its TEC mix from 75 per cent in 1973 to about 50 per cent today. This reduction is related more to its great desire to mitigate its energy vulnerability than due to the country's economic depression over the past decade.

Table 5: Proved Oil Reserves as of 1 January 2006 (billion barrels)

Brunei	1.350	Malaysia	3.000
Cambodia	—	Philippines	0.139
China	18.250	Myanmar	0.050
Indonesia	4.301	Singapore	—
Japan	0.059	Thailand	0.291
Korea	—	Vietnam	0.600
Laos	—		

Source: PennWell Corporation, *Oil & Gas Journal*, vol. 103, no. 47 (19 Dec. 2005) on EIA website.

Table 6: Estimated World Oil Reserves in 2025 (billion barrels)

United States	181.7
Canada	224.0
Mexico	87.1
Japan	0.5
Australia/New Zealand	12.1
Western Europe	72.1
Former Soviet Union	386.5
Eastern Europe	4.2
Central and South America	314.9
China	52.5
India	16.0
Other Developing Asia	49.5
Africa	285.2
Middle East	1,248.5
Total	2,934.8
OPEC	1,665.6
Non-OPEC	1,269.2

Source: IEA, *International Energy Outlook 2004*, p. 36. The IEA sourced the total figures from the *Oil and Gas Journal, World Petroleum Assessment 2000* and their own *International Energy Outlook 2002*.
Note: Resources include crude oil (including lease condensates) and natural gas plant liquids.

Table 7: Asia's Future Oil Demand and Supply (MMBD)

	2001	2010	2020	2025
Demand	21.2	27.2	34.8	39.1
Supply	8.0	8.6	8.5	8.3
Imports	13.2	18.6	26.3	30.8

Sources: IEA, *International Energy Outlook 2004*.
Note: This is not the ASEAN + 3 definition of Asia.

China, Japan and South Korea are the second, third, and seventh largest consuming countries in the world.[8] These three countries together have one-third of the world's supertanker fleet. ASEAN + 3 imports about one-third of the world's total oil trade including

[8] The US is the largest consumer in the world.

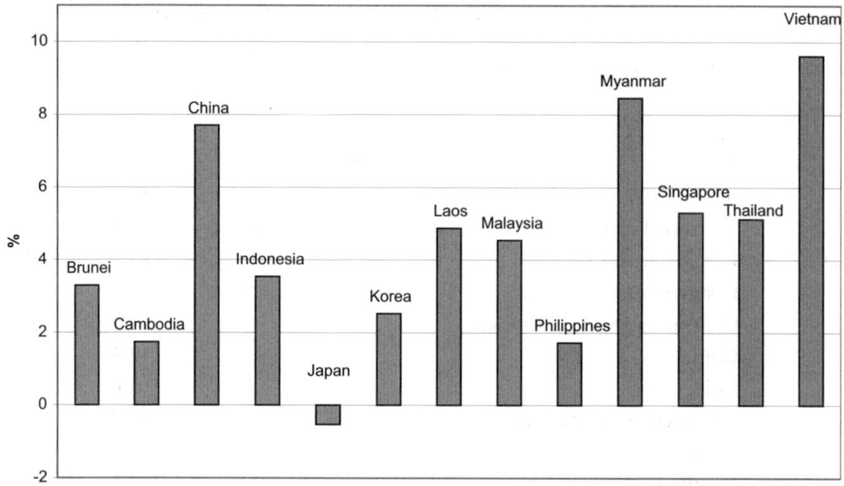

Figure 4: ASEAN Plus 3: Average Oil Consumption Growth, 1994-2004 (%)

Source: See Table 2.

refined products. South Korea exports large quantities of petroleum products to China and Japan.

China's demand obviously dominates the region's demand. Though it is the sixth largest producer of crude in the world, it is currently importing about 49 per cent of its requirements. Its imports will likely surpass those of Korea in less than five years. South Korea is the sixth largest importer of oil in the world. It has to import all of its requirements. Demand is still growing at this time, but is expected to taper and eventually begin to decrease in the near future.

As for the rest of the region, Indonesia, Malaysia, Brunei and Vietnam are net exporters of oil. However, there is no possibility of these countries ever meeting the full requirements of the region. Indonesia's and Malaysia's reserves are in fact declining. Indonesia's status as a member of OPEC became uncertain in 2004 when it became a net oil importer.

Close to 90 per cent of Malaysia's oil is exported to Japan, Thailand, Korea and Singapore.[9] Vietnam exports all of its crude,

[9] APERC and IEEJ, *APEC Energy Overview.* Singapore: Asia-Pacific Economic Cooperation Secretariat, 2003, p. 71.

sending it mainly to Japan, Singapore, the US and Korea. Brunei exports 90 per cent of its oil production, mainly to Japan, Korea, Singapore, Taiwan and Thailand, i.e., only 10 per cent of its production is used in the country.

Singapore, one of the largest refining centres in the world, imports all of its oil requirements and Thailand imports over 92 per cent. Cambodia still relies 85 per cent on non-commercial renewable energy (wood, charcoal and other biomass). It otherwise relies on imported petroleum products. Laos still relies 70 per cent on non-commercial renewable energy, and also imports petroleum products.

The region has a very heavy dependence on Middle Eastern oil. About 75 per cent of ASEAN + 3's crude presently comes from the Middle East. Half the oil to the region from the Middle East goes to China, Japan and Korea. Japan depends on the Middle East for 89 per cent of its oil, South Korea for 79 per cent and China, 51 per cent.

The IEEJ predicts that dependency on Middle Eastern oil by Northeast Asia (Japan, China and South Korea) will increase from 58 per cent in 2000 to about 70 per cent in 2020. If Siberia exported about 100 million tons to these countries per year, this figure could return to about 59 per cent.[10]

As for how the oil is used, transport use is the fastest growing sector though there is also strong demand from the industrial and petrochemical feedstocks sectors. It is not likely that any substitutes for the fossil fuels used in transport will be available on a large scale in the next 15 years. No large-scale conversion of vehicle engines to electricity, liquefied petroleum gas, compressed natural gas, hydrogen fuel cells, or ethanol is expected in this timeframe. The demand for private automobiles will not decline unless the cost of buying and operating them rises severalfold. Singapore has implemented several measures to discourage private car ownership with limited success.

Figure 5 presents oil balances for 1995 and 2003. The key message here is that the situation for Asia, especially China, has deteriorated

[10] IEEJ, *Asia/World Energy Outlook,* 385th Forum on Research Works, 10 Mar. 2004, p. 8.

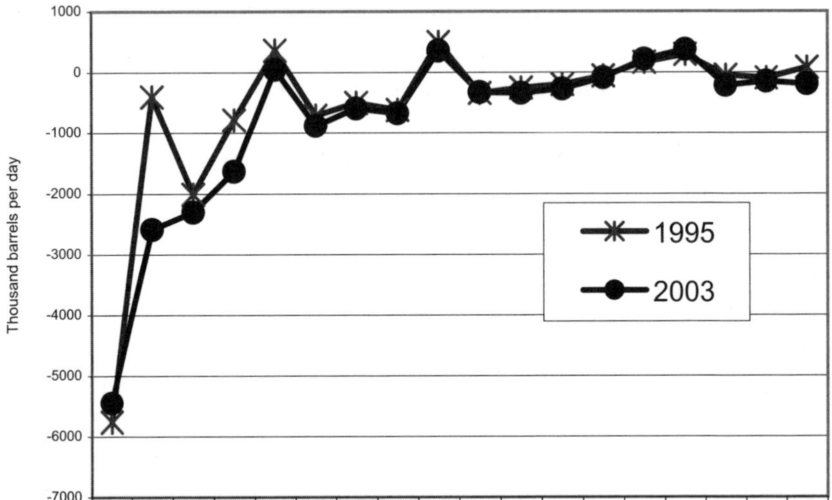

	Japan	China	South Korea	India	Indonesia	Taiwan	Thailand	Singapore	Malaysia	Philippines	Pakistan	Hong Kong	Bangladesh	Brunei	Vietnam	Australia	New Zealand	Other Asia Pacifi
1995	-5764	-401	-2009	-787	359	-713	-500	-617	494	-344	-255	-198	-59	165	292	-50	-95	76
2003	-5451	-2586	-2303	-1633	48	-880	-595	-672	356	-332	-342	-271	-87	214	372	-221	-149	-196

Country

Figure 5: Asian Oil Balance, 1995 and 2003

Source: Adapted from "Table 1.1 Oil Balances 1995" from Paul Horsnell, *Oil In Asia: Markets, Trading, Refining and Deregulation*, Oxford University Press, p. 6, which in turn was citing British Petroleum, *Statistical Review of World Energy*, 1996, *International Petroleum Encyclopaedia*, 1996 and various industry sources.The authors updated Figure 5 on the basis of British Petroleum, *Statistical Review of World Energy*, June 2004.

over this interval. China's dependency on imports has increased the most. South Korea and Indonesia have also become more reliant on imports. Japan, however, as noted above, has actually managed to reduce its consumption.

Oil Refining

Table 8 presents each country's oil refining capacity. China, Japan, Korea, Singapore and Indonesia have the largest capacities. Malaysia, Indonesia, Thailand and the Philippines have capacity but it is mostly for domestic supply. Singapore re-exports more than half of its oil imports as oil products.

Table 8: Crude Oil Distillation Capacity as of 1 January 2005 (thousand barrels per calendar day)

Brunei	9	Malaysia	545
Cambodia	0	Philippines	333
China	6,246	Myanmar	57
Indonesia	993	Singapore	1,337
Japan	4,672	Thailand	703
Korea	2,577	Vietnam	0
Laos	0		

Source: EIA Website.

As mentioned above, of the total increase in oil demand over the next twenty years, transport fuels are expected to account for the largest share. At the same time, supplies from the Middle East will likely become heavier and sourer. In other words, the region will have to adapt its refining equipment and technology to cope with crudes of different chemical constitutions compared to the past, and at the same time produce more transport products than before.

Oil Stocks

All IEA members are required to hold stocks equivalent to 90 days' supply. In this region, only Japan and Korea are members. Japan holds the equivalent of net import volume of 161 days. Korea plans to increase its strategic oil stocks from 44.5 days of net imports in 2002 (65.6 million barrels) to 60 days by 2006.[11]

In 2003, the Chinese Government after years of debate decided to build strategic petroleum reserves. However, they did not expect the price of oil to remain high for so long. Thus, at the time of writing, construction and filling are on hold. Several other countries in the region have discussed having their own stockpiles as well, notably Indonesia, Thailand and the Philippines. International relations specialists have suggested that large stockpiles be made available to the region as a whole at Singapore's Jurong Island, Philippines' Subic Bay and Thailand's Kra Isthmus. Industry analysts, however, believe that none of these could ever be feasible.

[11] *APEC Energy Overview*, p. 65.

Natural Gas and LNG

Natural gas is the cleanest of the fossil fuels and also a highly efficient and flexible fuel. Right now, governments can choose between building pipelines or shipping it in the form of LNG. In the future, however, there may be more use of compressed natural gas (CNG), gas-to-liquids (GTL) and gas-by-wire (electricity transmission).

Table 9 gives each country's reserves of natural gas. Indonesia, Malaysia and China have the largest gas fields, while Cambodia, Korea, Laos and Singapore have very little or none. Indonesia and Malaysia together account for about 70 per cent of Asia's gas trade. Indonesia is the world's largest exporter of gas. Seventy per cent goes to Japan, 20 per cent to South Korea and the rest to Taiwan. Some 90 per cent is exported in the form of LNG, 4 per cent as LPG and 6 per cent as piped gas.[12] Malaysia exports about 40 per cent of its gas, mostly in the form of LNG, to Japan, Korea and Taiwan.[13] A small percentage is piped to Singapore. In all, about 40 per cent of the world's supply of LNG originates in ASEAN. Production capacities at the end of 2004 for Australia (world's largest LNG producer in the world), Indonesia, Malaysia and Brunei were 11.7, 29.4, 22.7 and 7.2 million tons, respectively.[14] Most of Brunei's LNG is exported to Japan with a small amount sent to Korea.

Table 9: Natural Gas Reserves as of 1 January 2006 (trillion cubic feet)

Brunei	13.8	Malaysia	75.0
Cambodia	—	Philippines	3.96
China	53.33	Myanmar	10.0
Indonesia	97.79	Singapore	—
Japan	1.40	Thailand	14.75
Korea	—	Vietnam	6.80
Laos	—		

Source: PennWell Corporation, *Oil & Gas Journal,* vol. 103, no. 47 (19 Dec. 2005) on EIA website.

[12] *APEC Energy Overview,* p. 50.

[13] *APEC Energy Overview,* p. 72.

[14] "Companies Seek to Exploit LNG Growth", *Asia Gas & Power,* published by Argus-Energy iv, no. 24 (8 Dec. 2004): 4.

The demand growth for gas at this time is very high (see Figure 6). Demand growth in the Philippines was particularly strong due to the introduction of Malampaya gas in late 2001. Most governments are aiming to use more for power generation purposes. They seek to diversify their power capacity mix and also reduce pollution emissions resulting from the burning of coal and oil. Japan and Korea buy about two-thirds of the world's total gas exports. Korea has to import all it needs while Japan has to import 97 per cent.[15] At the time of writing, the main existing gas lines were these: Indonesia to Singapore (2), Indonesia to Malaysia, Malaysia to Singapore, Myanmar to Thailand (2) and China to Hong Kong. Lines in planning included: CAA (Commercial Agreement Area between Malaysia and Vietnam) to Malaysia, JDA (Malaysia-Thailand Joint Development Area) to Thailand, JDA to Thailand and Malaysia, Malaysia to the Philippines, Russia to Japan, Russia to China then Korea and China to Hong Kong.

Plans for a Trans ASEAN Gas Pipeline (TAGP) began in 1998.[16] There were already several point-to-point links underway and there are more

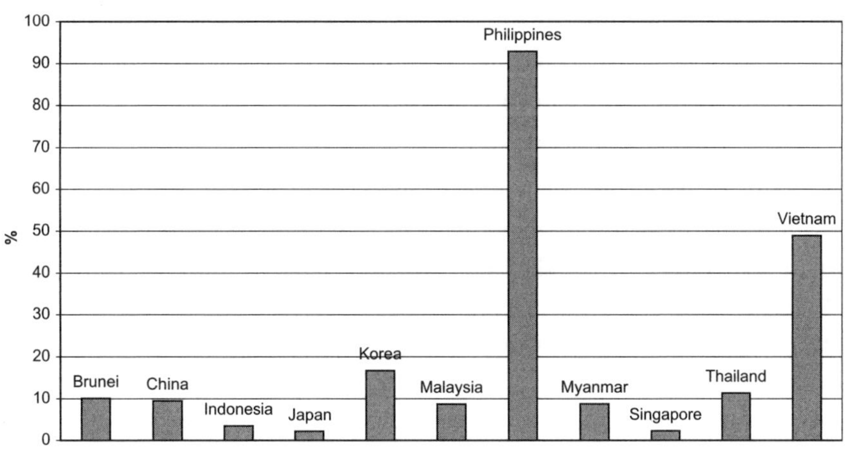

Figure 6: ASEAN Plus 3: Average Natural Gas Consumption Growth, 1994-2004 (%)

Source: See Table 2.

---

[15] *APEC Energy Review*, p. 57.

[16] Details of the plans for a TAGP can be found at the ASEAN website at http://www.aseansec.org/6578.htm [18 Mar. 2006].

planned. Little progress, however, has been made in terms of multi-country hook-ups. The main countries involved have done little to liberalise their energy markets and regulate anti-competitive behaviour. Thus, it would seem that realisation of the TAGP is yet some way off.

The IEA predicts that between 2002 and 2030, Asia's demand (including India) for gas will triple while gas imports will more than double.[17] If this happens, Asia may soon be depending as heavily on the Middle East for gas as it is now for oil. To prevent this, the Chinese, Japanese and Korean Governments would like very much to have pipelines built to their refineries from the Irkutsk region of Siberia.

By a large margin, Japan is the largest buyer of LNG (48 per cent of world imports in 2003) and South Korea is the second largest.[18] Since the early 1970s, Japan has pioneered the development of LNG in Southeast Asia and raised its share in Japan's TEC mix to 15 per cent. It imports about 30 per cent from Indonesia and 21 per cent from Malaysia.[19]

China will undoubtedly become a major importer in the near future. Three LNG terminals are under construction at this time and many more are planned. China, Japan and Korea are also hoping to be the beneficiaries of ongoing development of the gas fields near Sakhalin, off Russia's eastern seaboard.

Coal

From a terrorism point of view, coal is considered the most secure form of energy. It is also far cheaper to extract and transport. However, it is a dirty fuel to handle, transport and use. It can be made far less polluting through the use of clean coal technologies and coal liquefaction. However, these are extremely costly. Most of the coal in the region is used in thermal power plants and in heavy industries such as steel and cement.

Table 10 shows how China's reserves of coal dominate the region. Indeed they are third largest in the world and China is the

[17] IEA, *World Energy Outlook 2004,* Paris: OECD/IEA, 2004, p. 161.

[18] EIA, *The Global Liquefied Natural Gas Market: Status and Outlook,* Dec. 2003.

[19] *APEC Energy Demand,* p. 58.

Table 10: Coal Reserves, 2004 (million short tons)

Brunei	—	Malaysia	4
Cambodia	—	Philippines	366
China	114,500	Myanmar	2
Indonesia	4,968	Singapore	0
Japan	359	Thailand	1354
Korea	80	Vietnam	150
Laos	—		

Sources: *British Petroleum Review of World Energy*, June 2005 and EAI website.

largest producer of coal in the world. Indonesia has the second largest reserves in the region. This country became a major exporter of steam coal in the late 1980s and 1990s. It aims to double its coal production. Vietnam exports about 40 per cent of its coal production, mostly to Japan which has to import 99 per cent of its coal requirements.

Japan is the world's top steam coal and coking coal importer, though this is not reflected in Figure 7, which portrays the average annual growth rate of coal consumption in nine of the countries from 1994–2004, because it has been importing at a fairly steady rate for the past two decades. Imports amount to over 20 per cent of worldwide coal imports.[20] Half of it goes to the steel and pig iron industry which is second largest in the world after China's. Korea is the world's second largest importer of both steam and coking coal. It has to import 95 per cent of its coal supply.[21] The Philippines has to import almost all of its coal. China and Indonesia each provide about 40 per cent of the imports while another 4 per cent comes from Vietnam.

Electricity

The demand for electric power is soaring due to rapid economic growth, urbanisation and increasing living standards, including the exponential rate at which households in the region are purchasing

[20] *APEC Energy Overview*, p. 60.

[21] *APEC Energy Overview*, p. 63.

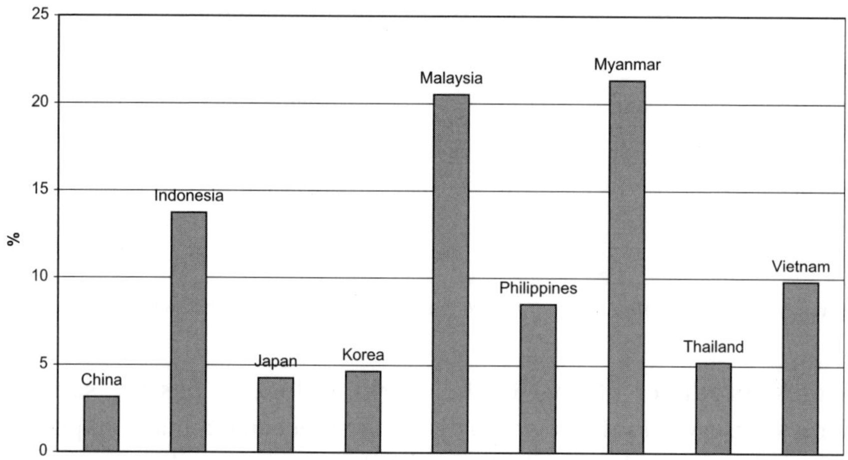

Figure 7: ASEAN Plus 3:Average Coal Consumption Growth, 1994-2004 (%)

Source: See Table 2.

Note:The other four countries have little or no coal reserves.

household appliances. Yet, several areas in the region suffer from electricity shortages, such as in rural China, Indonesia, the Philippines, Laos, Cambodia and Thailand. Only 15 per cent of Cambodia's population has access to electricity.

The highest growth in electricity consumption occurred in Laos (see Figure 8) due to massive construction of new power generating capacity for thousands of peasants who had no power at all until the 1990s. The same occurred in Vietnam. The growth rate for the rest of the region looks dwarfed by comparison. However, in actual fact it was very high, averaging around 8 per cent in all but Japan, a mature economy, which has had to add relatively little new power capacity over the past decade.

As mentioned above, Japan has worked hard at reducing its reliance on oil. In the power sector, 75 per cent of the electricity generated there in 1973 was oil-fired. Now, only 5 per cent is so derived. Its generation mix is the most diversified (see Figure 9). Brunei relies

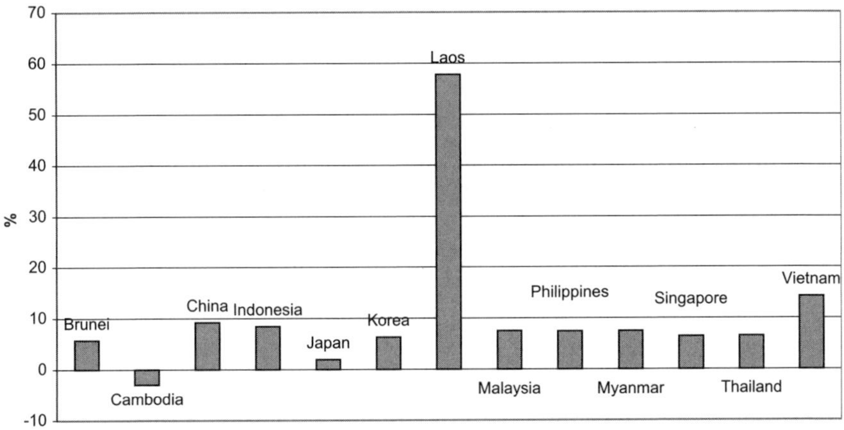

Figure 8: ASEAN Plus 3: Average Electricity Consumption Growth, 1994-2004 (%)

Source: See Table 2.

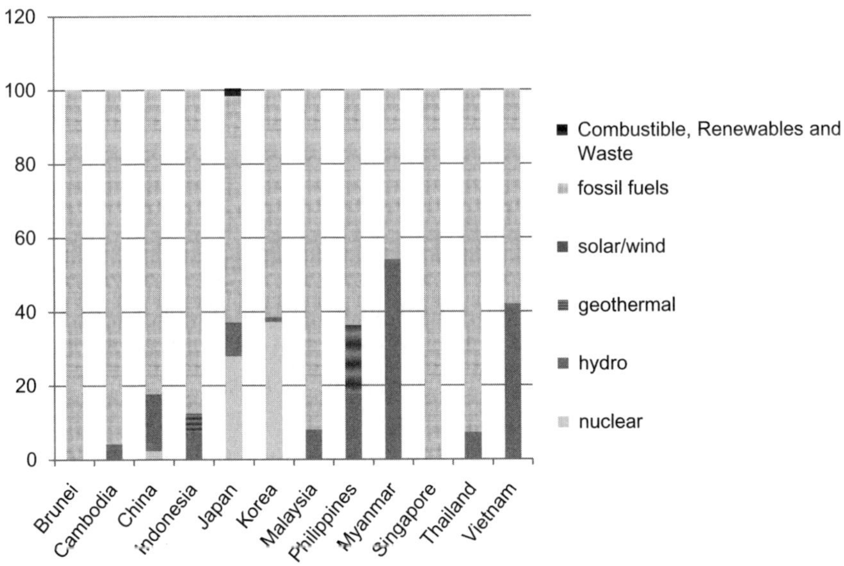

Figure 9: ASEAN Plus 3: Breakdown of Electricity Production, 2003 (%)

Source: IEA, *Electricity Information 2007*, Paris: OECD/IEA, 2007.

almost entirely on gas, Singapore relies on oil and gas, as does Thailand, while China relies very heavily on coal to generate electricity. The Philippines and Indonesia are the only producers of geothermal power.

Overall, ASEAN's electric power mix in 2002 was about: thermal at 70 per cent, hydropower at 22 per cent, nuclear at 6 per cent and geothermal at 2 per cent:[22] The IEEJ predicts that Asia's (including India) total electricity power breakdown by 2020 will change to: coal at 54 per cent, natural gas at 17 per cent, hydropower at 12 per cent, nuclear at 11 per cent, oil at 5 per cent, other renewables at 1.3 per cent and geothermal at 0.7 per cent.[23] They clearly believe, based on their low predictions concerning new oil and gas development in Asia, that some governments in this region will move towards nuclear power in a big way.

As for expansion of electric power capacity in the region, the most difficult issues to address are de-regulation and pricing. Most investors are in generation, not transmission and distribution which tend to remain monopolies. Investors need assurance of a good return, precisely-defined roles of government and energy bodies, as well as clear technical and market rules.

Hydropower

This region includes some of the world's largest potential hydropower sites, including the watersheds of the Salween (Nu Jiang in Mandarin) and other rivers in southern China and Myanmar, the Chao Praya River in Thailand and the Mekong. The Salween is mainland Southeast Asia's longest undammed river, originating from the Tibetan Plateau and traversing China, the Thai-Myanmar border, and emptying into the Indian Ocean in Myanmar.[24] Both Myanmar and China are discussing

---

[22] IEA, *Energy Balances for Non-OECD Countries* and for *Energy Balances for OECD Countries 2001/2002*. Paris: OECD/IEA, 2003.

[23] IEEJ, *Asia/World Energy Outlook,* 385th Forum on Research Works, 10 Mar. 2004, p. 8.

[24] Southeast Asia Rivers Network website at http://www.searin.org/salween_en.htm [19 Mar. 2006].

hydropower development of this River. The Chao Praya River has undergone considerable, harmful small-scale development in its tributaries. The Mekong River Commission (MRC) was founded in April 1995 by Cambodia, the Lao PDR, Thailand and Vietnam. In 2002, China and Myanmar became dialogue members. The MRC's Hydropower Development Strategy's objectives are "to identify the best options for sustainable hydropower development and to recommend criteria for prioritisation."[25]

The harnessing of rivers for hydropower generation anywhere typically leads to permanent ecological damage and the flooding of valuable land. Often irrigation and navigation are also deleteriously affected. Hydropower development inevitably requires detailed cost benefit analyses.

Nuclear Power

Nuclear power accounts for about 40 per cent of Korea's total electricity production, 30 per cent of Japan's and 2.3 of China's. Korea has 20 plants (total installed capacity of about 16.84 GW), Japan has 55 plants (installed capacity of 47.70 GW), and China has 9 plants (installed capacity of 6.587 GW). The Korean, Japanese and Chinese Governments have all endorsed plans for major nuclear power plant construction over the next 10–20 years.[26]

Indonesia has three research reactors while Thailand and Vietnam have one.[27] Indonesia plans to construct a plant capable of producing 6.00 GW by 2011, and Vietnam hopes to have one completed between 2015 and 2020 able to produce 1.40–4.00 GW.[28] Thailand,

---

[25] Mekong River Commission website at http://www.mrcmekong.org/programmes/hydropower.htm [19 Mar. 2006].

[26] These plans are discussed in detail on the World Nuclear Association Website at http://www.world-nuclear.org [Mar. 2006].

[27] World Nuclear Association Website at http://www.world-nuclear.org [Mar. 2006].

[28] Tatsuo Ikenaga, "Vietnam and Indonesia look to Nuclear Power", initially appearing in *Sekai Nippo* and reproduced in *World Peace Herald,* 30 Nov. 2005 at http://www.wpherald.com/storyview.php?StoryID=20051129-043843-3095r [18 Mar. 2006].

Myanmar and Malaysia have also recently begun to consider building nuclear power plants.

With increased domestic and international pressure to reduce carbon and sulphur emissions resulting from the burning of fossil fuels, nuclear power seems to be gaining favour despite the well-known safety issues, waste-disposal problems, high capital costs and long construction times.

Geothermal Power

The Philippines is aiming to become the world's largest producer and user of geothermal energy for power generation. It is now the world's second largest producer after the US. Indonesia is estimated to have 40 per cent (20,000 MW) of the world's geothermal resources.[29]

Renewable Energy and Cogeneration

The various forms of renewable energy (solar, wind, etc.) are recognised globally as being particularly useful in remote areas. However, generally speaking, they will continue to play a minimal role in Asia to 2020. An exception is China, which has set a remarkable target, namely, to have the various forms of renewable energy supply 15 per cent of total power consumption by 2020, up from 7 per cent in 2006.[30] ASEAN is developing cogeneration, namely, the production of steam and electricity from biomass.[31]

Air Pollution Created from the Burning of Fossil Fuels

The amount of carbon dioxide and sulphur dioxide emitted into the air is a function of the amount of fossil fuels burned and the efficiency with which they are burned. According to IEA data

---

[29] *APEC Energy Review,* p. 55.

[30] "Renewable Energy Quota Set for Power Companies", 17 Jan. 2006 from *Shenzhen Daily* at http://www.chinaview.cn [18 Mar. 2006].

[31] See COGEN 3 website at http://www.cogen3.net/ [18 Mar. 2006].

Table 11: Carbon Dioxide Emissions from the Consumption and Flaring of Fossil Fuels, 2003 (million metric tons of carbon dioxide)

Brunei	5.36	Malaysia	141.88
Cambodia	0.57	Philippines	72.02
China	3,540.97	Myanmar	9.78
Indonesia	318.35	Singapore	118.53
Japan	1,205.54	Thailand	195.25
Korea	469.53	Vietnam	61.12
Laos	1.01		

Source: EAI website.

(see Table 11), China's and Japan's carbon dioxide emissions in 2003 were 3,541 and 1,206 million metric tons respectively, or second and fourth highest in the world after those of the US (5,802) and Russia (1,606). Japan's emissions per unit of fuel consumed are lowest in the world because the average efficiency of the country's consuming equipment is generally the highest in the world. Laos and Cambodia's emissions, while minimal now simply because so little fuel is being used, will most certainly rise quickly over the next 15 years.

Table 12 gives each country's position vis-à-vis the Kyoto Protocol. Japan was the only country required to be obliged to the Kyoto Protocol. The others, by virtue of their lower economic development, could either ratify, accede or approve it. China approved it while Cambodia and Laos acceded to it. The rest ratified it.

FUTURE ENERGY CONSUMPTION

There are almost an infinite number of factors which determine energy consumption demand. Among the key ones which can be relatively easily quantified are GDP growth, GDP structure, population growth, climate, prices of energy, energy efficiency of buildings, automobiles, etc.

Over time, a country's energy consumption mix may change. For example, China will demonstrate the 'substitution effect' when it switches to more nuclear power, i.e., the total amount of energy

Table 12: Status of Each Country vis-à-vis the Kyoto Protocol

Brunei	—
Cambodia	Acceded
China	Approved
Indonesia	Ratified
Japan	Accepted
Korea	Ratified
Laos	Acceded
Malaysia	Ratified
Philippines	Ratified
Myanmar	Acceded
Singapore	Acceded
Thailand	Ratified
Vietnam	Ratified

Source: Official Kyoto Protocol Website http://unfccc.int/files/essential_background/kyoto_protocol/application/pdf/kpstats.pdf [17 Mar. 2006].

consumed will fall because the nuclear power (substituting the thermal power) is a much higher value form of energy.

Other variables are not so easily factored in such as the sudden introduction of a new technology which halves energy consumption requirements or a political disturbance which suddenly halves energy supplies. For example, energy supplies in this region would very likely be affected by a war on the Korean Peninsula, an attack on Taiwan, an attack on oil production facilities in the Middle East or their seizure by militants, or a related disruption in the sea lanes therefrom, a serious political crisis in China or a foundering of the Chinese or Japanese financial sector. This region's relations with the Middle East, Russia and the Caspian Region are also key variables. It is impossible to predict with any certainty where the oil will come from over the next 15 years. In short, energy demand and supply are both subject to political manipulation at either domestic or international levels, or both.

A critical factor affecting future demand for energy, particularly oil, is stockpiling. As mentioned above, the Chinese Government did not officially launch its stockpiling programme until 2003/4, though most analysts believe that the national oil companies had been stockpiling already. Indeed, most countries are not forthcoming about exactly how much crude oil and/or petroleum products they are storing or plan to store. China may amass oil on a regular basis or erratically depending on international fuel prices. The same is true of the other countries in the region considering building stockpiles.

We used a simple extrapolation method to create four different TEC scenarios from 2004 to 2020: the best-fit trend for actual consumption from 1980 and 2003 (hereafter called 'the trend'); a trajectory based on the TEC growth rate in 2003; a trajectory based on a constant growth rate of 10 per cent; and a trajectory based on a constant growth rate of 5 per cent (used as a sensitivity analysis for the 10 per cent trajectory).

These results (see the Appendix) are roughly comparable with those of the main energy agencies listed at the outset of this chapter. Direct comparisons are made difficult due to the fact that each organisation uses its own assumptions and units, date spans and definitions of Asia.

What can be said with certainty is that many of the poorer countries are expected to follow the pattern of manufacturing-led economic growth that the rest of the world has gone through. According to the IEA, 'Developing Asia' will account for 42 per cent of the increase in global total energy demand through 2030, compared to 34 per cent in the last three decades.[32] Over the next twenty years, there will be massive growth in energy consumption in Asia, but average per capita energy consumption by then will still be a fraction of what it is in North America.

CONCLUSIONS AND RECOMMENDATIONS

Over the next 15 years, the region will become less and less energy self-sufficient, and concomitantly, more vulnerable to the vicissitudes of international energy markets. Competition for all energy resources will increase. There are several factors which cannot be predicted with any

[32] IEA, *World Energy Outlook 2004.*

certainty — the rate at which the economies now in 'pre-takeoff' and 'takeoff' stages will continue to grow, the rates at which the populations will buy cars and deem nuclear power plants acceptable, consumption efficiencies will improve, solar and wind energy are harnessed, etc.

What can reasonably be affirmed is that over the next 15 years:

a. the demand for electric power can be met in the region using its own coal, hydropower and gas with support from Australia.
b. the region will become more dependent on imports of oil from the Middle East.
c. most of the demand for oil will be in the transport sectors, especially private motor vehicles.

It is imperative that the ASEAN + 3 countries cooperate on both the supply and demand sides. They must work towards increasing the amount of energy available and reducing and standardising the amount of energy needed for common consumption purposes. Assuming there is the political will, there is immense potential for making the region as a whole more secure energy supply-wise. Key, is for the countries to agree that working together is in their best strategic interests, i.e., to agree that it would be counter-productive for them to work independently either in terms of increasing supply and reducing demand within their own borders, or approaching international energy suppliers. The countries must agree to drop their reluctance to share information on how much resources they have, and how much of what type of refining capacity they have.

Before the fundamental physical and financial bridging work that is required for energy cooperation in the region to gain momentum, there must be considerable foundation work at the institutional level, i.e., the establishment of various groups of inter-government decision-makers which have the necessary functional mechanisms to deal with a great variety of issues. For example, governments need to ensure not only that they can share energy production and consumption data, but that they can generate *compatible* and *transparent* data. Also, for example, in the case of, say, a pipeline that traverses several countries, the various governments must have mechanisms to determine jointly who and how it would be maintained. At another level, government officials need a platform from which they can unite to negotiate with the major oil- and gas-producing countries in the Middle East, South America and with Russia.

This institutional bridging is fundamental in building the interest and confidence of potential private investors. The overcoming of the many differences and obstacles to energy cooperation in ASEAN + 3 at the institutional level, will more than anything else, demonstrate the 13 governments' serious long-term commitment to working together.

APPENDIX

TEC in ASEAN + 3 Countries from 1980 to 2020

The projections based on the trend yielded the lowest consumption increases over the period while those based on a 10 per cent growth yielded the highest (see Figure 10). Those based on the trend and 5 per cent growth were close, as were the projections based on the growth rate in 2003 and 10 per cent growth.

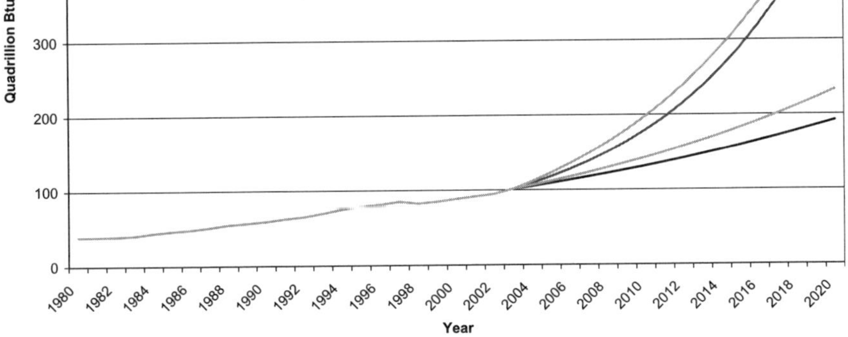

Figure 10: ASEAN Plus 3: Total Energy Consumption, 1980–2020

TEC in ASEAN + 3 Minus China from 1980 to 2020

Apart from the projection based on the 2003 growth rate, all the other projections were quite close (see Figure 11). This implies that TEC seems to grow at less than 5 per cent when TEC follows either the trend, or if the 2003 growth rate of TEC persists. For this group of countries, TEC growth in China would be higher than 10 per cent if the 2003 consumption growth rate were maintained.

TEC in ASEAN Countries from 1980 to 2020

Unlike the projections which include two or three East Asian countries (Japan, South Korea or China), those for ASEAN alone are lowest when based on the 2003 growth rate, followed by those based on the trend and 5 per cent growth (see Figure 12). This implies that TEC growth would be less than 5 per cent if the TEC growth in ASEAN countries maintained either the 2003 growth rate or followed the trend.

Trajectory Based on the 1980–2003 Trend

When the trend is used to project TEC for the ASEAN + 3 and ASEAN countries, the annual growth rate for ASEAN + 3 is higher from year

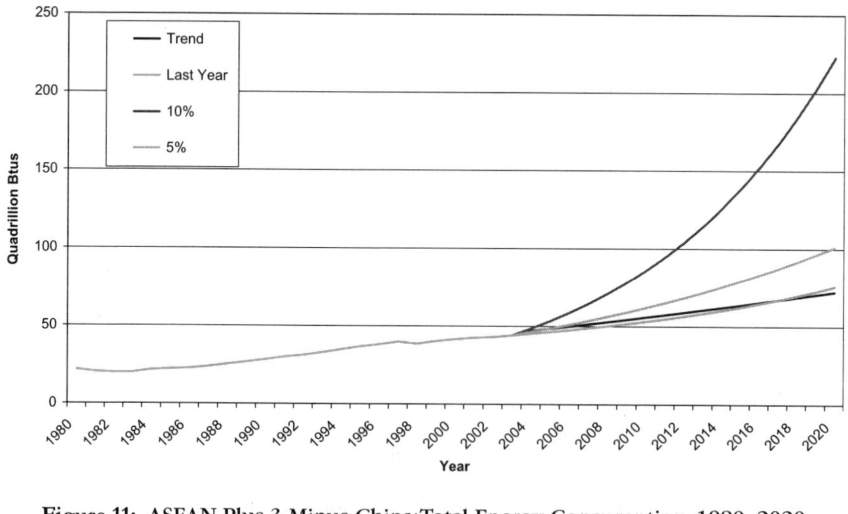

Figure 11: ASEAN Plus 3 Minus China: Total Energy Consumption, 1980–2020

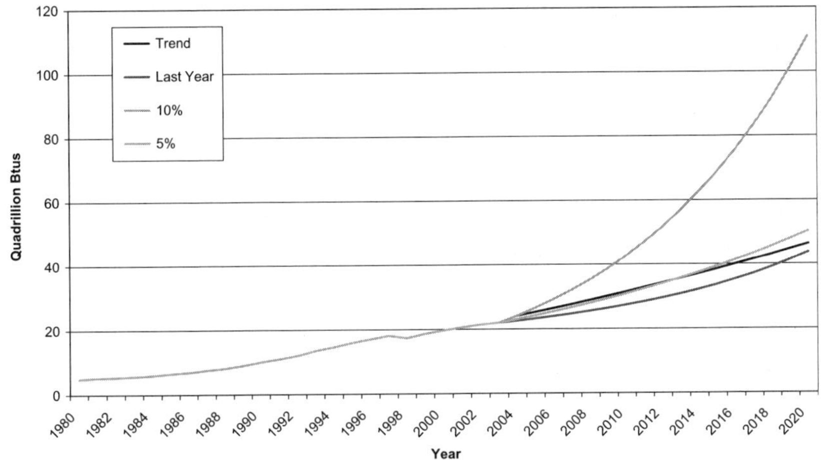

Figure 12: ASEAN Members Alone: Total Energy Consumption, 1980–2020

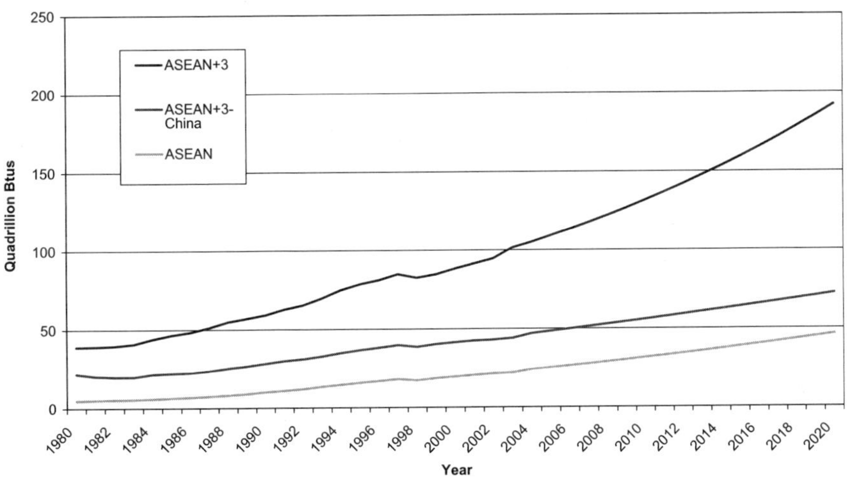

Figure 13: ASEAN Plus 3, ASEAN Plus 3 Minus China and ASEAN Members Alone: Continued Trend Lines

2017 than for ASEAN (see Figure 13). This implies that the annual growth rate will taper after a decade in the ASEAN countries and that their economies will become increasingly more energy-efficient and less energy-intensive. When the TECs of Japan and Korea are included, the annual growth rate becomes even smaller, implying that these economies are already energy efficient.

Trajectory Based on the 2003 Growth Rate

Unlike the downward-sloping annual growth rate projection based on the trend, the annual growth rate of TEC based on the 2003 growth rate produced an upward-sloping curve (see Figure 14). As expected, the annual growth rate for ASEAN + 3 minus China was the lowest, while that for ASEAN + 3 was the highest. The growth rate for ASEAN is in between, reflecting the fact that though the Japanese and Korean economies are large, they are relatively highly energy-efficient.

TEC Based on the Trend Identified from 1980 to 2003 and Projected for 2020

This scenario yields stark differences among the TEC trajectories from 2004 to 2020: stable growth for ASEAN plus Japan and Korea and for ASEAN alone at a lower trajectory while rapid growth for ASEAN plus Japan, Korea and China at a higher trajectory (see Figure 13). The main difference comes from the possible high con-

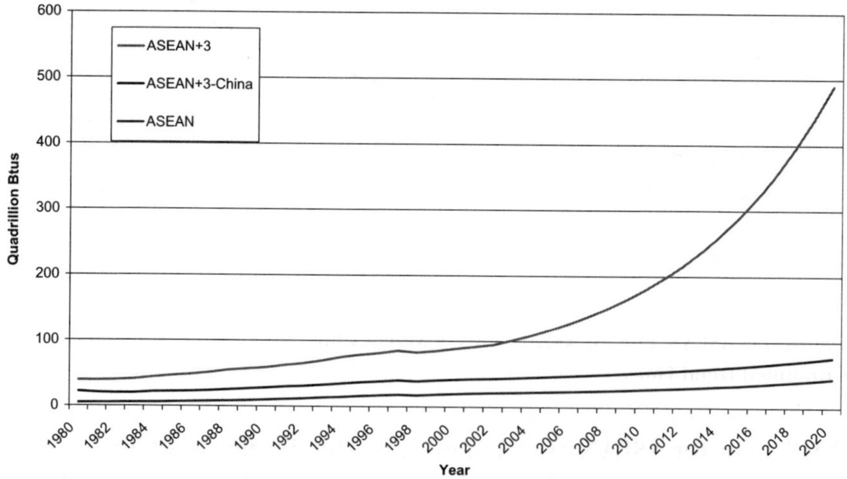

Figure 14: ASEAN Plus 3, ASEAN Plus 3 Minus China and ASEAN Members Alone: Trend Based on Last Year

sumption growth in China. The level of TEC would be far below 100 quadrillion BTUs for ASEAN + 3 minus China and ASEAN, a little less than 200 quadrillion BTUs for China, and the difference between ASEAN + 3 and ASEAN + 3 minus China would be about 150 quadrillion BTUs.

Trajectory Based on the 2003 Growth Rate

Using the 2003 growth rate as the basis for projection yields the widest differences in consumption growth between ASEAN + 3 and either ASEAN + 3 minus China or ASEAN (Figure 14). The difference between these categories is larger than that based on the trend, but the projected TEC for ASEAN + 3 minus China and ASEAN is still far below 100 quadrillion BTUs, while China's alone is close to 500 quadrillion BTUs in 2020.

TEC Based on a Growth Rate of 10 Per Cent per Year and Projected for 2020

This scenario produces the highest TEC growth (see Figure 15). TEC for ASEAN + 3 is more than 500 quadrillion BTUs in 2020. Even that for ASEAN would be more than 100 quadrillion BTUs in 2020. This case is therefore merely considered as an upper-bound case.

TEC Based on the Growth Rate of 5 Per Cent per Year and Projected for 2020

This scenario yields reasonably plausible TEC growth projections for ASEAN + 3, ASEAN + 3 minus China and ASEAN (Figure 16). The difference between ASEAN + 3 and ASEAN + 3 minus China is that TEC would be 200 quadrillion BTUs in 2020. TEC for ASEAN + 2 would be about 100 quadrillion BTUs while that for ASEAN would be 50 quadrillion BTUs. The main difference between this scenario and that of the trend is that the projected TEC for ASEAN + 3 minus China would be much less than 100 quadrillion BTUs. However, the projected TEC for ASEAN + 3 would be close to 200 quadrillion BTUs

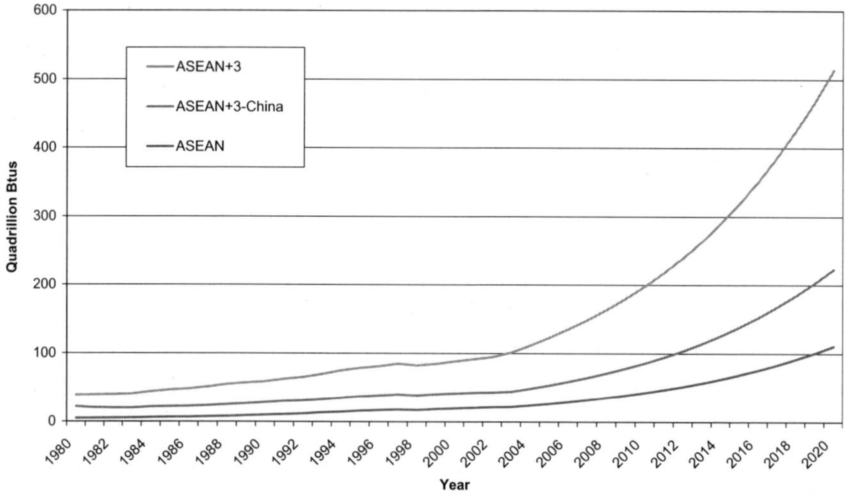

Figure 15: ASEAN Plus 3, ASEAN Plus 3 Minus China and ASEAN Members Alone: 10% Growth

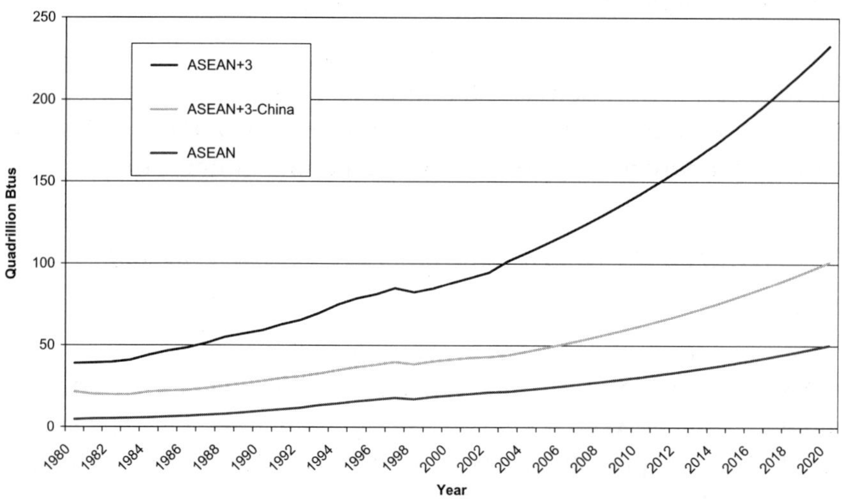

Figure 16: ASEAN Plus 3, ASEAN Plus 3 Minus China and ASEAN Members Alone: 5% Growth

in 2020, implying that China's share of TEC in 2020 is similar in the projections based on the trend and on 5 per cent growth.

Projections for Each Country

Brunei (Figure 17)

TEC in Brunei shows a cubic trend. The best-fit measure is not high ($R^2 = 0.6577$). It declined to a low in 1995 but thereafter increased at an increasing rate. Out of the four scenarios, the projection based on 5 per cent growth per annum yielded the lowest TEC trajectory followed by the 10 per cent growth per annum and trend. The highest TEC trajectory is given by the projections based on the 2003 growth rate, which is not surprising considering Brunei's growing economy and energy use pattern.

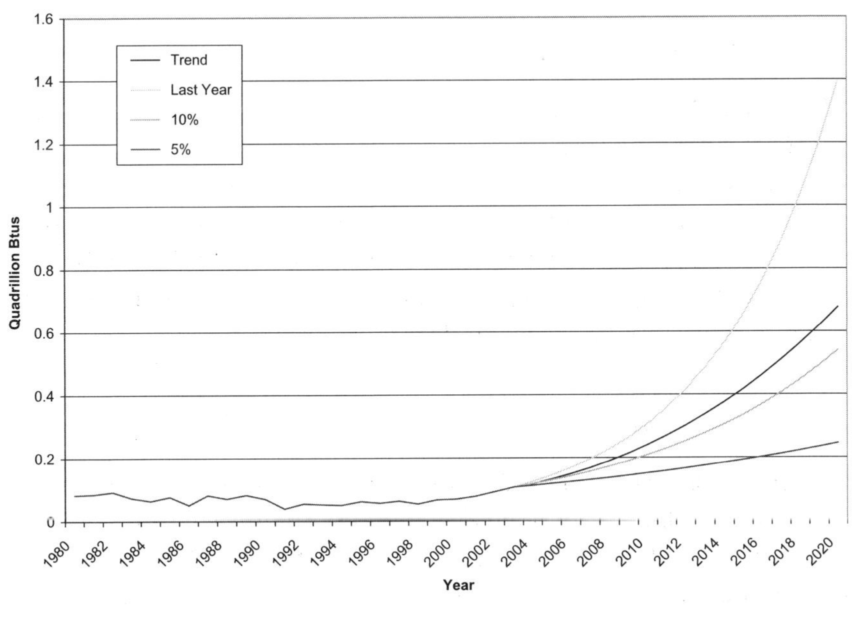

Figure 17: Brunei: Total Energy Consumption, 1980–2020

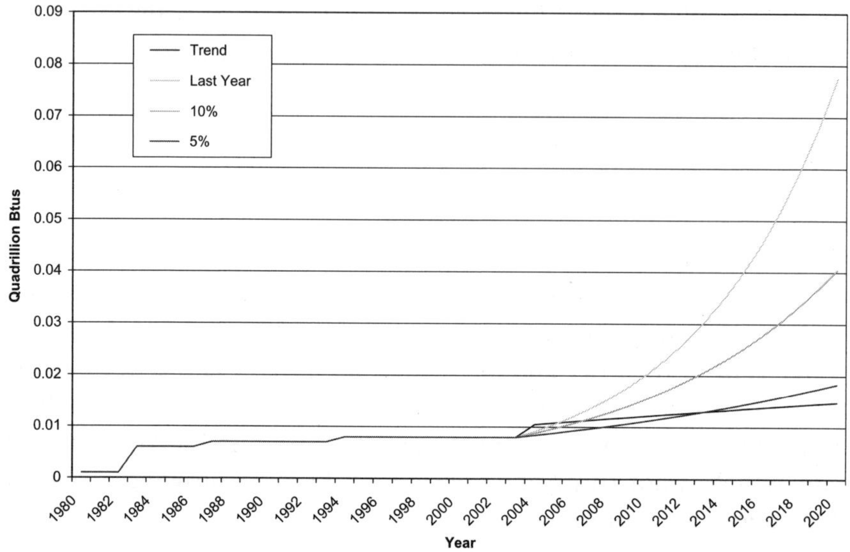

Figure 18: Cambodia: Total Energy Consumption, 1980–2020

Cambodia (Figure 18)

TEC in Cambodia rose sharply in 1983, but was generally quite low through 2003. The best-fit trend is a power function, the only one among ASEAN + 3 countries ($R^2 = 0.7385$). The projections based on the trend yield the lowest trajectory in TEC followed by that based on 5 per cent annual growth and 10 per cent annual growth. Like Brunei, the highest energy consumption trajectory is that based on the trend because there was no change in TEC for those years.

Indonesia (Figure 19)

TEC in Indonesia climbed fairly steadily from 1980 and the best-fit trend is a quadratic function ($R^2 = 0.9904$). The projections based on the 2003 growth rate give the lowest trajectory in TEC followed by the projections based on the trend and 5 per cent annual growth. These two projections are quite close, implying that the trend roughly follows a 5 per cent growth rate. The projection based on a

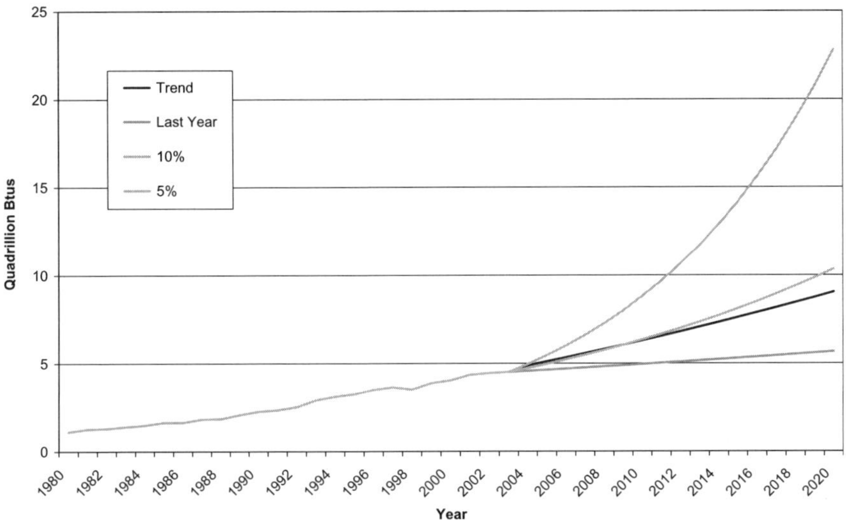

Figure 19: Indonesia: Total Energy Consumption, 1980–2003

10 per cent growth rate per annum yields the highest trajectory and is not credible given Indonesia's current economic status.

Laos (Figure 20)

TEC in Laos surged in 1998 and 1999 from below 0.015 quadrillion BTUs to over 0.04 BTUs and stayed at this level thereafter. Hence, the best-fit trend is a cubic function (R^2 = 0.9271). As expected, the projections based on the trend give the highest trajectory. The lowest trajectory is based on 5 per cent annual growth followed by that based on the 2003 growth rate and 10 per cent annual growth. The trajectory based on the trend is higher than that based on the 10 per cent growth, implying the former is far higher than 10 per cent annual growth. The four projections are noticeably diverse.

Malaysia (Figure 21)

As in Indonesia, TEC in Malaysia has also steadily risen and the best-fit trend is a quadratic function (R^2 = 0.9870). The lowest trajectory is

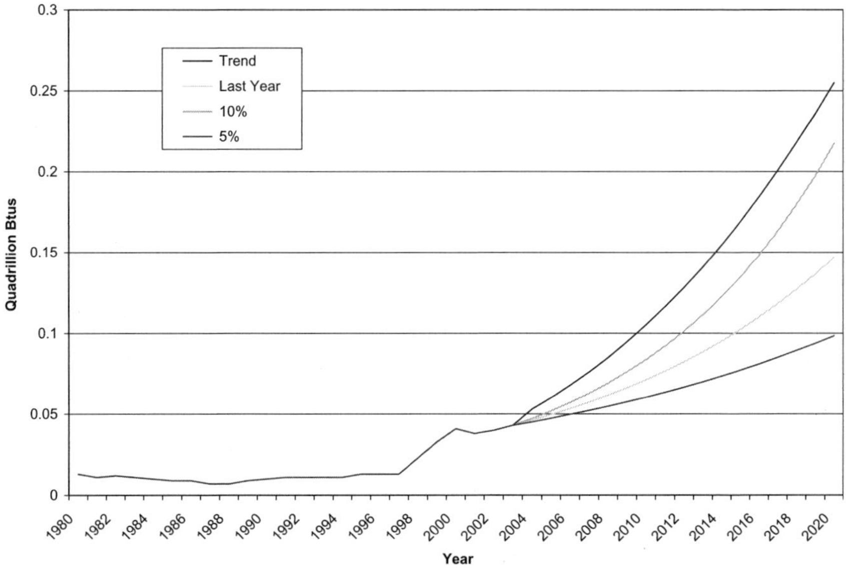

Figure 20: Laos: Total Energy Consumption, 1980–2020

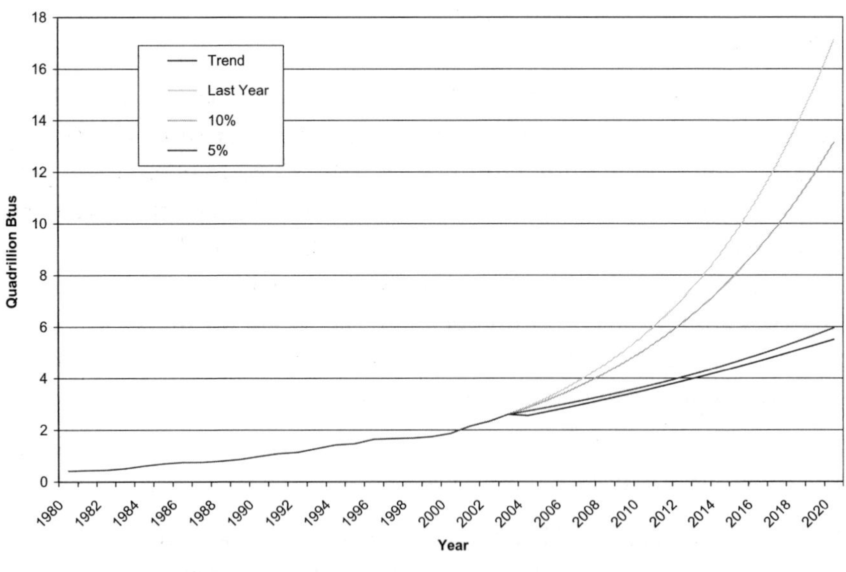

Figure 21: Malaysia: Total Energy Consumption, 1980–2020

given by the trend followed by that based on 5 per cent annual growth. These two projections are quite close implying that the one based on the trend is just under 5 per cent annual growth. The projections based on 10 per cent annual growth and the 2003 growth rate are fairly high, implying that the projection based on the latter is higher than either of the projections based on 10 per cent or 5 per cent annual growth.

Myanmar (Figure 22)

TEC in Myanmar has grown at a moderate pace and the best-fit trend is a quadratic function ($R^2 = 0.9426$). There was quite high growth until 1999 after which it stabilised at 0.16 quadrillion BTUs. The projection based on the trend gives the lowest trajectory and is closely followed by that based on 5 per cent annual growth. Between these is the projection based on the 2003 growth rate. The highest trajectory is that based on 10 per cent annual growth and is not realistic for Myanmar at this point in time.

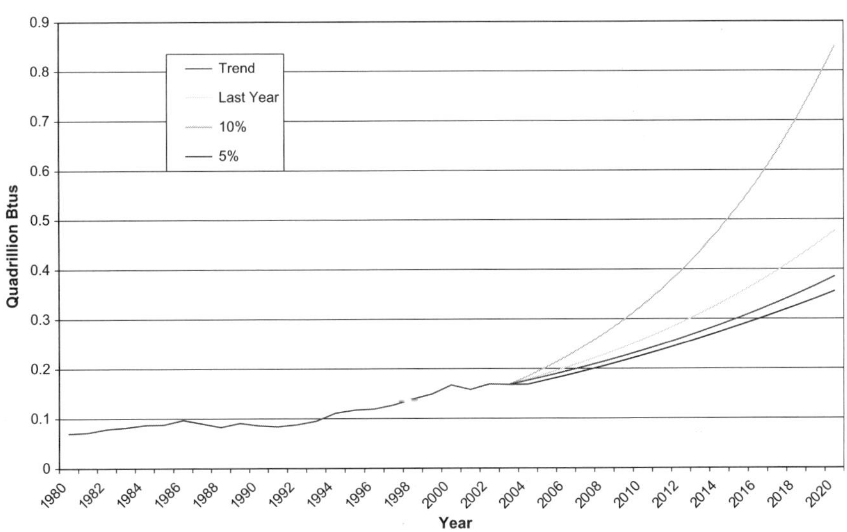

Figure 22: Myanmar: Total Energy Consumption, 1980–2020

Philippines (Figure 23)

TEC in the Philippines also grew steadily over the period and the best-fit trend is a quadratic function ($R^2 = 0.9468$). Since 2000, it stabilised at around at 1.2 quadrillion BTUs. Unlike the other ASEAN countries, the projection based on the 2003 growth rate yields the lowest trajectory. The projections based on the trend and 5 per cent annual growth cross at 2017 with the projection based on the trend falling below the other. The trajectory based on a 10 per cent annual growth is too high to be realistic for the Philippines.

Singapore (Figure 24)

TEC growth in Singapore was generally steady until 1998, but thereafter stabilised at 1.6 quadrillion BTUs. The best-fit trend is a cubic function ($R^2 = 0.9936$). The projections based on the 2003 growth rate yield a relatively flat trajectory which is also the lowest one. The projections based on the trend and 5 per cent annual growth are fairly close and cross at 2016. After that year, the projection based

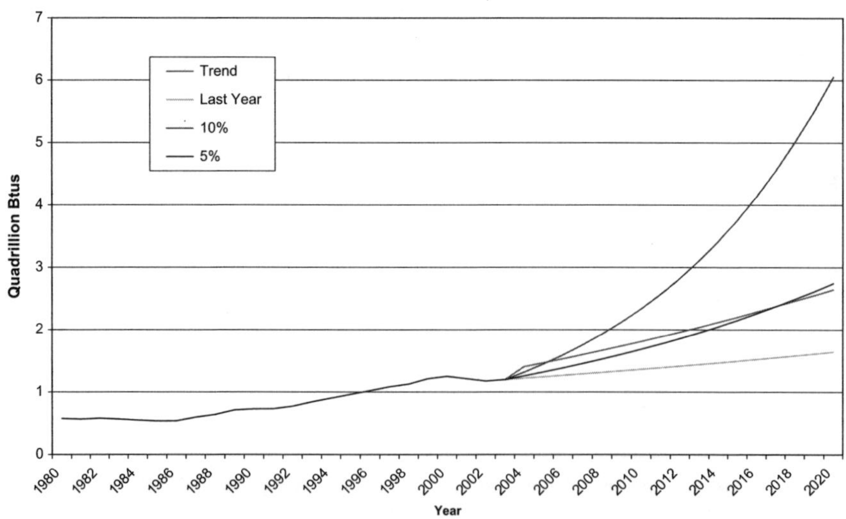

Figure 23: Philippines: Total Energy Consumption, 1980–2020

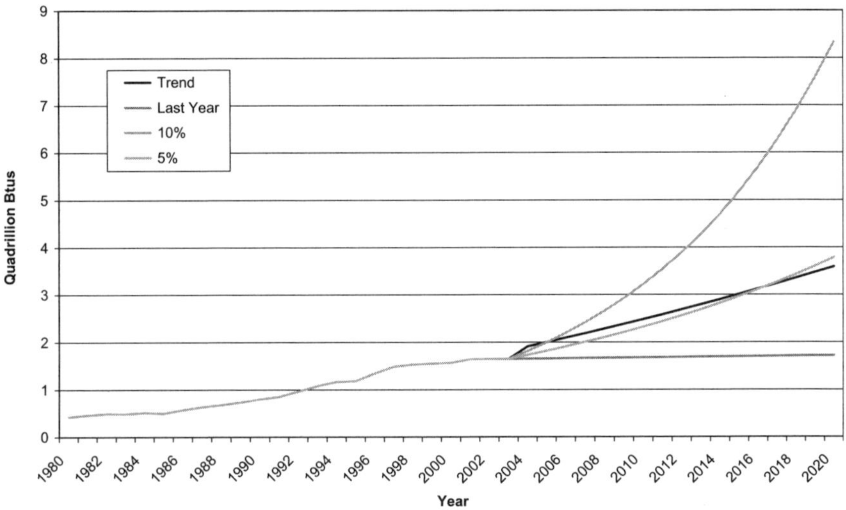

Figure 24: Singapore: Total Energy Consumption, 1980–2020

on the trend falls below the 5 per cent line. The projection based on 10 per cent annual growth is unrealistic for Singapore.

Thailand (Figure 25)

TEC in Thailand grew steeply from 1980 to 2002. However, from 2003 it appeared to stabilise at slightly over 3 quadrillion BTUs. The best-fit trend is a quadratic function (R^2 = 0.9767). As expected, the projection based on the 2003 trend yields the lowest trajectory and it is fairly flat, which is not likely for Thailand considering its booming economy and energy-intensive economic structure. The projection based on 5 per cent annual growth and the trend are more typical for Thailand. That based on 10 per cent annual growth is not ruled out in this case is due to Thailand's energy-intensive economic structure and fast-growing economy.

Vietnam (Figure 26)

As with other fast-growing economies, TEC in Vietnam has been growing steeply. The best-fit trend is a quadratic function (R^2 = 0.9745).

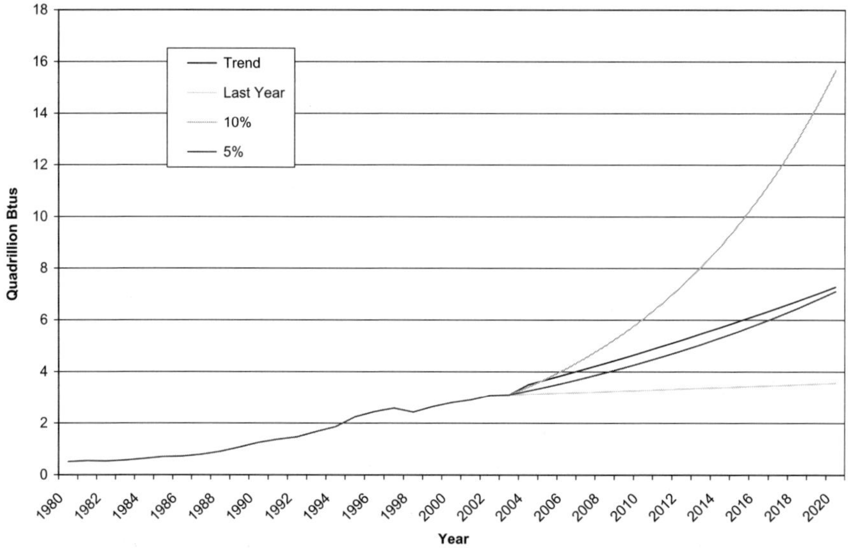

Figure 25: Thailand: Total Energy Consumption, 1980–2020

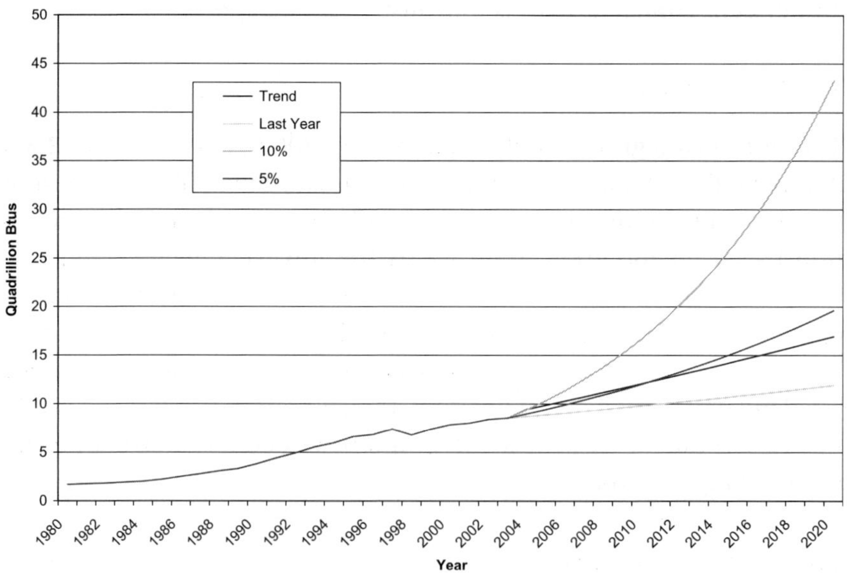

Figure 26: Vietnam: Total Energy Consumption, 1980–2020

Except in 1998, TEC increased every year from 1980. The projection based on the 2003 growth rate gives the lowest trajectory, but that based on the trend and 5 per cent annual growth are fairly close. The latter two trajectories cross at 2011 with the projection based on the trend falling below. The highest trajectory is given by the projection based on 10 per cent annual growth and is not ruled out considering Vietnam's fast-growing economy and energy-intensive economic structure.

Japan (Figure 27)

TEC in Japan generally declined over the period. Its best-fit trend is a quadratic function (R^2 = 0.9522). There was steady growth until 1998, but it thereafter stabilised at about 22 quadrillion BTUs. The projection based on the trend gives the lowest trajectory in TEC followed by that based on the 2003 growth rate and 5 per cent annual

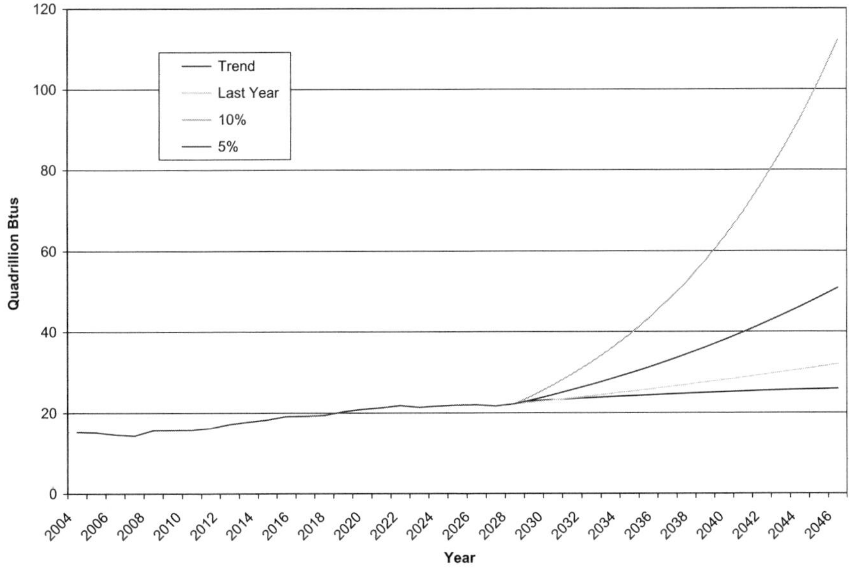

Figure 27: Japan: Total Energy Consumption, 1980–2020

growth. The projection based on 10 per cent annual growth is unrealistically high for Japan.

Korea (Figure 28)

TEC in Korea grew steadily over the period. The best-fit trend is a quadratic function ($R^2 = 0.9745$), implying that Korea has been a fast-growing economy and has had an energy-intensive economic structure. However, from 1998 the trend became flatter, implying its energy intensiveness improved and its growth rate slowed. The trajectory based on the 2003 growth rate is far lower than that based on the best-fit trend and based on 5 and 10 per cent annual growth. The next lowest trajectories are based on the trend and 5 per cent annual growth. These cross at 2011 and the trajectory based on the trend falls below. The projection based on 10 per cent annual growth is unrealistically high for Korea.

China (Figure 29)

TEC in China grew steadily over the period and was especially high in the last two years. The best-fit trend is a quadratic function

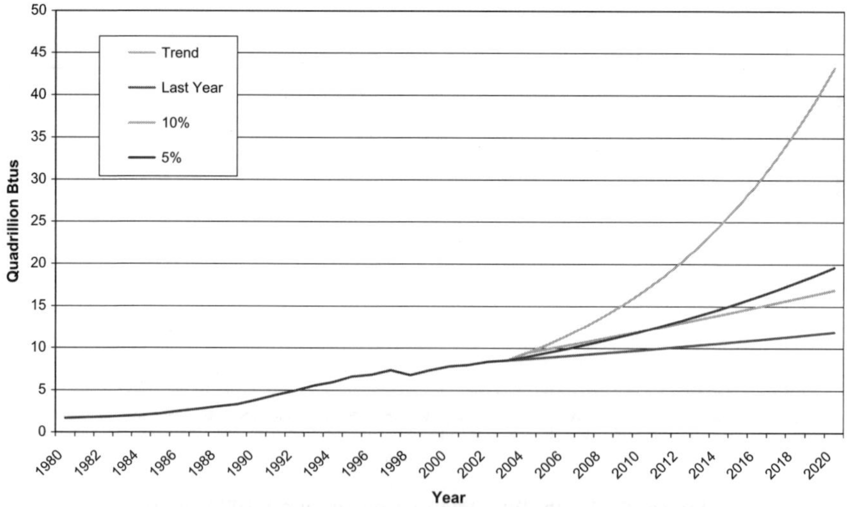

Figure 28: Korea: Total Energy Consumption, 1980–2020

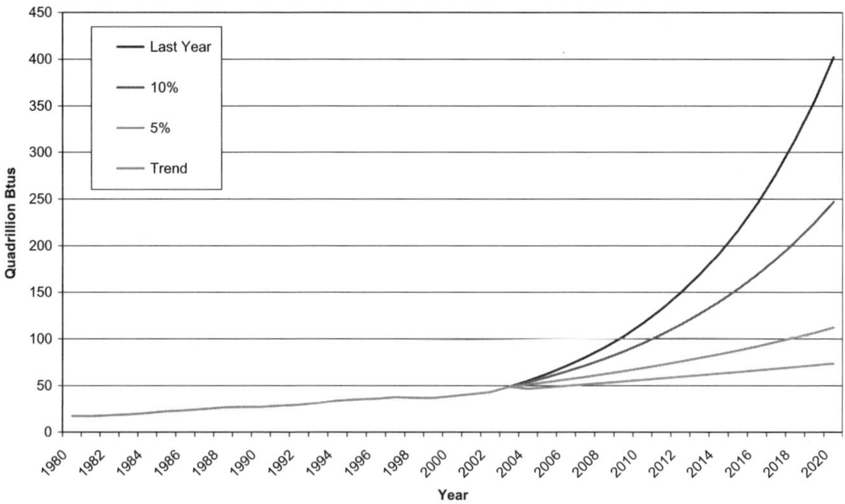

Figure 29: China: Total Energy Consumption, 1980–2020

(R^2 = 0.9811). The highest trajectory is based on the 2003 growth rate while the lowest is based on the trend. The former yields a result almost six times higher in TEC than the latter. In between are the projections based on 5 per cent and 10 per cent annual growth. The two highest projections are plausible for China considering it has the highest energy intensity among ASEAN + 3 countries and has experienced particularly rapid economic growth.

Chapter

2

The Potential for Energy Conservation in East Asia

Youngho Chang and Elspeth Thomson

INTRODUCTION

In 1973/4 the world experienced a historic economic shock. Due to conflict in the Middle East the price of oil quadrupled. Countries which had assumed comfortably that oil supplies and prices would remain stable well into the future scrambled to establish regulations which would force consumers to reduce their energy consumption quickly and improve their consumption efficiency. In the following years, out of crucial necessity, the output value of energy consumption in most countries improved dramatically. In 1980 there was a second, lesser though significant rise in oil prices. Again, most countries reacted abruptly and decisively.

Though not without its own energy resources, the ASEAN + 3 (ASEAN + China, Japan and Korea) region is by no means self-sufficient (see Chapter 1). Several countries, such as Japan, Korea, Singapore and the Philippines depend on imports to meet all or most of their energy requirements.

Thus, while efforts are being made to survey and harness all possible energy resources in the region, equal if not more effort must go into reducing demand, i.e., using energy, irrespective of its origin, as efficiently as possible. Ideally, there should be a clear, indubitable link between conservation and profit-making. Energy-saving activities are in fact normally much less capital-intensive than energy-production activities.

This chapter examines the successes in, and potential for energy conservation in the ASEAN + 3. The first section examines the energy intensity and income elasticity of energy consumption for each country. The second briefly describes the scope of the energy conservation programmes to date and the main problems encountered. The final section looks specifically at conservation in the sector most relevant to energy security in Asia at this time, namely, transport fuel.

ENERGY EFFICIENCY INDICATORS

There are two general indicators that measure the status and level of energy efficiency in an economy: energy intensity and elasticity of energy consumption. Energy intensity shows how much energy has been used to produce goods and services in an economy and is defined as the ratio of total energy consumption over GDP. It can be written as follows:

$$I = \frac{E}{GDP},$$

where I is the energy intensity, E is the energy consumption and GDP is the level of output in an economy. Changes in energy intensity may show that there are changes in the structure of an economy, changes in the mix of energy sources or changes in the efficiency of energy use in the economy. When examining an individual country's energy intensity trend, attention must be paid to what is the major force behind the changes in energy intensity. There are normally two key forces: the level of total energy consumption and GDP. If GDP growth is greater than total energy consumption growth, energy intensity decreases. This could imply that energy efficiency has improved. However, low energy intensity does not necessarily mean low income elasticity.

The income elasticity of energy consumption is an estimate of how much energy is required to produce an additional unit of GDP. In general, the lower the energy intensity and the further the elasticity is below 1, the more energy efficient the economy. Income elasticity is based on the assumption that total energy consumption can be expressed in terms of GDP and as a fraction of GDP. The derivation of income elasticity is:

$$E = K \cdot GDP^{\rho},$$
$$\log E = \log K + \rho \log GDP,$$
$$\frac{\Delta E}{E} = 0 + \frac{\Delta GDP}{GDP},$$
$$\rho = \frac{\Delta E / E}{\Delta GDP / GDP},$$

where E is the energy consumption, GDP is the level of output in an economy, K and ρ are parameters and considered constant. From the derivation, ρ is interpreted as income elasticity of energy consumption provided that total energy consumption is expressed as a fraction of GDP.

If the income elasticity of energy consumption is greater than 1, the economy's energy consumption is growing more rapidly than GDP. If it is less than 1, the economy's energy consumption is growing less rapidly than GDP. If it is 1, it is growing proportionally with GDP.

ENERGY INTENSITY

Figures 1–3 portray energy intensity from 1980 to 2003 for ASEAN + 3, and ASEAN + 3 minus China and the ASEAN member countries alone. China's energy intensity is so high that it causes the other countries' energy intensity levels in Figure 1 to appear low and relatively constant. The countries fall into four groupings: China and Vietnam have energy intensities of about 25,000 BTU/USD, while Malaysia and Indonesia have energy intensities of about 20,000 BTU/USD.[1] These four countries have the fastest growing economies of the region.

[1] All energy intensity figures here are in British Thermal Units per US dollar, in 1995 constant terms.

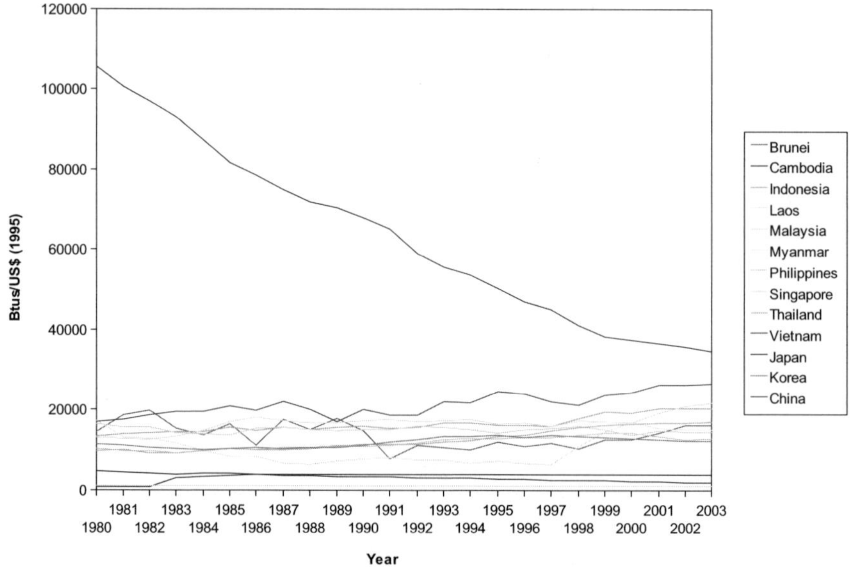

Figure 1: Energy Intensity in ASEAN+3, 1980–2003

Figure 2: Energy Intensity in ASEAN+3 Minus China, 1980–2003

Figure 3: Energy Intensity in ASEAN, 1980–2003

Thailand, Brunei, Laos, Singapore, the Philippines and Korea have energy intensities of about 15,000 BTU/USD. Of these, Korea is a developed economy in transition while the Philippines and Laos are growing economies with high energy consumption. Brunei and Singapore are borderline cases. They both have energy-intensive industrial structures. Thailand is a typical case of a fast-growing economy. Its energy intensity has been above 15,000 BTU/USD since 1998.

Japan, Cambodia and Myanmar have energy intensities of less than 5,000 BTU/USD. Japan is an example of a developed economy with high energy efficiency. Indeed, Japan, because it is so heavily dependent on energy imports, became one of the most energy efficient countries in the world following the world energy shocks.

ENERGY INTENSITIES OF EACH COUNTRY

Figures 4–16 show that the energy intensities of several of the countries were anything but steady.

Brunei: Energy intensity was lowest in 1991, thereafter it climbed, indicating that the economy was becoming more energy intensive.

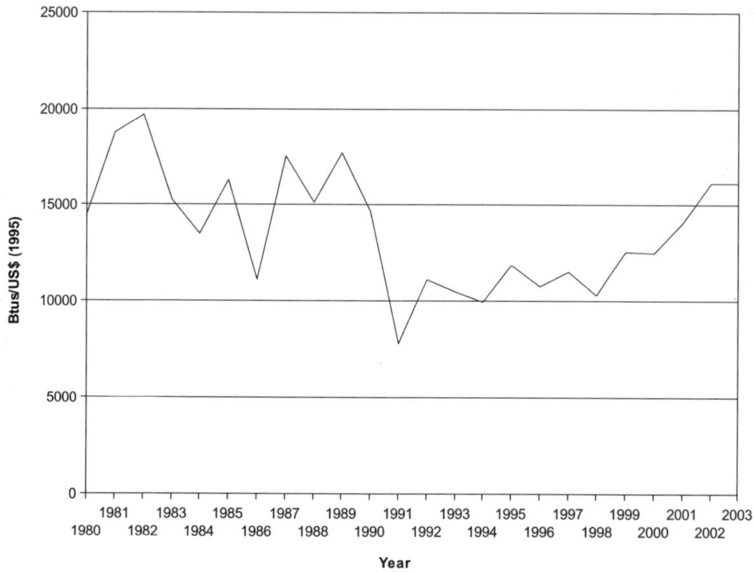

Figure 4: Energy Intensity in Brunei, 1980–2003

This is possibly due to energy inefficiency as well as reduced efforts to conserve energy. Its income elasticity in 2003 was also high (2.44).

Cambodia: Energy intensity decreased from 1986 onwards, reaching below 2,000 BTU/USD. However, the low energy intensity recorded for the early 1980s is dubious. It is not clear how it came to consume vast quantities of energy (and what kind of energy) from 1983 to 1986, soon after the second world oil shock. The income elasticity of energy consumption in 2003 was quite low (0.09) and therefore also suspect.

Indonesia: While the economy was growing rapidly apart from the years immediately following the Asian Financial Crisis which began in August 1997, its energy consumption was growing even faster. The income elasticity of energy consumption in 2003 was also high (1.70).

Laos: Energy intensity decreased steadily until 1998, but energy consumption grew rapidly from 1998 to 2000 as more energy became available, then stabilised. The income elasticity of energy consumption in 2003 was low (0.84).

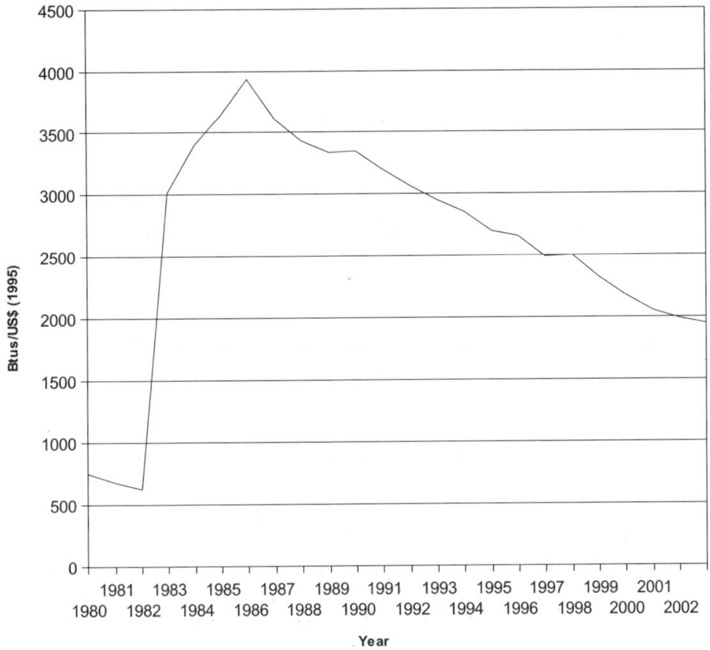

Figure 5: Energy Intensity in Cambodia, 1980–2003

Malaysia: Energy intensity increased from 1980 to 1986 but thereafter stabilised at around 17,000 BTU/USD. After 2000 it rose again, indicating that the economy became more energy intensive as it matured.The income elasticity of energy consumption in 2003 was high (1.81).

Myanmar: Energy intensity was generally low (below 1,100 BTU/USD) meaning the economy was not energy intensive (typical of an agricultural economy) and quite stable with a slightly increasing trend.The income elasticity of energy consumption was greater than 1 (1.37) indicating the economy was not very energy efficient.

Philippines: Energy intensity increased from 1983 to 1999 meaning the economy was using more energy to grow, but that energy efficiency was not exceptionally poor as the income elasticity of energy consumption in 2003 was less than 1 (0.97).

Singapore: Energy intensity was quite high in the early 1980s but dropped significantly thereafter. The country's energy efficiency was

Figure 6: Energy Intensity in Indonesia, 1980–2003

Figure 7: Energy Intensity in Laos, 1980–2003

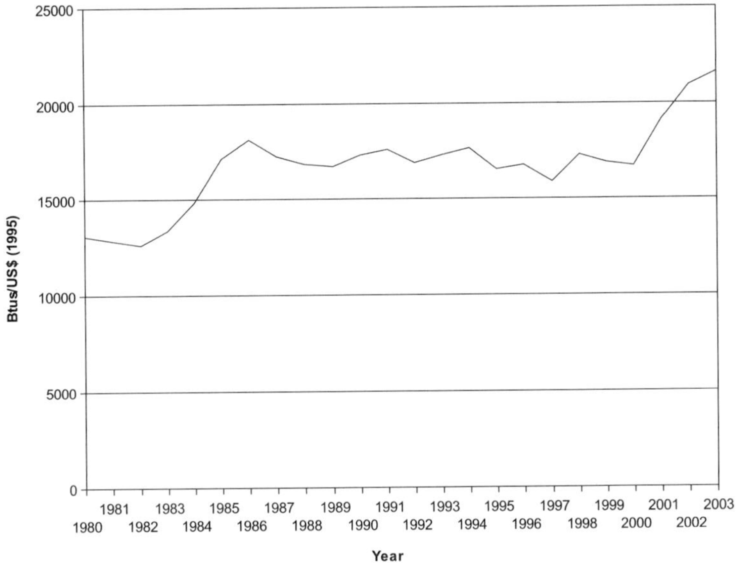

Figure 8: Energy Intensity in Malaysia, 1980–2003

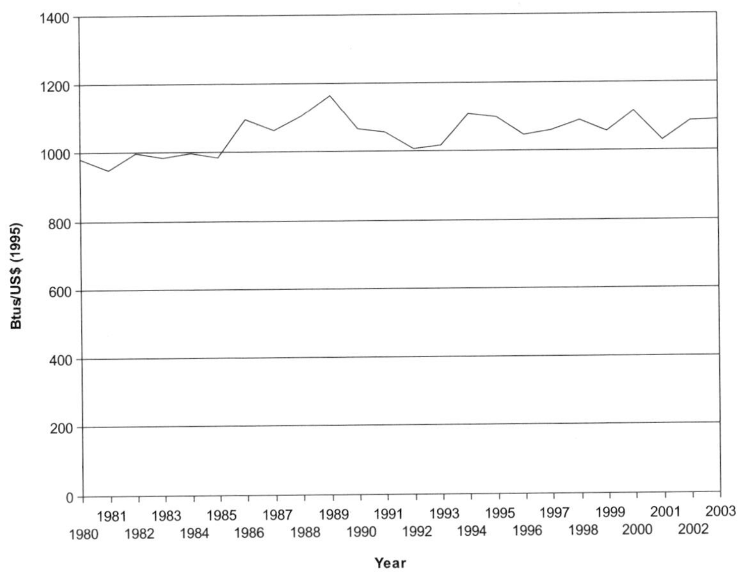

Figure 9: Energy Intensity in Myanmar, 1980–2003

Figure 10: Energy Intensity in Philippines, 1980–2003

Figure 11: Energy Intensity in Singapore, 1980–2003

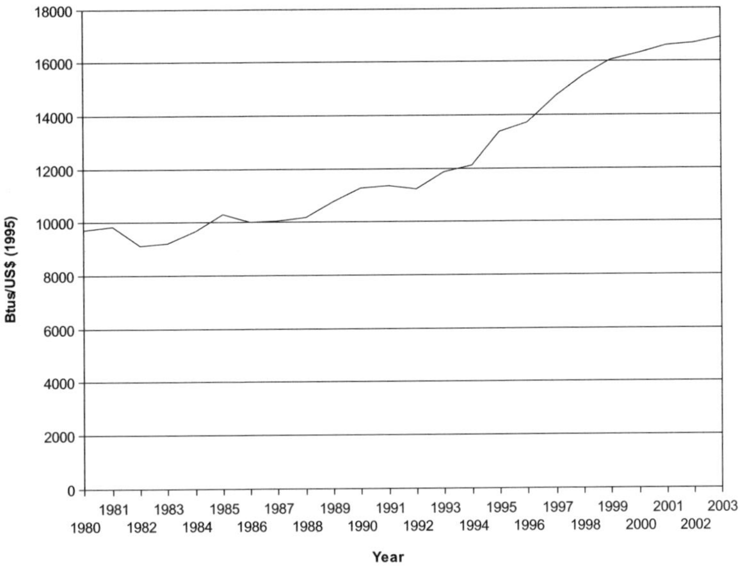

Figure 12: Energy Intensity in Thailand, 1980-2003

Figure 13: Energy Intensity in Vietnam, 1980-2003

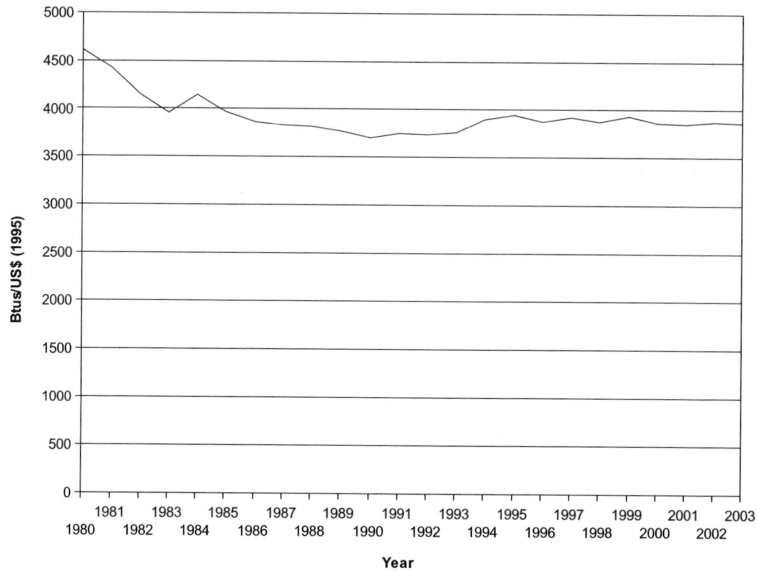

Figure 14: Energy Intensity in Japan, 1980–2003

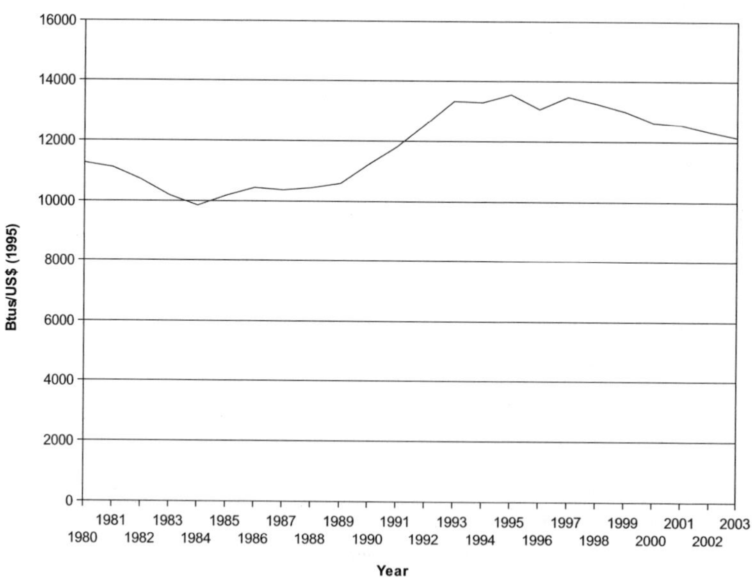

Figure 15: Energy Intensity in Korea, 1980–2003

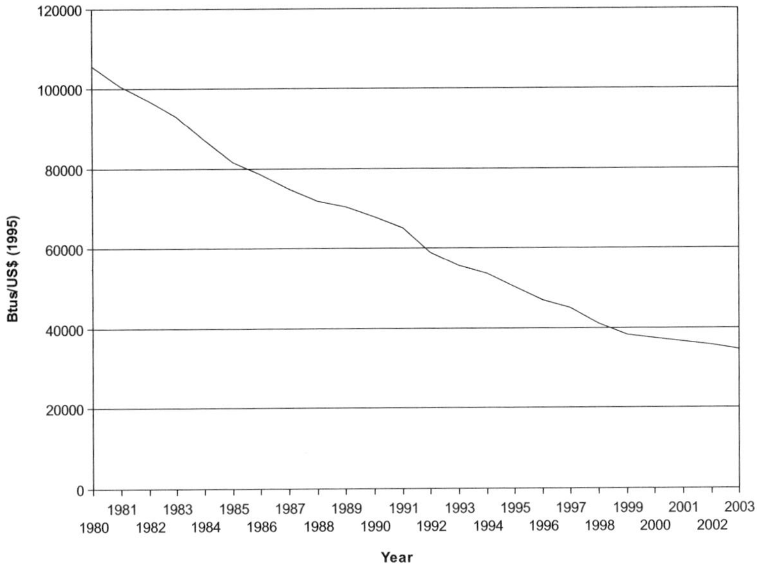

Figure 16: Energy Intensity in China, 1980–2003

reasonably good, as indicated by the income elasticity of energy consumption which in 2003 was lower than 1 (0.92). In 1986–7, 1993–4 and 1998, energy consumption surged reflecting perhaps the fact that more industries and oil refineries were introduced and more people bought private vehicles. In recent years it declined slightly.

Thailand: Energy intensity increased over the period by more than 60 per cent implying that the economy became more energy intensive and more people were buying home appliances. The economy's energy efficiency was poor as reflected in the income elasticity which in 2003 was close to 2 (1.84). In other words, energy consumption grew almost twice as fast as the economy.

Vietnam: Though energy intensity fluctuated around 20,000 BTU/USD, there was generally an increasing trend. Income elasticity of energy consumption was quite high, meaning that the economy required twice as much energy to fuel economic growth (2.01).

Japan: Energy intensity was low and stable at around 4,000 BTU/USD, indicating that the economy was very energy efficient as manifested by the low income elasticity of energy consumption (0.59). In 1990, energy intensity was at its lowest (3,703 BTU/USD). Thereafter it rose because more Japanese people used private cars.

Korea: Following the two world oil shocks, energy intensity decreased but industrial restructuring brought in more energy intensive manufacturing. This caused it to increase and peak around 1995 and 1997. From 1998, energy intensity decreased implying an improvement in energy efficiency due to energy conservation efforts. However, more energy was required for economic growth as income elasticity of energy consumption in 2003 was larger than 1 (1.13).

China: Energy intensity here in 2003 was the highest of the ASEAN + 3 countries at 34,602 BTU/USD. However, income elasticity of energy consumption was the lowest (0.26), apart from the doubtful case of Cambodia. For over 20 years, energy consumption doubled while economic growth increased more than seven times. Energy efficiency improved as a result of major structural changes in the economy which reduced the emphasis on heavy industry. The government has also had some success in its energy conservation policies. Much further work is required.

INCOME ELASTICITY

If energy consumption growth is higher than GDP growth, the income elasticity of energy consumption is larger than 1. The countries with high income elasticity of energy consumption in 2003 were Brunei, Indonesia, Malaysia, Myanmar, Thailand and Vietnam (see Figure 17). This means that they need more energy to grow and indirectly implies that there is considerable room for them to improve their energy efficiency. Those with lower income elasticity of energy consumption were Cambodia, Japan and China, but China is a special case as its GDP growth has been far higher than its energy consumption growth. The data for Cambodia is problematic, indicating that energy consumption did not change much from 1987 to 2003.

Countries such as Laos, the Philippines, Singapore and Korea had close to unitary elasticity, implying that almost the same proportion

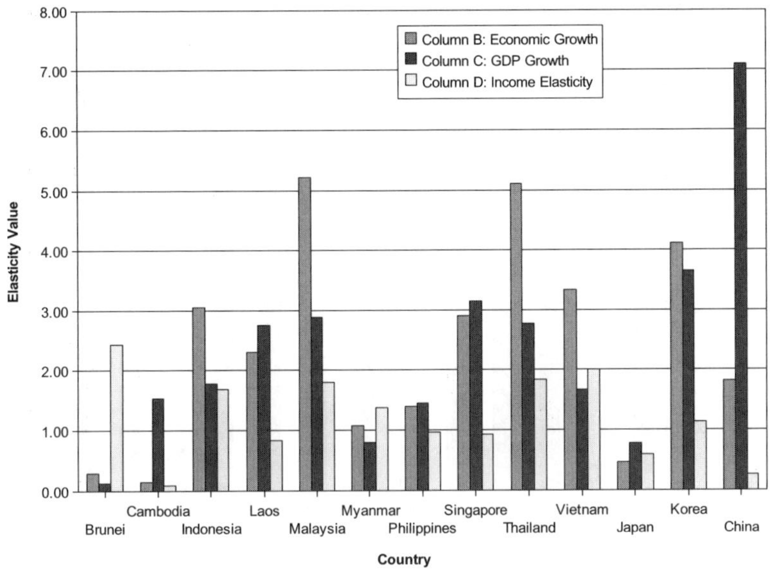

Figure 17: Energy, GDP and Income Elasticity of Energy Consumption, 1980–2003

of energy was required to increase one unit of GDP. Cambodia, Japan and China had far lower than unitary elasticity, meaning that they needed less energy to increase one unit of output. However, unlike the higher income elasticity of energy consumption case, it does not necessarily mean the economy is energy efficient. For Cambodia or Laos, it could mean that the economy is at a very low level in terms of using energy or that it does not require much energy as an input for its economy. Japan, on the other hand, is extremely energy-efficient. For countries such as the Philippines, Singapore or China, this measure gives mixed signals in terms of energy efficiency.

THE RELATIONSHIP BETWEEN ENERGY INTENSITY AND INCOME ELASTICITY

There was seemingly a relationship between energy intensity and income elasticity of energy consumption in 2003: the higher the energy intensity, the higher the income elasticity of energy consumption. However, there were a few exceptions (see Figure 18). Myanmar's energy intensity (1,090 BTU/USD) was quite low but its income

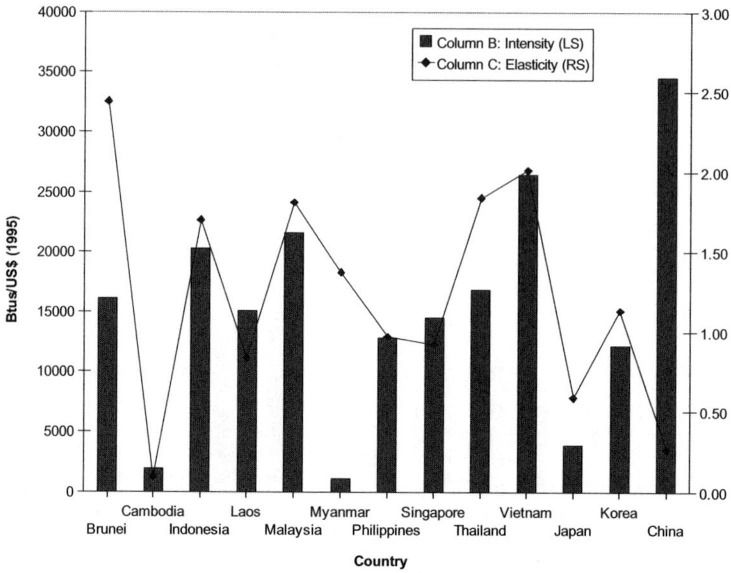

Figure 18: Energy Intensity (2003) and Income Elasticity of Energy Consumption in ASEAN + 3

elasticity of energy consumption was quite high (1.37). This could be interpreted to mean that the economy had low energy efficiency and was simultaneously energy-intensive.

China was another extreme case. Its energy intensity (34,602 BTU/USD) was the largest among the 13 countries but its income elasticity of energy consumption was the lowest (0.26). For perspective, the energy intensities for the same year for the US, UK and Australia were 10,575, 7,039 and 11,936, respectively.

Generally speaking, the countries with high energy intensity in 2003 also had high income elasticity of energy consumption. Brunei, Indonesia, Laos, Malaysia, the Philippines and Singapore for example, had both high energy intensity and high income elasticity. Korea is clearly a high energy-intensive economy.

Cambodia and Japan had both low energy intensity and low income elasticity, but these two countries represent two extreme cases. Japan is a highly energy-efficient economy while Cambodia is a case where the level of energy demand is simply low because the economy is not yet using much commercial energy.

China is an abnormal case where energy intensity in 2003 was high but the economy showed low income elasticity. Its economic growth overwhelmed its energy consumption growth, and its energy intensity had a decreasing trend. This does not, however, indicate that China is energy efficient, though it is possible that its efficiency has improved.

GOALS AND SCOPE OF ENERGY PROGRAMMES TO DATE: A BRIEF OVERVIEW

Dozens of conservation programmes have been launched in Asia over the past 30 years. For each country there are huge tomes of information available detailing when and how these programmes were launched and their relative successes. There have been many attempts to summarise these.[2] Basically, these programmes have together aimed to:

a. reduce energy consumption,
b. improve energy efficiency,
c. decrease carbon emissions.

Progress is now being made in:

a. Constructing a basic set of energy efficiency indicators for all countries.
b. Constructing specific indicators for specific sectors (industrial, residential, services and transport).
c. Developing a network of energy efficiency experts and agencies.

Table 1 lists the main energy conservation products and services available in the world. The products are the various types of industrial, commercial and residential equipment which have been specifically designed to use as little energy as possible. The services are

---

[2] For example, Peter Rumsey and Ted Flanigan, *Compendium: Asian Energy Efficiency Success Stories*, published for the International Institute for Energy Conservation by Global Energy Efficiency Initiative, Washington, DC, 1995; UN, *Energy Efficiency: Compendium of Energy Conservation Legislation in Countries of the Asia and Pacific Region,* New York: UN, 1999; *Energy Efficiency Indicators: A Study of Energy Efficiency Indicators in APEC Economies,* published by the Asia Pacific Energy Research Center and Institute of Energy Economics, 2001; Ming Yang and Peter Rumsey, "Energy Conservation in Typical Asian Countries", *Energy Sources* 19, no. 5 (1997): 507–21.

Table 1: Main Energy Conservation Products and Services

Products	Services
Motors, pumps, drives, compressors	Demand-side management consulting
Insulation and building materials	Building and system design software
Heating, ventilation, air conditioning	Energy performance contracting
Lighting equipment	Process improvement engineering
Office equipment	Demonstration projects
Cogeneration/combined heat and power	Labelling programmes
District heating systems	Financial incentives
Windows	Energy audits
Household appliances	MEPS (minimum energy performance standards)
Sensors and control systems	Building codes
Distribution transformers	
Control and metering equipment	
Industrial process equipment	
High-efficiency transport technologies	

Source: Adapted from International Institute for Energy Conservation with the Export Council for Energy Efficiency, *Developing and Financing Energy Efficiency Projects and Ventures in Emerging Markets.* IIEC Publications: Washington, DC, 1998, p. 2.

management packages which enhance energy consumption efficiency in businesses, industry and the home.

As clearly evident from the previous section, Japan's efforts at conserving energy have, by a wide margin, been the most successful. Indeed, the Japanese economy is highly respected as one of the most energy efficient in the world. Korea and Singapore are the next most energy efficient economies in the region. The historical, geographical, economic, etc. factors determining these countries' relatively high energy efficiency are obvious and do not need to be discussed here.

THE MAIN PROBLEMS FACED IN ASEAN + 3

Much has been written about the problems faced in introducing energy conservation programmes in Asia. Many government officials have rightly complained that while there has been a great deal of discussion and research about energy conservation, little consistent and effective action has been taken. There has been generally plenty of good intent in launching the various energy conservation programmes, but problems have been encountered almost immediately in their implementation. These include limited capital for start-up, maintenance and enforcement of the programmes. Moreover, the projects are often too small, become bogged down in bureaucracy, and the legal and regulatory frameworks are often not compatible with energy efficiency investments. The widely varying conditions (social, financial, physical, language, etc.), among the Asian countries also complicates progress.[3]

Energy consumption behaviour is greatly influenced by price. Fundamental in energy conservation work is pricing the energy at its real value. However, in Asia, fuel prices have long been subsidised. As almost anywhere in the world, price reform is a very complex problem, inextricable from a nation's price/wage structure. No government is keen to face the inevitable anger of consumers when fuel subsidies are lifted.[4]

In the lesser developed areas of Asia, the programmes have been further affected by the fact that low income groups spend a disproportionate share of their income on energy. These people are generally not in a position to choose which type of energy they use or how they use it. Their attention is focused on simply surviving, and their governments are unable to assist as their tax revenues are continuously fully stretched merely trying to provide the most basic human needs of food and water, and to avert civil strife, epidemics, etc.

On the bright side, however, there is tremendous potential for the poorer members and areas of ASEAN + 3 to learn from the experiences and mistakes of the richer areas within the region and from the west. There is also enormous potential for them to 'leap-frog' technologically speaking, i.e., to employ the latest energy conservation technologies developed by other countries instead of using valuable resources to develop their own prototypes.

---

[3] In many parts of Asia, about half of the businesses and industries are small-scale family-run enterprises, while the rest are huge government-run enterprises. However, most energy conservation programmes are most easily introduced, managed and monitored by medium-sized enterprises.

[4] For example, in Indonesia see Urip Hudiono, "US$966 Million Lost in Subsidised Fuels to Smuggling", *Jakarta Post,* 11 Oct. 2005 and "Easing the Chocks", *The Economist* (1–7 Oct. 2005): 28. For China, see, Hugh Dent, "China: Where Has All the Oil Gone? IEA Wonders", at Petroleumworld.com, 12 Aug. 05 [13 Oct. 05].

THE CRUX IS PETROL SAVING

It is difficult to suggest what ought to be the priorities in energy conservation work in the ASEAN + 3 region. From the calculations discussed above, the wide gap between the high energy efficiencies achieved by Japan and the rest of Asia is clear. Thus, all conservation initiatives are important and worthwhile if Asian countries are ever to use fuel as efficiently as does Japan, a world leader in this area. As the main theme of this book is energy security, the authors decided that instead of examining Asia's energy inefficiency in a range of industries such as cement, steel, etc., it would be most useful to address one of the key energy problems, namely the region's inefficient consumption of transport fuels.

As stated in Chapter 1, the demand for electric power can be met in the region using its own coal, hydropower and gas with support from Australia. The region, cannot however, and likely never will, be able to produce the amounts of petroleum it requires. Asia is already importing a large share of the world's oil production (37.3 per cent of global crude imports and 37.0 per cent of global oil product imports) and this share is expected to rise severalfold and very quickly in the near future.[5]

Most of the region's petroleum is now and will continue to be used for transport. Table 2 indicates that in Asia as a whole, 47.2 per cent of all petroleum products were used in transport in 2002/2003, compared to a world average of 58.0 per cent, this despite very high proportions in Brunei (82.9), Malaysia (67.2) and Myanmar (77.6). These differences are due to differences in economic structure.

From Table 3 it can be seen that the use of petroleum products for transport use has increased sharply in Asia. Between 1970 and 2003 the increase for all of Asia was 473.5 per cent compared to 206.1 per cent

---

[5] These figures, for China, Japan and the Asia Pacific, were calculated from *British Petroleum Statistical Review of World Energy*, June 2005.

Table 2: Use of Petroleum Products for Transport Use in 2002/2003 (thousand tons oil equivalent)

	Total Final Consumption (TFC)	Total Transport	% (TFC)	Road Transport	% (TFC)	Air Transport	% (TFC)	Rail Transport	% (TFC)
Brunei (2002)	461	382	82.9	298	64.6	84	18.2	—	
Indonesia (2003)	47,369	23,979	50.6	20,683	43.7	867	1.8	—	
Malaysia (2003)	21,172	14,224	67.2	12,371	58.4	1,854	8.8	—	
Myanmar (2002)	1,484	1,151	77.6	1,054	71.0	75	5.1	—	
Philippines (2003)	13,609	8,902	65.4	7,085	52.1	878	6.5	306	2.2
Singapore (2002)	9,633	4,947	51.4	2,010	20.9	2,937	30.5	—	
Thailand (2003)	31,586	19,914	63.0	16,731	53.0	3,005	9.5	106	0.3
Vietnam (2002)	8,756	4,943	56.5	4,441	50.7	283	3.2	—	
China (2003)	215,176	83,709	38.9	57,079	26.5	6,271	2.9	10,779	5.0
Japan* (2003)	217.50	91.64	42.1	76.66	35.2	10.69	4.9	0.22	
Korea* (2003)	87.62	33.80	38.6	25.91e	29.6	3.32	3.8	0.33	
Asia* (excluding China) (2002)	290.92	137.23	47.2	113.97	39.2	16.83	5.8	—	
OECD* (2003)	1,975.35	1,231.00	62.3	1,034.88	52.4	156.18	7.9	16.33	0.8
Non-OECD* (2003)	1,123.08	566.85	50.5	465.35	41.4	63.41	5.6	—	—
World* (2002)	3,098.43	1,797.85	58.0	1,500.23	48.4	219.59	7.1	—	

Sources: IEA, *Energy Balances of Non-OECD Countries 2002–2003* and *Energy Balances of OECD Countries 2002–2003*. Paris: OECD/IEA, 2005.

Note: * Unit is millions of tons oil equivalent.

Table 3: Consumption of Oil in Transport, 1971/3 and 2003 (million tons oil equivalent)

	1971	2003	% Increase
Brunei	0.03	0.39	1,200.0
Indonesia	2.72	23.98	781.6
Malaysia	1.50	14.22	848.0
Myanmar	0.52	1.36	161.5
Philippines	2.40	8.90	270.8
Singapore	0.74	4.96	570.3
Thailand	2.33	19.91	754.5
Vietnam	3.32	5.56	67.5
China	7.11	83.71	1,077.4
Japan	41.26*	91.64	122.1
Korea	2.58*	33.80	1,210.1
Asia	24.67	141.49	473.5
Non-OECD	184.90	566.86	206.1
OECD	610.95	1,231.00	101.5
World	795.85	1,797.86	125.8

Sources: IEA, *Energy Balances of Non-OECD Countries, 2002–2003.* Paris: OECD/IEA, 2005, p. II.299–300; *Energy Balances of OECD Countries, 2002–2003.* Paris: OECD/IEA, 2005, p. II.194.
Note: * Data is for 1973.

for non-OECD countries, 125.8 per cent for the world and just 101.5 per cent for the OECD countries.

Asia is not likely to see complete conversion to non- or hybrid-petrol-fuelled cars (powered by electricity, natural gas, ethanol, hydrogen, etc.) in the next 15 years. As per capita incomes in this part of the world steadily rise, and more and more Asians aspire to the 'middle class' life of the west, complete with a family vehicle or two, it is imperative for the governments of Asia to enforce the strictest energy conservation and environment-friendly practices now. As the western countries have already been so polluting in their modernisation and maturity, the consequences of Asia's buying cars at the same rates as in North America and Europe are staggering from both energy supply and environmental points of view.

MUCH CAN BE DONE TO REDUCE CONSUMPTION OF ROAD TRANSPORT FUELS[6]

The use of transportation fuels differ greatly in Asia from that in North America or Europe simply because a much smaller proportion of the population in Asia can afford to buy and maintain a private vehicle. The relative cost of doing so is many times higher. The majority of adults living in Asia at this time will never own a private vehicle in their lifetime. This is not to say, however, that absolute numbers of private vehicles are not already high as can be seen from Table 4 which gives comparative car ownership statistics for ASEAN + 3. China currently has about 17 million private cars, Japan has about 42.6 million and Korea has about 10.3 million. For the sake of comparison, the UK has about 25.7 million and Australia, 10.6 million.[7]

Car ownership per 1,000 population is obviously highest in the mature economies of Japan and Korea at 335 and 216, respectively. Malaysia is a distant third with 212 cars per 1,000 population. For perspective, in the US, Australia and the UK it was 533, 464 and 433, respectively. It would most certainly be higher in Singapore if the country's small geographical size did not make it necessary for the government to limit car ownership severely through 'certificates of entitlement' and heavy taxation.

In 2005, the IEA published a report on how fuel used for road transport could be reduced very quickly in the event of either a large-scale oil supply disruption or a small localised one.[8] It estimates how various types of demand restraint measures could reduce transport fuel consumption quickly for each IEA region: Japan and Korea, IEA Europe, US and Canada, and Australia and New Zealand (there were 26 IEA members at the time of writing). It also provides the mathematical formulae for calculating the effectiveness and cost of each (see Table 5 which ranks the various measures in terms of cost).

---

[6] Road transport fuels include: motor gasoline (unleaded and leaded), and diesel.

[7] For sources, see Table 4 and "China Encouraging Smaller Cars and Improved Fuel Efficienrcy", *World Watch*, 27 Jan. 2006 at http://www.evworld.com/view.cfm?section=communique&newsid=10833 [21 Mar. 2006].

[8] International Energy Agency, *Saving Oil in a Hurry*. Paris: OECD/IEA, 2005.

Table 4: Vehicle Numbers in Each Country

Brunei (registered vehicles, 2001)
Private cars = 188,720
Private cars per 1,000 population = 0.55
Goods vehicles = 17,828
Motorcycles and scooters = 7,162
Buses and taxis = 2,267
Others = 4,470

Cambodia (estimated number of motor vehicles in use, 2002)
Passenger cars = 209,128
Private cars per 1,000 population = 15.1
Buses and coaches = 3,196
Trucks = 29,968
Other vehicles = 421
Motorcycles and mopeds = 586,278

China (2003)
Passenger cars and buses = 14,788,100
Private cars per 1,000 population = 11.3
Goods vehicles = 8,535,100

Indonesia (2000)
Passenger cars = 3,038,913
Private cars per 1,000 population = 14.4.
Lorries and trucks = 1,707,134
Buses and coaches = 666,280
Motorcycles = 13,563,017

Japan (2002)
Passenger cars = 42,655,000
Private cars per 1,000 population = 334.6
Buses and coaches = 233,000
Trucks including trailers = 7,666,000
Special use vehicles = 1,720,000
Heavy use vehicles = n.a.
Light 2-wheeled vehicles = 23,266,000

Korea (2003)
Passenger cars = 10,278,923
Private cars per 1,000 population = 215.5
Goods vehicles = 3,016,407
Buses and coaches = 1,246,629
Motorcycles and mopeds = 1,730,193

(*Continued*)

Table 4: (*Continued*)

Laos (1996)
Passenger cars = 16,320
Private cars per 1,000 population = 3.4
Buses and coaches = n.a.
Lorries and vans = 4,200
Motorcycles and mopeds = 231,000

Malaysia (2002)
Passenger cars = 5,069,412
Private cars per 1,000 population = 211.5
Buses and coaches = 51,158
Lorries and vans = 713,148
Road tractors = 345,604
Motorcycles and mopeds = 5,842,617

Myanmar
Passenger cars = 175,400 (2001)
Private cars per 1,000 population = 3.6
Trucks = 42,828 (1996)
Buses = 15,639 (1996)
Motorcycles = 85,821 (1996)
Others = 6,611
Commercial vehicles = 98,900

Philippines (2004)
Passenger cars = 798,160
Private cars per 1,000 population = 31.7
Utility vehicles = 1,647,524
SUVs = 141,447
Buses = 35,003
Trucks = 267,977
Motorcycles and mopeds = 1,847, 361
Trailers = 23,121

Singapore (2004)
Cars (including private, company, tuition and private care hires) = 417,103
Private cars per 1,000 population = 98.0
Motorcycles and scooters = 136,122
Motor buses = 12,892
Taxis = 20,407
Goods vehicles (including private) = 126,709
Others = 14,162

(*Continued*)

Table 4: (*Continued*)

Thailand (2001)
Passenger cars = 2,281,000
Private cars per 1,000 population = 37.1
Buses and trucks = 804,000
Vans and pickups = 3,341,000
Motorcycles = 15,236,000

Vietnam (2003)
Incomparable categories

Sources: *The Europa World Yearbook 2005*. London: Routledge, 2005; IMF, *International Financial Statistics Yearbook* (IMF): Washington, DC, 2005.

Table 5: Ranking of the Costs of the Measures

Cost-Effectiveness Range	Measure	Oil Savings
Very Inexpensive (less than USD1 per barrel saved)	Car-pooling, driving ban, telecommuting, compressed workweek, ecodriving	Moderate to very large
Inexpensive (less than USD15 per barrel saved)	Reduced speed limits, driving ban	Large
Moderate cost (less than USD50 per barrel saved)	Bus-only lanes	Small
Expensive (more than USD100 per barrel saved)	Telecommuting (including purchase of computers for 50 per cent of participants), free public transit or 50 per cent fare reduction, increased weekend and off-peak transit service and increased transit peak service frequency by 10 per cent	Moderate to large

Source: Adapted from International Energy Agency, *Saving Oil in a Hurry*, Table E-2, p. 25.

Table 6 summarises the expected results of each of the various demand restraint measures for the four IEA regions. The calculations are based on the assumed occurrence of a 90-day disruption. It gives the savings in terms of billions of barrels per day, as a per cent

Table 6: Estimated Fuel Savings for Each IEA Region

	Japan/ RK	IEA Europe	US/ Canada	Aus/ NZ	Japan/ RK	IEA Europe	US/ Canada	Aus/ NZ	Japan/ RK	IEA Europe	US/ Canada	Aus/ NZ
Current road transport fuel consumption (2001), thousand bbls per day	2,101	5,643	11,816	528								
Current road petroleum fuel consumption (2001), thousand bbls per day	3,760	8,882	15,428	743								
	Fuel Savings (thousand bbls per day)				% of Road Transport Fuel Saved				% of Total Petroleum Fuel Saved			
Public transit:												
50% fare reduction	64	172	42	3	3.1	3.0	0.4	0.5	1.7	1.9	0.3	0.3
100% fare reduction	128	344	85	6	6.1	6.1	0.7	1.2	3.4	3.9	0.6	0.8
Off-peak service	59	96	31	3	2.8	1.7	0.3	0.5	1.6	1.1	0.2	0.3
Peak and off-peak service	74	117	38	3	3.5	2.1	0.3	0.5	2.0	1.3	0.3	0.3
Bus and HOV enhancement	3	11	4	0.2	0.12	0.19	0.03	0.05	0.07	0.12	0.02	0.03

(Continued)

Table 6: (*Continued*)

	Japan/ RK	IEA Europe	US/ Canada	Aus/ NZ	Japan/ RK	IEA Europe	US/ Canada	Aus/ NZ	Japan/ RK	IEA Europe	US/ Canada	Aus/ NZ
Bus and HOV expansion	5	21	7	0.5	0.24	0.38	0.06	0.09	0.14	0.24	0.05	0.07
Car-pooling infrastructure and programme	125	277	800	38	6.0	4.9	6.8	7.2	3.3	3.1	5.2	5.1
Car-pooling programme	13	41	112	6	0.6	0.7	1.0	1.1	0.3	0.5	0.7	0.8
Telecommuting	88	102	523	21	4.2	1.8	4.4	4.0	2.3	1.2	3.4	2.9
Compressed four-day workweek	61	71	367	15	2.9	1.3	3.1	2.8	1.6	0.8	2.4	2.0
Odd/even day driving ban	516	1,992	1,467	109	24.5	35.3	12.4	20.7	13.7	22.4	9.5	14.7
One day in ten driving ban	73	284	110	19	3.5	5.0	0.9	3.6	2.0	3.2	0.7	2.5
Speed limits at 90 kph	15	184	363	7	0.7	3.3	3.2	1.4	0.4	2.1	2.4	1.0
Ecodriving campaign	102	275	574	26	5.0	5.0	5.0	5.0	2.7	3.1	3.7	3.4

Source: International Energy Agency, *Saving Oil in a Hurry*. Paris: OECD/IEA, 2005, p. 116.

Note: bbls is billion barrels; HOV is high occupant vehicle.

of road transport fuel consumption and of total petroleum consumption.

According to the IEA's calculations, total road transport fuel consumption for all the IEA members in 2001 was 20,088,000 barrels per day and total petroleum consumption was 28,813,000 barrels per day.[9] They estimate that transport fuel savings resulting from the implementation of all of these transport demand restraint measures would amount to 10,092,700 barrels per day or 50.2 per cent of total daily transport fuel consumption. Moreover, besides saving energy, the demand restraint measures would reduce greenhouse gases (GHG). Every litre of gasoline consumed produces 2.4 kilograms of carbon dioxide.[10]

Not surprisingly, by far the most effective measure in all the IEA regions was the odd/even day driving ban (described below), i.e., completely curtailing driving for periods of time. The second, third and fourth ranking measures aimed at limiting traffic differed considerably among the IEA regions due to varying levels of fuel taxation, geographical factors, social factors, etc.

APPLICATION OF THESE METHODS TO THE ASEAN + 3 REGION

A Closer Look at the Results for Japan and Korea

From Table 6, it can be seen that total potential road transport fuel savings in 2001 for Japan and Korea alone were 1,326,000 barrels, or 63.1 per cent of their total road transport fuel saved. The potential savings for Japan and Korea alone represent 6.6 per cent of total IEA members' road transport fuel consumption.

Obviously, if these two countries alone, and more importantly, two countries which are already very fuel efficiency conscious, could potentially save such a large amount of fuel,

---

[9] These figures are the sums of the first two rows in Table 6.

[10] See Canadian Ministry of Natural Resources, Office of Energy Efficiency Website at http://oee.nrcan.gc.ca/communities-government/transportation/municipal-communities/articles/idling-tips.cfm?attr=28 [17 Oct. 2005].

the savings in the rest of Asia could also be expected to be very significant.

Applicability of These Methods to the Rest of the ASEAN + 3 Region

In the absence of the data required to carry out the precise fuel savings as measured by the IEA, it is nonetheless useful to speculate here on the utility of each of the measures:[11]

a. Public transit

50 per cent fare reduction
100 per cent fare reduction
Off-peak service
Peak and off-peak service
Bus and HOV enhancement
Bus and HOV expansion

Reducing public transit fares by 50 or 100 per cent, providing off-peak service and creating dedicated lanes for buses and high occupant vehicles (HOV) are not likely to result in much saving of fuel in much of Asia simply because the vast majority of people rely on public transport or their own leg power (walking or pedal bicycles). Most Asians have little choice but to take public transport or use their own leg power. As noted above, this is not to say that in absolute terms there is not already a large number of passenger vehicles on the roads. It must also be recognised that the ratio of public transit commuters to vehicle commuters is rapidly changing. The governments, ideally, need to keep public transit fares to the lowest possible. This is discussed further in the conclusion.

[11] These speculations are for the years 2005–2006. In the ensuing years, it is expected that social conditions, including population growth, per capita income, etc., will change drastically, and concomitantly greatly affect fuel consumption behaviour.

b. Car-pooling

A car-pooling programme involves creating a system for matching potential car-poolers, and also possibly dedicated car-pooling lanes and parking. Such a programme would not likely make a large impact on Asia's transport fuel consumption at this time because the number of private vehicles available for participation in car-pooling schemes relative to the number of people who do not have cars is small. The only exceptions where it might have an effect are Singapore, and possibly around the region's mega-cities such as Bangkok, Kuala Lumpur, Jakarta, Manila, etc.

c. Telecommuting

Telecommuting, i.e., having people who work at computers all day work at home for part of the week, would likely have a measurable impact in Singapore especially because it is already one of the most 'wired' countries in the world. Telecommuting would also result in considerable fuel savings in and around the region's mega-cities such as Bangkok, Kuala Lumpur, Jakarta, Manila, etc. The proportion of people working all day at terminals will likely continue to increase rapidly over the next decade.

d. Compressed four-day workweek

Reducing the number of commutes to work to only eight rush-hours per week instead of ten or twelve (by private vehicles, taxies, buses, motorcycles, tuktuks or jeepneys)[12] by having people work one day less at their offices, factories, etc., would definitely result in transport fuel savings. It would also have a positive impact in terms of pollution reduction.[13]

---

[12] The term 'tuktuk' is used in several Asian countries. It refers to taxi-type transport using a 3-wheeled vehicle which looks like a motorcycle with a covered wagon on the back. In the Philippines, 'jeepneys' serve the same purpose to carry 20–30 passengers. These are converted US jeeps.

[13] In some parts of Asia, some workforces are still required to work half days on Saturdays, hence twelve rush-hours per week.

e. Odd/even day driving ban

From Table 6, it can be seen that in all the IEA country groupings, reducing consumption of transport fuel by enforcing driving bans, i.e., by plate number, odd/even days, alternate Sundays, etc., was by far the most effective measure in terms of total fuel saved. The complete prohibition of driving of private vehicles taxis, buses, motorcycles and tuktuks for certain periods is similarly expected to have a measurable impact on the ASEAN + 3's total consumption of transport fuels.

f. One day in 10 driving ban

As for items (d) and (e) above, the expected fuel savings would be high.

g. Speed limits set at 90 kph

Setting speed limits at 90 kph would not greatly reduce fuel consumption in Asia because the number of private cars vis-à-vis public transport is still so low. It would have little effect on fuel consumption by taxis, tuktuks and buses because they do not normally travel at great speed anyway due to heavy traffic congestion. Setting speed limits for motorcycles, however, would likely have a measurable impact.

h. Ecodriving campaign

'Ecodriving' refers to the enforcement of efficient driving and maintenance habits. Of all the fuel-reduction measures listed in Table 6, apart from outright driving bans, this would likely be the most effective in Asia at this time. Japan and Singapore are well-known for their tight laws on driving and maintenance, but in most of the rest of Asia, driving instruction is poor, i.e., people are not taught and continuously urged (via the pricing mechanism) to drive with minimal fuel use, i.e., in acceleration/deceleration and idling, maintaining minimal weight in the vehicle, maintaining a clean air filter, etc. The regulations for maintaining private vehicles, taxis, buses, motorcycles and tuktuks range from minimal to none at all and several governments in Asia have provided large fuel subsidies which actually *encourage* drivers to be nonchalant or unaware of their low fuel consumption efficiency.

A poorly maintained vehicle can cause consumption to be up to 50 per cent higher than normal.[14] For example, a clogged air filter can increase fuel consumption by ten per cent. Having one tyre under-inflated by 6 pounds per square inch (40 kiloPascals) can increase consumption by three per cent. A poorly-tuned engine consumes up to 15 per cent more fuel when idling than a well-tuned one.

The IEA estimated that in 2001 a total of 21,656 million litres of fuel could be saved in the IEA member countries from a tyre inflation campaign, i.e., having people keep the tyres of their vehicles (including buses and heavy goods vehicles) at the optimum pressure level. In Japan and Korea, the total was 2,638 million litres.[15] If such a campaign were applied across Asia, the results would likely be quite significant, especially if motorcycles, public buses, taxis and tuktuks, were also given equal, if not more attention.[16]

Other Possible Transport Fuel-Saving Measures Which Would Likely be Effective in Asia

a. Road pricing policy

Placing a direct fee based on vehicle-kilometres travelled by private cars, motorcycles and tuktuks could result in large fuel savings. However, the cost of installing a tamper-proof gauge on all such vehicles would be costly.

b. Ban on imports of old, fuel-inefficient cars, buses, motorcycles and tuktuks

Banning the importation from foreign countries of older models of vehicles would greatly reduce fuel consumption in Asia. The older models now shipped from Japan and Singapore to the rest of Asia are

[14] These statistics are from Natural Resources Canada, Office of Energy Efficiency at http://oee.nrcan.gc.ca/transportation/personal/index.cfm?attr=0 [17 Oct. 2005].

[15] International Energy Agency, *Saving Oil in a Hurry*, Table A-19, p. 156.

[16] A survey carried out in 2003 revealed that over two-thirds of Canada's personal vehicles had at least one tyre that was either under- or over-inflated. See http://oee.nrcan.gc.ca/transportation/personal/index.cfm?attr=0 [Dec. 2007].

usually much less energy-efficient than the newer ones replacing them. Technologically outdated equipment purchased at low prices, and thought to be a bargain, has frequently turned out, ultimately, to be very expensive in terms of fuel use and air pollution.

c. Park and ride lots

The construction of park and ride lots at this time would have an effect only in the wealthier parts of Asia, such as Singapore and around the above-mentioned mega-cities. However, it would be useful to construct these now throughout Asia so as to stem fuel consumption use as more and more people acquire cars.

d. Raise price of petrol

It is fact that those places in the world with the highest fuel prices have both the most efficient vehicles and most efficient vehicle use. At the time of writing, several Asian governments were facing public demonstrations against measures to lift fuel subsidies.

e. Restrict parking

Same as for item (c).

f. Congestion pricing

Restricting traffic in certain areas or closing it off entirely on some days or hours, would reduce private vehicle, taxi, motorcycle and tuk-tuk traffic, and would consequently likely be an effective way to reduce fuel consumption.

g. Encouraging walking and cycling

Walking or cycling to work is possible only in the more northern areas of Asia and in the winter seasons of the other areas simply because it is too hot. Before China's reforms, its citizens had long used bicycles as the main form of transport. In recent years, city roads have increasingly been designed more for motor vehicles instead of bicycles. If the Chinese Government is serious about saving fuel, it should ensure, as in countries such as Holland and Canada, that bicycle lanes are not taken over completely by cars. Japan and Korea ought to build more bicycle lanes. The added social advantage of bicycles besides less pollution is citizen's improved physical fitness.

h. Improving the quality of the roads

Fuel transport consumption efficiency is much higher on smooth roads than on poorly surfaced roads. In Asia, considerable fuel could be saved if all roads were paved. There are some 2.1 million kilometres of unpaved roads in the region (see Table 7).[17] Fuel consumption anywhere, is also higher on roads where there are many traffic lights and/or stop signs. Idling engines consume large quantities of fuel as do repeated accelerations from a dead stop.[18] Driving in Asia's

Table 7: Unpaved Highways in ASEAN + 3 (kilometers)

Brunei	0.00 (2000)
Cambodia	10,327 (2000e)
Indonesia	73,048 (1998)
Laos	12,052 (1999e)
Malaysia	15,942 (1999)
Myanmar	24,760 (1996)
Philippines	159,575 (2000)
Singapore	0.00 (1999)
Thailand	1,615 (1999e)
Vietnam	69,882 (1999e)
China	1,088,494 (2000)
Japan	627,423 (1999)
Korea	22,182 (1999e)
Total	2,105,300

Source: CIA, *World Factbook 2005*. Washington, DC: CIA, Supt. of Docs, USGPO, 2006.

[17] Calculated from CIA, *World Factbook*, 2004.

[18] For example, in Canada, a single car's idling for ten minutes per day consumes an average of 100 litres of gasoline per year. See Natural Resources Canada, Office of Energy Efficiency at http://oee.nrcan.gc.ca/transportation/idling/issues/why-idling-problem.cfm?attr=8#wastes [17 Oct. 2005].

mega-cities typically involves considerable stopping and starting because many lights and stop signs have been installed to allow pedestrians and cyclists to cross. The solution is to separate vehicular traffic from the other traffic as much as possible and to construct more overpasses (perhaps with escalators in the very hot climates) so that it is not necessary for vehicles to stop and start so frequently.

CONCLUSION

It is apparent from the analysis of the region's energy intensities that there is tremendous potential for energy conservation. Fortunately, Japan is a world leader in energy conservation products and management and is keen to help the rest of Asia raise its energy consumption efficiencies. The region's energy security would also be considerably raised if Japan could take a greater role in introducing energy conservation programmes in urban and rural areas alike. Moreover, international environmental concerns over global warming caused by excessive and inefficient burning of fossil fuels are quickly forcing governments to reduce energy consumption and the emissions resulting from consuming fossil fuels.[19]

Recommended measures for general conservation of energy:

Regional Public Education: More must be done to educate workforces and the public, in the ASEAN + 3 region as a whole, about the extent of the wastage and how to reduce it. Workers at all levels, not just management, must be made aware of their enterprise's consumption performance compared to the best and worst performers in their industry nationwide, and internationally. Similarly, residents must be made aware of the costs of leaving air conditioners and extra lights on, keeping refrigerator temperatures unnecessarily low, etc.

Regional Strict Enforcement of Energy Conservation Measures Already in Place: A fundamental priority for all levels of government

[19] For example, the Montreal Climate Summit, held in December 2005.

in ASEAN + 3 as a whole, over the next decade must be the rigorous enforcement of the energy conservation measures already passed.The penalties for non-compliance should be made more severe and the incentives for improved efficiency more attractive. More officials should be appointed to monitor energy use, and be empowered to set not merely guidelines but absolute standards and deadlines for improvement.

Regional Removal of Subsidies and Pricing Reform: As energy has often been heavily subsidised by governments in the region, its use has generally been highly wasteful. People may be quite unaware how wasteful their consumption behaviour is. Energy pricing reform is often a socio-politically painful and complicated step.[20] Ideally, it should be done simultaneously in the ASEAN + 3 as a whole. The demand for energy, including transport fuels, tends to be very price inelastic in the short run, i.e., increases in price do not alter consumption behaviours. It is only when the increase(s) are large and sustained that action is taken to drastically reduce consumption.

Recommended measures for conservation of transport fuels, in particular:

a. Carry out a study for the ASEAN + 3 region as a whole, to measure potential transport fuel savings on the basis of the IEA's work discussed here, or a comparable study.
b. Keep public transit fares as low as possible in order to make intercity trips considerably cheaper via this mode than by private automobiles or taxies.
c. Put measures into place which not only reduce consumption now, but stem the expected increases in consumption, i.e., more investment now in public transit and park and ride.

---

[20] There are indeed many people in Asia whose livelihood very much depends on their use of vehicles of some sort, either for transport of goods or people, or simply to make a long commute each day. Thus raising fuel prices often hurts the most vulnerable sectors of society.

d. Assist bus and taxi companies to convert their fleets to natural gas.
e. Carry out urban planning at the regional level with simultaneous planning for housing, transport, commerce, industry and parks, instead of separately as has been the case to date for many of Asia's cities.

As noted, there are already large numbers of private vehicles in Asia. It is difficult for governments to persuade people already driving cars to work and for leisure purposes to take public transport instead. Asian city governments, as do many North American and European city governments, must focus their efforts on measures aimed at encouraging people to commute to work via public transport, not private cars.

Indeed governments ought to subsidise public transport and improve the services in terms of scheduling and routes and otherwise make driving to work less desirable, thus dissuading people from buying cars and using them for the work commute. Throughout the world, public transport rarely earns profits, however, the costs of subsidising the construction and operation of public transport in the ASEAN + 3 region, in both the short and long runs, will be lower than the costs incurred in tacitly encouraging more and more people to commute by private vehicle. Like the metropolises of the west, such as London, Paris and New York, public transport in the ASEAN + 3 must be made sufficiently convenient location- and frequency-wise, as well as inexpensive so as to constitute a viable, if not a superior alternative to driving private vehicles.

Chapter

3

Asian Energy Partnership: Opportunities and Obstacles

Yasuo Tanabe

WHAT IS 'ENERGY SECURITY' AND HOW CAN WE ACHIEVE IT?

When we identify energy challenges, the notion of 'energy security' should be defined broadly in the modern context as the need to secure a stable energy supply at a reasonable cost in an environmentally friendly manner. Energy security is not just about ensuring a physical supply of oil, as envisaged after the two 'oil shocks' in the 1970s and early 1980s. Since the mid-1980s, international oil markets have evolved and oil supply decisions have been made largely on the basis of price signals in the spot and future markets. Prices rise in a tight market and fall in a glutted market. Compared with the pre-market era (up to the mid 1980s), price volatility has increased in recent years and has also become a target of energy security policy. Also, there has been growing awareness and concern, globally, regionally, and nationally about the environmental impact of energy consumption, especially air pollution from SO_x

and NO_x emissions and climate change from CO_2 emissions. The environmental aspect, often called sustainability, has also become an important objective of energy security policy. The modern concept of energy security, therefore, should be defined in the broader terms of physical, economic and environmental aspects of energy supply.

There is a widely shared view among energy experts that energy security can be maintained through increasing the 'resiliency' of economies and the international market to changes in the terms of energy supply which can mean anything from changes in price, physical availability or the quality of the commodity.[1] 'Resiliency' is the virtue that allows a system to rebound, adapt or adjust itself in response to pressures from the external world. It implies not only a degree of strength, but also the ability to evolve in response to changing conditions. Governments have a role in designing the market and monitoring the market players, such as regulation for safety and the environment. Those with good political leadership can overcome parochial nationalism, laying the groundwork for an international market where business players can act on economic terms to promote international cooperation.

WHY AN ENERGY PARTNERSHIP IN ASIA?

As background to the Asian energy challenge, we must first acknowledge the fundamental fact that economic integration has progressed on a market basis in Asia, particularly in East Asia. Although energy experts tend to view energy issues from energy sector viewpoints, the most important approach in considering energy issues is to integrate them into the broader economic (and sometimes political) context. Thus Asian energy issues in the 21st century should be viewed from the standpoint of Asian economic integration which has evolved dramatically in recent years.

---

[1] FACTS Inc./EWCI Ltd, "Paths to Achieving ASEAN + 3 Energy Cooperation", Unpublished Report Commissioned by the Institute of Energy Economics Japan, 2004.

Interdependence through trade and investment among Asian economies has led to the build-up of a highly sophisticated network of production and distribution in the region. No one can deny that the shared goal for East Asian economies is to sustain and strengthen this network towards sustainable economic growth and prosperity.

East Asia's economic development has historically progressed via the 'flying geese' model of trade and investment linkages among developed economies, in particular, Japan and developing economies — first with NIEs, then ASEAN and China. South Korea enjoyed high economic performance through the 1960s, 1970s and 1980s following a similar pattern to the Japanese economic model. After the appreciation of the yen that resulted from the Plaza Accord in 1985, Japanese overseas investment rushed to the NIEs and ASEAN countries, leading to the current production networks prevalent throughout East Asia. South Korean investment has also surged into ASEAN. China started its 'reform and opening policy' in 1978 and attracted massive foreign direct investment especially from the 1990s. To date, China has recorded high economic growth comparable to Japan's in the 1950s and 1960s and is now often called the 'factory of the world'. However, judging from the widespread production networks in the region today, it is more accurate to say that East Asia, as a whole, is the factory of the world.

As a result of trade and investment-linked economic development in Asia, in particular East Asia, economic integration has progressed to a level of intraregional trade dependence of around 50 per cent, comparable to that of Europe in the 1970s and 1980s (see Figure 1). This business-led de facto economic integration is being followed by government-level efforts aimed at creating institutional economic integration through arrangements such as free trade agreements (FTAs). An FTA is often called an Economic Partnership Agreement (EPA) when a broader cooperation and institutional harmonisation agenda is included in addition to a trade liberalisation agenda. Government efforts are also fundamental in functional cooperation mechanisms, such as the 'Chiang Mai Initiative' (May 2000) in the currency field, especially after the Asian Financial Crisis of 1997–1998. Thus, it is

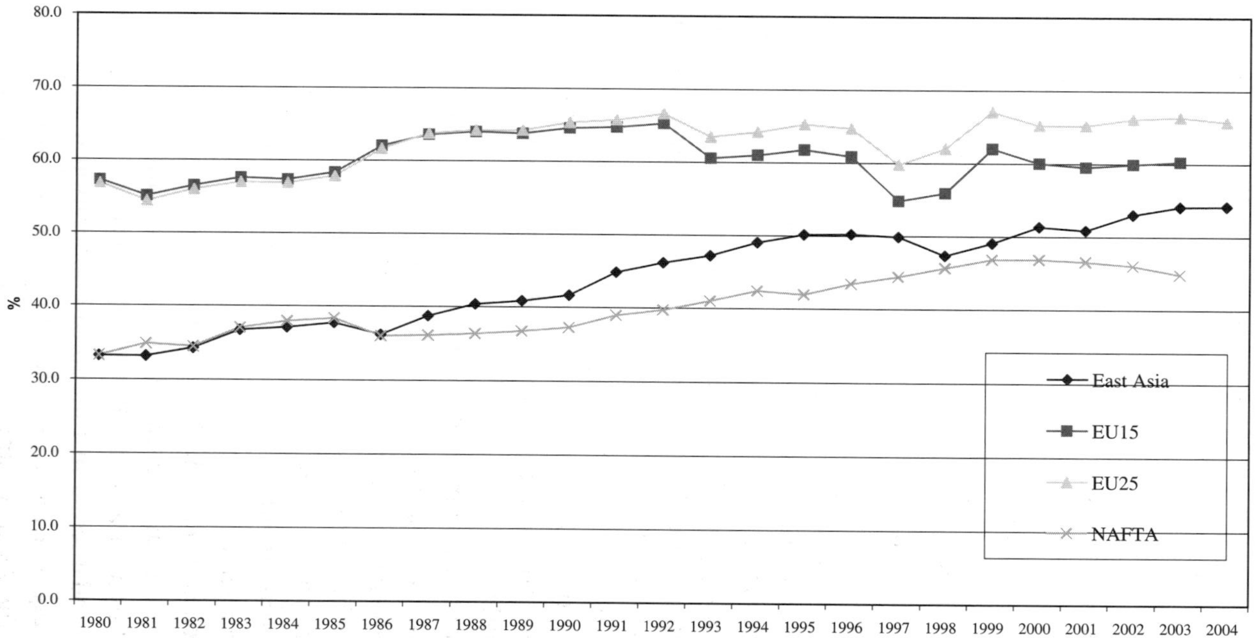

Notes: East Asia includes Japan, China, Korea, Hong Kong, Taiwan and ASEAN 10. The export and import data for each country and region based on Taiwan is for 1989–2003. However, the export and import data for Taiwan based on each country and region is for 1983–2003.
Source: DOT (IMF), *Trade Statistics* (Board of Foreign Trade, Taiwan, Chinese Taipei at http://eweb.trade.gov.tw/default.asp [3 Mar. 2006].

Figure 1: Intra-regional Trade Ratio of Each Region

imperative to look at this economic reality and the institutional arrangements in Asia when considering energy issues in Asia.

Against the above-mentioned background, Asia is the fastest growing and largest region in the world in terms of energy consumption. According to forecasts made by various international and national institutions, demand for energy in Asia will grow substantially. According to the International Energy Agency (IEA), energy demand in Asia will increase by 2.8 billion tons of oil equivalent (toe) from 2002 to 2030. This figure is larger than total current energy consumption in the US. Asia's share of world energy demand will increase from 30 per cent in 2002 to 36 per cent in 2030 (see Figure 2). There are other forecasts which predict even larger increases for Asia, especially China, based on higher economic growth and lower energy efficiency projections than those used by the IEA.[2] Supply for this growing energy demand cannot be secured without adequate investment in supply and transportation capacity within the supply chain from resource development to delivery to end-users. Among the energy sources, oil will be the major source of energy demand growth in Asia due to industrialisation and motorisation. As most of the world's oil reserves are held in the Middle East area, Asia will have to depend on production in and imports from the Middle East. The IEA forecasts that Asia's import dependence share of oil will increase from 62 per cent in 2002 to 83 per cent in 2030, most of which will be from the Middle East. This is the first basis of the energy security challenge for Asia.

Furthermore, as a second challenge, the environmental impact of the growing energy consumption will continue to increase. Air pollution and climate change are becoming serious policy challenges in the region. There are two causal factors. The first is poor energy efficiency except in Japan which has amongst the highest energy efficiencies in the world. Energy intensity measured as energy consumption per unit of GDP is a good indicator of energy efficiency.

---

[2] Li Zhidong, "Chugoku no EnerugiiKankyo no Doko" (Trends in Chinese Energy and Environment), in Yasuo Tanabe, *Ajia Enerugii Paatonaashippu* (Asian Energy Partnership). Tokyo: Energy Forum, 2004, pp. 42–45.

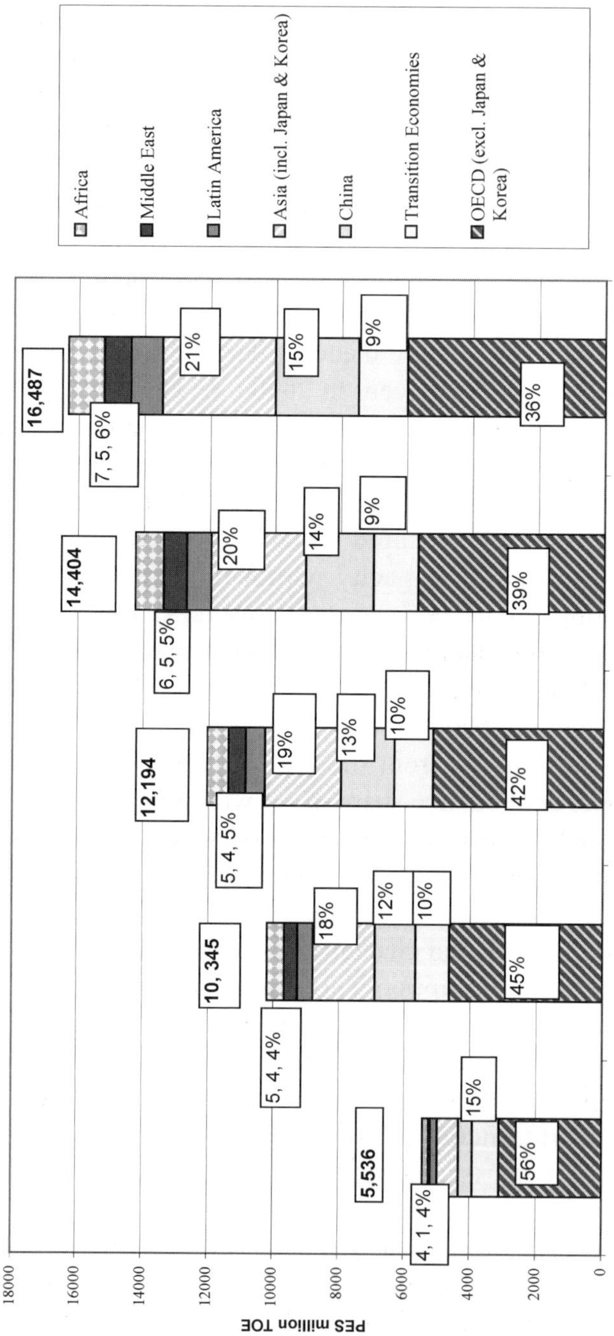

Figure 2: Asia's Position in the World Energy Demand Growth (PES million TOE)

Source: IEA, World Energy Outlook, 2004

Compared to Japan, which is the world leader, energy intensity of China is about nine times higher, ASEAN's is about four times higher and South Korea's about three times (see Figure 3). Higher efficiency would contribute to sustained economic growth, and also reduce the environmental impact of fossil fuel consumption. Therefore, energy conservation and energy efficiency should be the most important agenda items for energy security. The second causal factor is the use of coal. Although it is an economical and secure energy source, the environmental impact of its use is much larger than that of the other fossil fuels. Yet, certain Asian countries are expected to depend on coal for a significant share of their energy mix.

The third fundamental energy security challenge in Asia is to strengthen the market function centred on a price mechanism. As noted earlier, energy economics tell us that the supply-demand balance is determined by price movement. This market mechanism is best at meeting supply and demand. Poor market function causes misalignment of supply and demand. In much of Asia, the market function is weaker than in North America and Europe. Therefore, market reform is an important policy challenge. Economic efficiency will be best achieved through strengthened market function.

The fourth point to keep in mind is the relation between energy and economy, that is, the fact that a stable supply and rational demand for energy are important prerequisites for sustainable economic growth. Reasonably-priced energy is an indispensable input for all economic activity. Looking back at Japan's history, it can be said that the country's rapid economic growth and energy supply were inseparable. Lacking energy resources of its own, Japan's rapid growth in the 1950s and 1960s would not have been possible without a stable supply of low-priced energy from overseas. Economic growth stagnated during the two oil crises in the 1970s, and Japan's economic structure was transformed after these crises. As a result, Japan enjoyed medium growth through the 1980s.

Fifth, is the need to recognise the fact that energy markets are linked regionally and globally. Since the mid-1980s, international oil markets have developed to their current level where futures trading of West Texas Intermediate (WTI) in the New York Mercantile

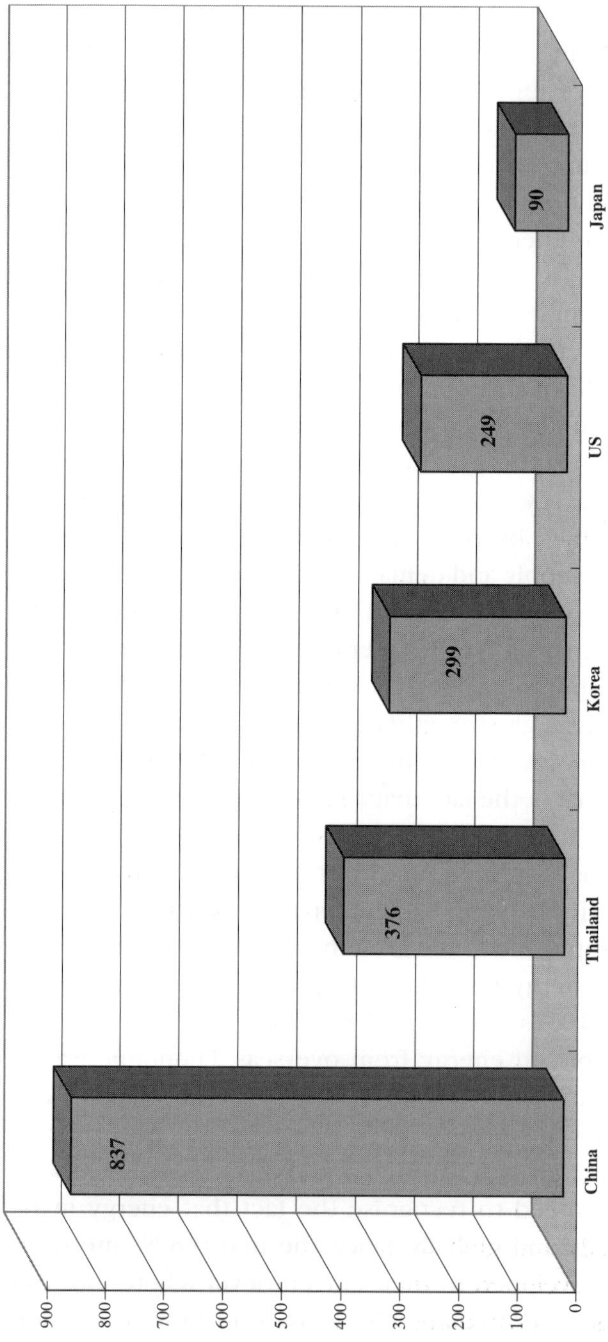

Figure 3: Energy Intensity of Key Countries (toe/million US$, 1995)

Source: IEA.

Exchange (NYMEX) records several to more than ten times the real global demand. The WTI prices are used directly or indirectly as the benchmark for almost all trade of crude oil and oil products in the world. The hegemony in global oil trading has shifted from the 'Seven Sisters' in the 1950s and 1960s, through OPEC in the 1970s and to the market in place since the mid-1980s. All markets — North America, Europe and Asia — are linked, though there is weaker linkage between Asia and the others compared with the Atlantic linkages. In these linked international markets, the effect of terrorism in the Middle East, hurricanes in the US, technical problems of production facilities in the North Sea, suspension of nuclear power plants in Japan, etc., are quickly transmitted regionally and globally. In the recent tight energy markets, a key factor behind the higher energy prices has been demand growth in Asia, particularly China. Asia is already the largest market, i.e., consumption region, in the world, surpassing North America and Europe. The Asian market will grow further and become even more significant in the years to come.

However, Asia need not trigger an energy crisis in the globally linked markets. Japan, in particular, as a developed and mature economy with a plausible energy policy performance, should not be passive about the recent energy situation causing energy security concerns in Asia. Energy security should not only be considered by each country, but by and for the region. A coordinated policy approach has good potential to perform more effectively and efficiently than an individual country approach in response to Asia's energy challenge because of complementarity and synergy. Ideally, Asia should demonstrate rational cooperative policy-making and implementation at a time when the markets are sensitive not only to fundamentals of supply and demand but also massive speculative movement.

Sixth, is the fact that there remain political tensions in Asia, particularly Northeast Asia due to historical issues and growing nationalism in each country. Needless to say, political stability is a prerequisite for sustained prosperity in Asia. The Japanese and East Asian 'economic miracles' were possible with political stability in the region underpinned by the strong US presence, both militarily and economically. Since the 1990s, globalisation has proliferated remarkably throughout

Asia, epitomised by the Asian Financial Crisis of 1997–1998. There was some doubt about American leadership in tackling this crisis. During this time the Asian countries became conscious of their egos, leading to the rise of various types of nationalism. As a result, events have occurred which could jeopardise political stability in the region. Disputes over energy resource sovereignty, for example, often relate to politics. Examples include the rivalries over gas deposits in the South China and East China Seas. Such matters tend to raise international political tension. At the same time, though, this means that it is possible to overcome political confrontation through solving energy resource problems, thereby building political stability as a basis for economic and social development.

To sum up, based on the above six points, it is of vital interest for all Asian countries to work together to sustain economic growth by solving energy security concerns. There is good potential for joint and/or coordinated efforts in the region, recognising the nature of the Asian- and internationally-linked markets. Overcoming parochial nationalism, will in turn, enhance political stability in the region, and act as a solid basis for further prosperity. This regional energy cooperation, will be embodied in a framework to be called an 'Asian Energy Partnership' under which various kinds of projects and businesses in the energy field will be guided.

HISTORY OF ENERGY COOPERATION IN ASIA

Beginning in the 1990s, energy experts warned that Asia's energy demand growth coupled with its high economic growth and dependence on oil imports from the Middle East would become key issues in need of tackling through international/regional cooperation. For example, Valencia and Dorian called for the need for regional cooperation based on the concept of energy security.[3] China's becoming a

[3] Mark J. Valencia and James P. Dorian, "Multilateral Cooperation in Northeast Asia's Energy Sector: Possibilities and Problems", in *Energy and Security in Northeast Asia: Supply and Demand; Conflict and Cooperation*, ed. Michael Stankiewicz, eScholarship Repository, University of California, 1998, p. 54.

net oil importer in 1993 was a major turning point. Also, territorial/resource disputes began to intensify in the South China and East China Seas. These events put a political dimension on the energy security debate and worried those in foreign policy circles. Calder discussed the impact of Asia, especially, China's energy demand growth and warned that the newly significant insecurity exacerbated strains ranging from Chinese territorial disputes and the North Korean nuclear programme to fears the region would draw too close to Iraq and Iran.[4] He urged US cooperation with the Pacific powers to help ensure energy supplies for Asia.

Regional government cooperation with respect to energy began in the APEC (Asia Pacific Economic Cooperation Conference). It started in 1989 as a ministerial conference to discuss economy and trade issues in the Asia-Pacific region. Establishment of the energy working group (EWG) was proposed by Australia and Japan. The first energy ministerial meeting was held in Sydney in 1996, then in Edmonton in 1997, Okinawa in 1998, San Diego in 2000, Mexico City in 2002, Manila in 2004, Pusan in 2005 and Darwin in 2007. To serve the EWG, the APERC (Asia Pacific Energy Research Center) was established in Tokyo pursuant to the Osaka Summit declaration in 1995. APEC/EWG works with energy policy experts exchanging views on common energy issues including energy data, clean fossil fuel, energy conservation/energy efficiency and new and renewable energy. All of these fall under the '3E' goals of energy security, environmental protection and economic growth.

In the 1990s, a regional cooperation framework was pursued to consider the construction of natural gas cross-border pipelines. Natural gas, a relatively non-polluting form of energy, is indigenous in Asia and the eastern part of Russia. Ideally, cross-border pipelines, i.e., those connecting neighbouring countries could help contribute to political stability. Japanese companies and politicians promoted the idea of an 'Asia-Pacific Energy Community'. This grand scheme envisioned a 42,500-kilometre pipeline grid extending from Yakutsk in Siberia to Dampier in northwestern Australia, connecting China,

[4] Kent E. Calder, "Asia's Empty Tank", *Foreign Affairs* 75, no. 2 (Mar./Apr. 1996): 55–69.

South Korea, Japan, Taiwan and six ASEAN countries.[5] Though discussion of these kinds of ideas has continually taken place between various groups over the past 15 years, any tangible outcome is yet to be seen. Momentum was lost in the late 1990s due to the Asian Financial Crisis and oil price slump. The economic growth of the major five ASEAN countries fell to minus 8 per cent and minus 6.7 per cent for South Korea in 1998. The WTI price fell from USD22.9 in January 1997 to USD9.75 in January 1999. Not surprisingly, interest in Asian energy investment, both from outside and inside the region decreased.

By 2000, however, discussions about energy investment and security resumed. Then, following the terrorist attacks in the US on 11 September 2001, fear spread over the world. As economic growth in the Asian countries that had been hit by the Asian Financial Crisis recovered, concerns over broader security issues came to the fore, including energy supply, destabilisation of the Middle East and possible terrorist attacks on Middle Eastern and Asian energy facilities. In 2001, the Comprehensive Resources and Energy Research Committee, which is the advisory council for the Japanese Minister of Trade, Economy and Industry (METI), announced a report from its Energy Security Working Group stating that Asian regional response in the form of an 'Asian Petroleum Security Initiative' was necessary because the emergency response system was not well-established and external import dependence would only increase as the Asian energy demand grew.[6]

The first ASEAN + 3 (China, Japan and Korea) Senior Officials Meeting on Energy (SOME) was held in Bali in August 2002 and the 8th International Energy Forum (IEF) was held in Osaka in September. Launched in 1991, the IEF is a dialogue meeting of producers and consumers of oil designed to enhance mutual understanding. The main focus of the meeting in Osaka, at which ministerial-level

[5] Mark J. Valencia and James P. Dorian, "Multilateral Cooperation in Northeast Asia's Energy Sector", pp. 36–39.

[6] METI, "Report of the Comprehensive Resources and Energy Research Committee 2001", 3 Mar. 2006 at http://www.meti.go.jp/report/downloadfiles/g10628bj.pdf [1 Mar. 2006].

participants gathered from 65 countries and 10 organisations, was Asian energy issues. On the sidelines of the meeting, an ASEAN + 3 informal meeting was held upon the proposal of then METI Minister Takeo Hiranuma. This was the first, though informal, ministerial-level meeting for ASEAN + 3 policy-makers to gather and discuss energy issues. Minister Hiranuma proposed a Japan-China-South Korea-ASEAN Energy Cooperation plan (Osaka Initiative) consisting of establishment of an emergency network, an oil stockpile initiative, oil market study, natural gas development and promotion of energy efficiency and renewable energy. He instructed all senior officials to follow up on these initiatives. At that time, oil prices were rising to about USD25–30 per barrel in fear of a US attack on Iraq and many Asian countries were nervous about the oil market situation. Thus, the meeting and the initiatives reflected concerns widely shared by Asian countries.[7]

During the course of 2003, Asia's energy security concerns increased. Oil prices rose in advance of the Iraq War in March 2003 with the WTI price reaching USD39 in February 2003. Before the outbreak of war it fell, but soon regained strength breaking record high levels month-by-month over 2004 and 2005. In June 2003 the second SOME + 3 decided to establish an Energy Policy Governing Group (EPGG) of ASEAN + 3 to discuss the common energy issues in line with the 2002 Osaka initiatives. In August 2003, the first EPGG meeting was held and it was decided to form five fora to discuss: (1) energy security; (2) the oil markets; (3) oil stockpiles; (4) natural gas; and (5) renewable energy. In June 2004, at the first official ASEAN + 3 Energy Ministers Meeting held in Manila, the joint declaration, Forging Closer ASEAN + 3 Energy Partnership, was signed.[8] Thus, the ASEAN + 3 framework for energy partnership was established and the ministerial meeting was regularised. The second ASEAN + 3 Energy Ministers Meeting was held in Siem Reap, Cambodia in July 2005. The third Ministerial Meeting was held in Bienchan, Laos in July 2006.

---

[7] METI, *Sekiyu Shinseiki* (Oil New Century). Tokyo: Energy Forum, 2003, pp. 74–78.

[8] Joint Ministerial Statement, "Forging Closer ASEAN + 3 Energy Partnership", 3 Mar. 2006 at http://www.eppo.go.th/inter/phil2004/ASEAN-AMEM + 3-JMS.html.

Japan and ASEAN have been the engines for this ASEAN + 3 framework. The first initiative was made by METI Minister Takeo Hiranuma in Osaka in 2002. The ball then was passed to the ASEAN-centred mechanism of the SOME + 3. It was established by the SOME + 3 as the main vehicle to discuss energy cooperation among ASEAN + 3. The EPGG designated the ASEAN Centre for Energy (ACE) as coordinator for ASEAN, and METI as coordinator for Japan, China and South Korea. An international strategy for Japan was discussed at the Comprehensive Resources and Energy Research Committee in 2004. This Committee subsequently adopted a report which stated that Japan should lead the multi-layered international cooperation frameworks with an emphasis on Asian consumer cooperation centred around ASEAN + 3. In 2004 the Philippines had the rotating presidency for the ASEAN energy group. Secretary Vicente Perez of the Philippine Department of Energy and METI Minister Shoichi Nakagawa worked closely together in preparation for the 1st ASEAN + 3 Energy Ministers' Meeting as did their bureaucrats.

This ASEAN + 3 framework has been formalised, combining the existing ASEAN framework plus the Northeast Asian countries. This platform has a certain rationality for Northeast Asian countries, because without it, head-to-head confrontation might occur among the Northeast Asian countries. Indeed, a fierce argument over natural gas development in the East China Sea took place in the bilateral meeting between Minister Nakagawa and Vice Chairman Zhan Guobao of the National Development and Reform Committee (NDRC) at the margin of the first ASEAN + 3 Energy Ministers Meeting in Manila in 2004. The regional forum has the merit of providing opportunities to have bilateral meetings that otherwise could not happen.

The ASEAN + 3 framework in general has evolved since the 1997 summit meeting and provides a good basis for regional energy cooperation. It launched discussions on financial cooperation in response to the 1997 Asian Financial Crisis and thereafter, the subjects dealt with by this grouping have expanded to include ministerial meetings on foreign affairs, finance, economy, agriculture, etc.

The finance ministers' meetings produced agreement on the Chiang Mai Initiative, which created a currency swap mechanism for emergency situations, and the Asian Bond Market Initiative which aims at developing bond markets to utilise more Asian money for investment in Asia in 2003. The Agricultural Ministers Meeting in 2001 agreed on a feasibility study for rice stockpiling in East Asia. Thus, ASEAN + 3 has evolved as a platform able to nurture functional cooperation in various fields in East Asia, and the time is now ripe for further cooperation in energy.

The East Asia Summit, consisting of ASEAN + 3 + 3 (India, Australia and New Zealand), started in December 2005. The second East Asia Summit, held in January 2007 in Cebu, Philippines, adopted the Cebu Declaration on East Asia Energy Security. This East Asia Summit process has also become the forum for Asian energy cooperation.

WHAT KIND OF ENERGY PARTNERSHIP IN ASIA?

Basic Principles

The 'Asian Energy Partnership' is a regional energy cooperation framework designed to cope with a common Asian challenge. It should be an institutional framework, with policy cooperation at the state level, which promotes and justifies individual projects and businesses. There is consensus that the goals of the partnership should be the 3Es of energy security, environmental protection and economic growth. In the author's view, there should be three important principles shared by the partnership members if meaningful cooperation is to take place.

First, the Asian region should share a common interest of net oil importer position. The biggest challenge shared by the Asian countries is energy security, in particular oil security, stemming from its net importer position and high dependence on Middle East oil. Economic and environmental challenges occur as a result of the growing energy consumption. Asian countries should maximise their collective power in addressing these energy challenges. On the supply side, they should promote joint development and the development of alternative energy supplies as much as possible. Consumer cooperation in Asia, in particular ASEAN + 3, should be modelled on the IEA in terms of policy-oriented international cooperation among developed and consumer nations.

Secondly, regional cooperation should be based on self-determination on the one hand, but shared responsibility on the other. As energy policy is primarily a national policy for national security, any international or regional cooperation should be pursued on the grounds of self-help and self-determination, though seeking common interests. Oil security measures include emergency response measures, such as oil stockpiling, diversification of energy sources other than oil and energy conservation/efficiency. Each country should commit fully to these policy measures and seek cooperation, including technical cooperation and joint projects, with neighbouring countries based on shared responsibility. The form of cooperation could vary from one project to another.

Thirdly, the regional framework must have market function as its basis. As discussed earlier, the basic solution for energy security is enhanced 'resiliency' in market function. In a well-functioning market, price signals determine the response of both the supply and demand sides leading to equilibrium in both the short and medium-to-long terms. Market function ought to apply to both the domestic and international markets. In order to enhance market resiliency, impediments or price subsidies which distort the market function should be lifted. Of course, it is justifiable for individual governments to tax on oil used for building and for maintenance of stockpiles for the sake of national security, and to set regulations on product specification for environment purposes. For the international market, however, it is important to eliminate trade and investment barriers and to work towards building an integrated common Asian market where energy commodities are traded freely. A common oil products market in Asia would promote free trade movement and balance between excess demand and excess supply. There should also be harmonisation of product specification, for example sulphur content. An integrated market is additionally necessary for crude oil and LNG. In order for crude oil and LNG to move freely in the market, restrictive business practices, such as destination limitations which suppliers impose on buyers, should be eliminated.

Elements of Policy Cooperation

What kind of policy elements should the regional policy cooperation framework cover? There are both demand- and supply-side approaches. Policy-makers should be aware that the demand-side approach, namely energy conservation and raising efficiencies, must be the first priority of the Asian Energy Partnership agenda. Energy efficiency in Asia, apart from Japan is very poor. Compared to Japan, China consumes nine times, ASEAN four times and South Korea three times as much energy per unit of GDP (see Figure 3). There is clear potential for most of Asia to improve energy efficiency. According to the IEA, the primary energy supply of Asia in 2030 in the reference scenario will rise to 5,967 million toe from 3,140 million toe in 2002. In one of IEA's alternative scenarios, which depicts a more efficient and environment-friendly future based on policies and technology, the primary energy supply in Asia in 2030 will be around 12 per cent lower than that in the reference scenario. This means that about 700 million toe (roughly equivalent to the present total energy consumption in Japan and South Korea combined) can be saved with proper policies and technology (see Figure 4).

The main way to drastically reduce energy consumption is to implement energy efficiency measures for industrial processes, appliances and lighting. Japan has well-advanced experience, technology and know-how in these areas and would be pleased to share it. According to the market function principle, investment in plants and equipment, as well as human resources, should be rewarded a fair return. Thus, it is imperative that energy prices be set on a market basis. Subsidies, if necessary, should be provided to investment rather than to energy prices. Technology should also be priced properly or

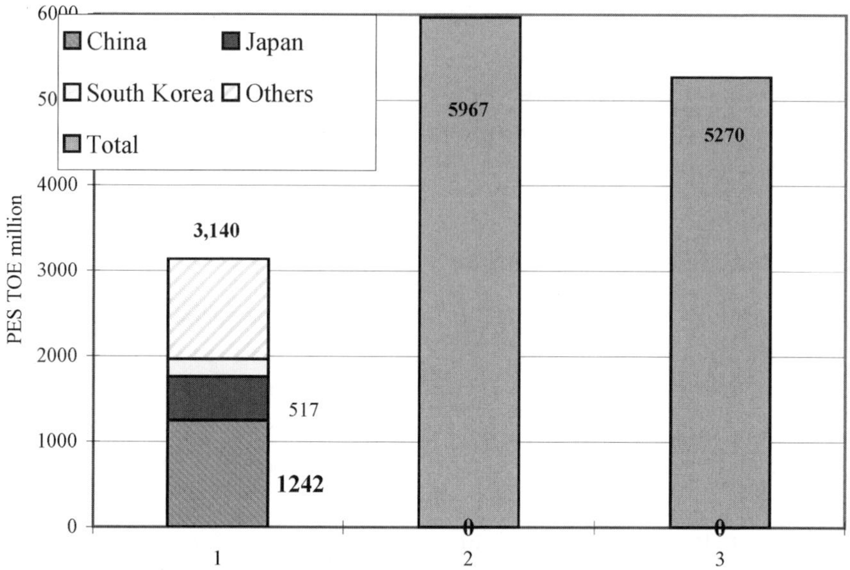

Notes: 1 = year 2002; 2= 2030 Reference Scenario; 3= 2030 Alternative Scenario
Source: IEA World Energy Outlook 2004 modified by Tanabe.

Figure 4: Energy Saving Potential in Asia (PES TOE million)

subsidised, if necessary, through certain mechanisms such as licensing contracts and the 'Clean Development Mechanism' (CDM).[9]

Energy conservation makes energy available which would otherwise be consumed. As most of the world's additional oil supply capacity will be in the politically fraught Middle East, energy conservation makes no less sense than investment in expansion of resource supply. A framework should be established whereby the energy conservation and energy efficiency know-how of Japan can be smoothly transferred to developing Asian countries via both government/institutional and

[9] The Clean Development Mechanism (CDM) is an arrangement under the Kyoto Protocol allowing industrialised countries with a greenhouse gas reduction commitment (so-called Annex 1 countries) to invest in emission reducing projects in developing countries as an alternative to what is generally considered more costly emission reductions in their own countries. From *Wikipedia* at http://en.wikipedia.org/wiki/Clean_Development_Mechanism] [3 Mar. 2006].

business platforms. If the energy efficiency of Asia as a whole could reach Japan's level, there would be less political tension arising over competition for energy resources.

The energy efficiency framework should involve both the public and private sectors combined with certain governmental intervention in the form of regulation and/or incentives. One possibility would be to adopt some kind of common benchmark or standard of efficiency to be pursued as a common policy goal by the Asian countries.

As for policy cooperation on the supply side, a long-term policy goal must be to achieve the best energy mix. Each Asian country should aim to have an energy mix which is not overly dependent on oil and that is based on country-specific conditions such as economic development stage, social peculiarity and energy resource endowment. Natural gas, a clean form of energy readily available in the region, should be more greatly utilised. Thus, Asian countries should cooperate in the development of infrastructure for supply and transportation of natural gas, promotion of more flexible trading practices including destination free contracts, and development and diffusion of gas-related technology.

Coal is an economical and indigenous energy resource in Asia which is expected to continue to play a major role in the future. Asia should promote cleaner use of coal with less sulphur and lower CO_2 emissions through the use of clean coal technologies (CCT). The most relevant indigenous and sustainable energy in Asia is renewable energy, including not only hydro but also solar, wind, geothermal and biomass. The general disadvantage of renewable energy, however, is its higher costs. Therefore, it is necessary to introduce incentives and regulations such as renewable portfolio standards (RPS) as well as further technological development.[10] There is much experience to draw upon, not only from Japan but also other Asian countries. It is worth sharing this experience in order to identify best practices.

[10] RPS is a policy that requires those who sell electricity to have a certain percentage of renewable power in their mix.

It cannot be denied that oil is indispensable despite our best efforts to diversify energy resources. Indeed, energy security has been, and will likely continue to be largely *oil* security. There are two ways to enhance oil security. Short-term security, or emergency response, should be based on oil stockpiles to prepare for supply disruptions. The motivation to establish the IEA in 1974 was creation of oil stockpiles to be shared by the member countries in response to the Arab oil embargo. It is not likely that Saudi Arabian or other oil producers will launch an oil embargo in the near future. However, the international oil markets may occasionally encounter supply disruptions due to events such as terrorism in the Middle East or in certain sea-lanes such as the Malacca Strait. The arguments for oil stockpiling have not changed. In Asia, only Japan and South Korea have oil stockpiles up to IEA standard (90 days of net imports), though China is building national oil stockpile facilities and some ASEAN countries are planning to build stockpiles in relation with the existing ASEAN petroleum security agreement (APSA). Construction of oil stockpiles in Asia should be encouraged with the support of Japan and South Korea who can share their experience in the financing and maintaining of the stockpiles. The mechanism determining when and how to release the oil should be coordinated among the Asian countries and the IEA members with its CERM (Coordinated Emergence Response Measures) programme.[11] Ideally, Asia's oil stockpile should supplement the IEA's.

Long-term oil security in Asia will be enhanced if there is sufficient investment in resource development and transportation. This is why dialogue between producers and consumers is so important, i.e., producer-consumer dialogue and cooperation, consumer-consumer (C-C) dialogue and cooperation. The C-C cooperation can take the form of joint investment and joint procurement commitment. The joint procurement of oil by Asian refineries and traders is relevant for bargaining power in the Asian trading market which has historically been biased towards Middle Eastern oil producers.

[11] CERM is the IEA's rapid and flexible system of response to actual or imminent oil supply disruptions including the release of oil stocks, restraining of demand, switching to other fuels or increasing of domestic production.

The 'Asian Premium' epitomises the producers' dominance over consumers in Asian markets, i.e., Asian buyers have been paying roughly USD1 more than European and American buyers.[12] The Asian market function should be strengthened for oil products as well in view of recent refining capacity bottlenecks. The regulation of trade (both imports and exports) in crude and oil products should be eliminated to strengthen the international market function in the Asian region. Compared with the pricing systems of North America and Europe, Asia's is relatively weak. The Asian countries must work together to rationalise the supply and demand for crude and products, through strengthened market function with more open and transparent pricing at the spot and futures trading market.

Vital for all of the policy areas identified above is the availability of reliable data and statistics. There is an urgent need to improve the quality of energy data and statistics in Asian countries. Unsound policies and quirky market reactions are inevitable when using faulty data and statistics. It has been noticed that there are 'missing barrels' of oil and strange miscellaneous items in the oil statistics. 'Joint Oil Data Initiative' (JODI) is a collaboration of six organisations including APERC to improve oil statistics, in the context of the International Energy Forum. Hopefully, this will rectify some of the problems. Members are required to disclose detailed oil statistics, such as exports and imports and inventory level. The quality of data and statistics for total energy and CO_2 emissions must be further improved.

EXPERIENCES OF THE EU AND IEA

Regional and international energy cooperation is not new. The EU practises regional cooperation and the IEA coordinates international cooperation among developed consumer countries.

---

[12] Yoshiki Ogawa, "Ajia no Sekiyu Mondai" (Asian Petroleum Problems), in Yasuo Tanabe, *Ajia Enerugii Paatonaashippu* (Asian Energy Partnership). Tokyo: Energy Forum, 2004, pp. 72–76. It is said that the Asian premium mainly reflects the fact that Asia is more dependent on Middle Eastern oil than is Europe and North America.

EU Experience

The European integration process started with the formation of the European Community of Steel and Coal (ECSC) in 1951 by France, West Germany, Italy, Belgium, the Netherlands and Luxembourg. The focus was joint management of the highly strategic coal and steel production resources. It was the first case of national governments commissioning power to a supranational body. Over several decades, the European integration process evolved to a single market and common currency. In the energy area also, the process of market integration and policy cooperation and unification have been continuously evolving. Although it took more years than originally thought necessary, the idea of a single energy market, initially announced in 1987, finally bore fruit in 2003 when it was agreed there ought to be full liberalisation of all electricity and gas markets by 2007. A notable recent development in the EU energy market has been the European Commission's intervention in the removal of the destination restriction clause for natural gas contracts between EU buyers and external suppliers such as Nigerian LNG and Gazprom. The EC has been investigating territorial sales restrictions in supply contracts between gas producers and European wholesalers for some time. The clauses prevent wholesalers from reselling the gas outside the countries, which, in the EC's view, represents a breach of European competition law and undermines the on-going creation of a European gas market. In December 2002, the EC settled a case concerning the Nigerian gas producer Nigeria LNG Ltd and in October 2003, it settled the Gazprom/ENI case, removing the restrictive clauses. This is a good example of how a dominant power — the single European market — can exert influence over energy suppliers. It used its leverage of competition law for the EU integrated market. Another aspect is the offering of differentiated treatment to newly participating, less developed members such as Portugal and Greece. The EU has a scheme to offer assistance from its 'Structural Fund' for infrastructure such as natural gas pipelines in those countries. In the energy market integration process, the EU also offers Greece exemption from liberalisation obligations, because the Greek market is not interconnected with the rest of the EU. Thus, the EU has a flexible technique to offer exemptions from the common policy to more recent member countries and those in different development stages.

Though of course, Europe differs considerably from Asia, its experience provides many lessons. European integration started with western European countries of similar cultures, economic development levels and political regimes, while Asia is composed of countries of a wide range of economic development levels, cultures, religions and political regimes. However, despite the differences, the EU proves that economic integration and energy cooperation can be pursued in a mutually supporting manner, that energy policy is centred on market integration, that differentiated treatment may be offered to countries in different stages of development, and that energy security is best considered in collective, rather than individual country terms.

IEA Experience

The IEA was founded in 1974 on the basis of a proposal by then Secretary of State, Dr. Henry Kissinger, in response to the first oil crisis triggered by the embargo of oil exports from members of the Organization of Arab Petroleum Exporting Countries (OAPEC). It was a counter measure instigated by the Organization of Economic Cooperation and Development (OECD) countries to the OAPEC. Legally, the IEA is an implementing body of the International Energy Programme which was established by the OECD.

The IEA has two functions. The founding purpose was to have emergency measures for oil supply disruption. The member countries are required to maintain emergency oil reserves equivalent to at least 90 days of net oil imports. Other measures include demand restraint, fuel-switching and surge oil production. In the event of an oil supply disruption, at the decision of the Governing Board of the IEA, member governments are requested to coordinate their response measures including their stock draw. There have been two cases of coordinated stock draw in the past: during the Gulf Crisis in 1991, and when Hurricane Katrina hit the US in 2005.

IEA's other important function is long-term policy coordination, including energy conservation and the development of alternative energies to reduce the dependence on imported oil. In response to the development of international energy markets and the shift in policy objectives and means, the goals of the IEA were refined as 'Shared Goals' at the Ministerial Governing Board in 1993. It was at this time that the '3E' goals of energy security, environmental protection and economic growth were identified. The way the IEA achieves policy goals is through peer review, i.e., a country policy review process. Every four to five years each IEA member country undergoes an intensive policy review by an IEA team consisting of officials of member countries and the country has a moral obligation to perform the policy recommendations. This is a very effective policy coordination process which aims to improve nations' policies and build a sense of solidarity among the members for the shared goals of the IEA. These twofold functions of the IEA should provide lessons for Asian economic policy cooperation.

The 'Asian Energy Partnership' should have an oil stockpile programme among its emergency response measures. The reason why the IEA was a legitimate and effective countermeasure by the oil consumer group against the OAPEC was because the primary energy supply of the OECD countries accounted for 62 per cent of the world total in 1973. Today, the OECD countries account for 51 per cent of the world primary energy supply. Of the rest, 25 per cent lies in non-IEA member countries in Asia. Asia as a whole should have an effective oil stockpile programme which can supplement the IEA programme and stabilise the effects of an emergency. If the IEA and Asia work together, oil consumers all over the world will gain. The 'Asian Energy Partnership' should not only oversee the construction of oil stockpiles, but also seek policy coordination. If the Asian countries adopt and implement a set of energy policies in line with the 'Shared Goals' of the IEA through a peer review process, energy security in Asia will be dramatically enhanced.

The first option is for the Asian countries, other than Japan and South Korea (which are already members) to join the IEA. However, there are strict eligibility requirements to become a member of the

IEA, such as membership in the OECD, a club of developed countries. South Korea joined the OECD in 1996 and the IEA in 2002, but it is not realistic to imagine countries such as China, Thailand and the Philippines joining the OECD and IEA in the near future. Therefore, the second option, a more realistic one, is to have a similar regional cooperation framework and strong relations with the IEA. The 'Asian Economic Partnership', therefore, may be called the 'Asian IEA', signifying an Asian group to function in line and in cooperation with the IEA, where systematic policy cooperation is to be made.

FUTURE OUTLOOK OF THE 'ASIAN ENERGY PARTNERSHIP'

How will the Asian Energy Partnership grow? Will it disintegrate due to confrontation and rivalry among the Asian countries? Will it be another 'talk shop' destined to fizzle without any results, or will it evolve as a relevant regional cooperation regime? There are obstacles and opportunities.

The diversity among the Asian countries could hinder meaningful cooperation. Asian countries exhibit broad diversity not only in their economic development, political regimes, social and cultural backgrounds, but in their energy supply and demand situations as well. For example, Japan is a mature, high- income, democratic country with a population of 127 million living on narrow islands, while China is a rapidly developing, but low-income, autocratic country of 1.3 billion population living on a large continent. Energy consumption per capita in China is one-fifth that in Japan. There is a 70-fold difference in per capita income between Singapore and Cambodia. Singapore's energy consumption per capita is more than 10 times that of Cambodia. Some people doubt the possibility of meaningful regional cooperation among such diversified countries.

Another obstacle is nationalism among the Asian countries. In particular, there is conspicuous nationalistic sentiment growing in Northeast Asia which accounts for 90 per cent of total East Asian GDP. The Japanese public became inward-looking in the 'lost decade' of recession since the burst of the bubble economy in the early 1990s.

Meanwhile in China, the economy has grown tremendously and the Japanese have noticed that China's anti-Japan movement has become larger and bolder. This in turn gives rise to anti-Chinese feelings and nationalistic sentiment within the Japanese public and political discourse. In China, there is growing patriotic nationalistic sentiment stemming from self-confidence as a result of economic growth and the effect of patriotic education. The rapid increase in internet communications has only served to fuel the fury as seen clearly in the anti-Japan demonstrations which took place in the spring of 2005.

In South Korea, nationalism has grown from a movement in search of Korean identity. It has quickly taken hold under the current left-wing government and the sense of being liberated from a repressed atmosphere under military governments until the 1980s. All bilateral relations at the time of writing between Japan, China and South Korea were tense and these cast a shadow over potential energy cooperation. A typical example is the dispute between Japan and China over the development of a natural gas field in the East China Sea by CNOOC. The Chinese side argues that they are working on the Chinese side of the median line which Japan claims is the border for the Exclusive Economic Zone (EEZ) but China does not accept. The Japanese side argues that the project will take gas resources that lie on the Japanese side of the median line. They have requested provision of data and information on the gas field and suspension of the development. This is a typical dispute when resources are spread over the borders of two countries. In many similar cases, the solution has been joint development with shelving the solution of sovereignty issues. Both China and Japan have agreed so far, in principle, on joint development. However, their concurrent nationalism makes an amicable solution difficult. Given that Japan and China together account for 80 per cent of East Asia's GDP, one may conclude that an 'Asian Energy Partnership' will have difficulty achieving any success.

However, to make the Asian Energy Partnership successful, the obstacles should and can be overcome by the policy-makers and people of Asia. One of the most important elements for the success of the regional cooperation framework is the sharing of common interests and challenges. The Asian region shares common interests based on its position as a net importer of energy, in particular oil. The hike in oil prices due to the tight market situation is negatively affecting all economies in the region. According to the IEA, a USD10 increase in the price of oil would have an impact on Japan of minus 0.4 per cent of its GDP and on the rest of Asia of minus 0.8 per cent of its GDP. Further, as discussed earlier, there has been the 'Asian Premium' which is peculiar to the Asian region. Therefore, Asian countries share a common interest in energy security defined as a stable supply at a reasonable price for sustainable economic growth.

A relevant approach to the common challenge is cooperation in energy efficiency. It is encouraging to note that Japan and China have engaged in high level promotion activities for energy efficiency. In April 2007, Premier Weng of China visited Tokyo and issued a joint press announcement with Prime Minister Abe stating that both countries would strengthen their cooperation on energy efficiency.

The 'Asian Energy Partnership' may not necessarily involve all countries in Asia. Countries could join on an opt-in basis, in other words, countries with specific shared interests can form specific partnerships. This way the diversities in political systems, economic development and social conditions, which are typical in the region will not hinder rational, practical forms of partnership. As mentioned earlier, Asia should carefully study the experiences of the EU which offers differentiated treatment to developing and/or newly participating member countries.

Another important catalyst for cooperation is recognition of the international and regional nature of the energy markets and community. In the international market, a change in the supply-demand situation of one country affects other countries through the price effect. It has been China's demand growth which has been the main determining factor in the oil price hike in recent years and hurt the other Asian economies. The suspension of nuclear power plants in Japan in 2003 squeezed the LNG and light sweet crude markets in the region as a result of irregular sourcing by Japanese electricity companies. Thus, international markets can be compared to a pool, where pouring from any source loosens the general conditions, while draining from any leakage tightens the general conditions. Japan, China, South Korea and ASEAN are all in the same pool. Of course the market is not perfect and the governments should intervene to rectify market failure via policy means such as setting rules and providing information. The word 'pool' or 'market' can be replaced by the word 'community'. Japan, China, South Korea and ASEAN all belong to the same community. They should all refrain from acting in their own narrow-minded self-interests, which might negatively affect the other members. Recognition of the sense of community, in a general way, is expected to grow, as movement of persons, goods, services, money and information is intensified in Asia. The sense of an *energy* community should also grow if the evolution of an integrated Asian energy market is accelerated through policy cooperation. Ideally, a sense of community and cooperation should mutually support and intensify.

As in political science theory, the fledgling regional framework for energy cooperation can be viewed as the early stages of an international regime defined as 'sets of implicit or explicit principles, norms, rules and decision-making procedures around which actors' expectations converge in given areas of international relations'. An international regime is created or evolved based on its key function, which is to stabilise international relations in a certain issue area. In the ordinary 'prisoners' dilemma' game theory, a player chooses to defect from the other player, believing that he or she will be better off than choosing to cooperate while anticipating the other will defect. This is true for a one-time only game, or in short-term decision-making. If the games are repeated in the future, or during long-term decision-making, a different solution emerges. A player will choose to cooperate with the other, since he or she will expect the other to cooperate with him or her in future games. This is the 'iterated prisoners' dilemma'.

In the case of Asian players, for instance Japan and China, it is clear that they will 'play many games' in the future. As the theory of the iterated prisoners' dilemma predicts, it is expected that there will be cooperation among the Asian players as a rational consequence of anticipating iterated games in future. The improvement in Japan-China relations in 2007 centred on energy efficiency cooperation and shows the cooperative nature of iterated prisoners' dilemma.

The concept of the international regime is also explained by using the concept of public goods from the field of economics. Historically, public goods in international relations were thought to be maintained by a hegemon country according to the 'hegemonic stability' theory. This theory, first espoused by Charles Kindleberger in the 1970s, argued that the overwhelming dominance of one country, such as the UK in the 19th century and the US in the 20th century, was necessary for the existence of an open and stable world economy. Such a hegemon served to coordinate and discipline other countries so each could feel secure enough to open its markets and avoid beggar-thy-neighbour policies. Conversely, the theory asserted that the decline of a hegemon tends to be associated with international economic instability.

In the 1980s, Robert Keohane argued that even with the decline of hegemony, an international regime can emerge and be maintained. The oil regime was used as the best example. An issue-differentiated theory of hegemonic stability, rather than crude theory of hegemonic stability, makes a contribution to the understanding of changes in the regimes. Although the US was not a hegemon in the international oil system during the 1973 oil embargo, the US, Europe and Japan successfully founded the IEA as an insurance regime by establishing international procedures to share the costs of an oil supply disruption and reduce the risks faced by consumer countries. It was a result of rational judgement on the part of major oil consumers, i.e., the US, Europe and Japan, based on common interest. Today, in Asia, Japan, South Korea and China account for about 90 per cent of the GDP and energy consumption in the East Asia. Japan and South Korea, as members of the IEA, fully commit to maintaining an oil stockpile of more than 90 days of net oil imports and having an energy policy based on the '3E' goals. China is also committed to becoming a law-abiding, responsible member of the international economic order, as seen in its accession to the WTO. These three Northeast Asian countries together with ASEAN will form a natural centre for the regional framework, or regional regime, of energy cooperation.

CONCLUSION AND POLICY RECOMMENDATIONS

Economic integration in Asia is progressing to the level where the intra-regional trade ratio exceeds 50 per cent. Energy security, that is, a stable supply of energy at a reasonable price in an environmentally-friendly manner, is an indispensable condition for continued progression towards Asian economic integration. To attain these energy policy goals, Asian countries must strengthen their policy cooperation in the form of an 'Asian Energy Partnership'.

Energy policy cooperation should be based on the central principles of consumer status, self-determination and market function. In policy cooperation, the first priority should be on the demand side, namely, energy conservation and efficiency. On the supply side, the long term goal should be formation of a diversified energy mix which is not overly dependent on oil, i.e., there should be development and further utilisation of natural gas and renewable energy, and cleaner usage of coal. Oil security should be addressed by emergency response measures, in particular, an oil stockpile for short-term risk, and dialogue and cooperation with oil producing countries to create a better environment for investment in oil production and transport.

Among the various Asian energy security efforts to date, the most recent is the energy partnership of ASEAN + 3 (Japan, China and South Korea). The Manila Declaration announced by ASEAN + 3's Energy Ministers in 2004 is a good basis. Another is the East Asia Summit, consisting of ASEAN + 3 + 3 (India, Australia and New Zealand). The 'Asian Energy Partnership' should examine the examples of the EU and IEA. Although Asian conditions are very different from those of Europe, the European integration process provides lessons about the interaction of market integration and energy cooperation based on the common interests of energy consumers. The 'Asian Energy Partnership' should emulate the functions of the IEA, that is, concerted emergency response measures based on an oil stockpile and peer review of energy policy for 3E goals of member governments.

There are obstacles for the regional policy cooperation framework, namely the diversity in political, economic, social and resource endowment conditions among the Asian nations which could hinder meaningful cooperation. The nationalism and the political tensions which have been prevalent in the northeast Asian countries in recent years are also problematic.

However, those obstacles can be overcome. There are clear opportunities. Asian countries share a common interest in energy security for the sustained growth of their economies. Wise political leadership and skilled technocracy, backed by public support, will hopefully nurture a sound policy cooperation framework, under which private sector business interests can pursue rational cooperation in unobstructed markets. It is already apparent that the once sour Japan-China relationship has improved with cooperation in energy efficiency. Dialogue geared towards mutual understanding and confidence-building among the various government and non-government actors is the key for a successful 'Asian Energy Partnership', which leads to community building in Asia.

Chapter

4

Energy Conservation Policy Development in Brunei

Jianjun Tu

INTRODUCTION

With a land area of 5,765 km^2 and a population of 383,000 in 2006, Negara Brunei Darussalam, or Brunei Abode of Peace (hereafter Brunei), is a small sultanate located on the north-west coast of Borneo Island in Southeast Asia. As the world's fourth-largest producer of liquefied natural gas (LNG) and the fifth-largest producer of crude oil in the Southeast Asia, Brunei is an economy heavily supported by petroleum exports. For instance, in 2005, the revenues from the hydrocarbon sector alone accounted for 66 per cent of GDP at current prices, and yielded 94 per cent of merchandise exports and 92 per cent of government revenues.[1]

The Asian Financial Crisis in 1997 and 1998, coupled with a large decline in the world price of crude oil and the 1998 collapse of the Amedeo Development Corporation, Brunei's largest construction

[1] JPKE, *Brunei Economic Bulletin*. Bandar Seri Begawan, Department of Economic Planning and Development, Prime Minister's Office, vol. 4, issues 3 and 4, 2006.

firm, caused the country to slip into a mild recession, but the economy rebounded after the recent oil price spike, with real GDP growing 3 per cent annually between 2000 and 2005.[2] Government revenue and grants, excluding income from international reserves, amounted to USD5.1 billion in 2005, whereas Government expenditure totalled USD3.1 billion, leaving a surplus of USD2.0 billion. The per capita GDP of USD16,800 in 2005 places Brunei among the World Bank's high-income non-OECD group. Brunei's oil wealth, and its position in the Association of Southeast Asian Nations (ASEAN), gives it a disproportionately large influence in the world in relation to its size.[3]

Oil and gas are the main sources of energy in all economic sectors in Brunei, contributing about 19 per cent and 81 per cent of all primary energy consumption between 1980 and 2004, respectively. From 1980 to 1981, Brunei's primary energy consumption dropped near 30 per cent, and fluctuated widely thereafter but showed a general upward trend, reaching 2.7 million tons of oil equivalent (mtoe) in 2004, which was 40 per cent higher than in 1981. Brunei's energy intensity was 0.49 mtoe per billion USD in 2004 while the similar intensity of the world was only 0.25 mtoe per billion USD.[4] Brunei's significantly higher than world average energy intensity indicates a potential for energy efficiency improvement and conservation.

In this chapter, the recent and projected energy supply and demand patterns in Brunei between 1980 and 2030 are outlined to provide an overview of Brunei's energy sector. Then the Brunei Government's programmes to reduce energy wastage and improve energy efficiency are discussed and evaluated. Recommendations are

[2] United Nations, *National Accounts Statistics Database*. Economic Statistics Branch of UN Statistics Division, 2007.

[3] Asian Development Bank, *Key Indicators 2006: Measuring Policy Effectiveness in Health and Education.* Manila: Asian Development Bank, 2006; Central Intelligence Agency, *The World Factbook 2006*; Economist Intelligent Unit, *Country Profile 2006: Brunei*. London: Economist Intelligent Unit, 2006.

[4] International Energy Agency (IEA), *IEA Energy Statistics*, 2006. United Nations, *National Accounts Statistics Database*. Economic Statistics Branch of UN Statistics Division, 2007.

made with regard to further tapping the potential of energy conservation in Brunei.

ENERGY SUPPLY AND DEMAND IN BRUNEI: 1980–2030

The Asia Pacific Energy Research Centre (APERC), the International Energy Agency (IEA) and the Brunei Government all publish Brunei's energy statistics on a regular basis. In this study, the IEA statistics were used whenever possible and supplemented with data from the APERC, British Petroleum (BP) and the Brunei Government.

Brunei's energy supply and demand between 1980 and 2030 in this study serve as the benchmark for understanding the historic path and possible trajectory under which Brunei's energy conservation potential could be tapped. While the energy forecast of this study was based on many simplified macroeconomic assumptions defined below, the uncertainties associated with the fluctuations in the price of oil, additions to Brunei's petroleum reserves, potential changes in the country's national development plan and the migration of foreign workers cannot be fully incorporated. Thus, the projected energy supply and demand trends highlighted below must be viewed with caution, especially for the later years.

The Energy Drivers — Macroeconomic Indicators

Brunei had a population of 370,100 in 2005, 78 per cent of which was categorised as urban. GDP stood at 4.76 billion (in 1990 constant USD).[5] Figure 1 presents the general economic trends underlying the energy service demand projections. The population and urbanisation projections are based on *World Population Prospects and World*

[5] United Nations, *World Urbanization Prospects: The 2003 Revision*. New York: Population Division, Department of Economic and Social Affairs, UN, 2004; Department of Statistics, *Brunei Sarussalam Key Indicators 2006*. Bandar Seri Begawan: Prime Minister's Office, 2006; UN, *National Accounts Statistics Database*. Economic Statistics Branch of UN Statistics Division, 2007.

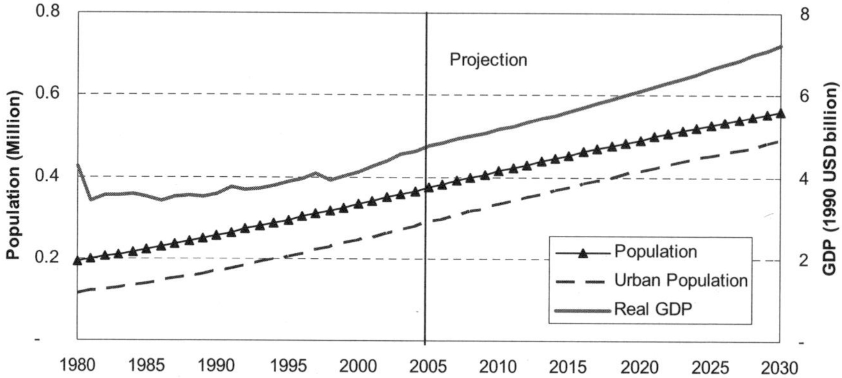

Figure 1: GDP and Population in Brunei, 1980–2030

Sources: APERC, *APEC Energy Demand and Supply Outlook 2006*. Tokyo: Asian Pacific Energy Research Centre, 2006; UN, *World Urbanization Prospects: The 2003 Revision*. New York: Population Division, Department of Economic and Social Affairs, UN, 2004; UN, *World Population Prospects: The 2004 Revision*. New York: Population Division, Department of Economic and Social Affairs, UN, 2005; UN, National Accounts Statistics Database, the Economic Statistics Branch of the UN Statistics Division, 2007.

Urbanization Prospects, and represent the medium variant forecast.[6] The GDP projections are based on the growth rate defined in the *APEC Energy Supply and Demand Outlook*.[7] During the 25-year forecast period, Brunei's population is forecast to increase by 52 per cent, but the proportion of urban population will grow more significantly. In 2030, 488,000 people, or 87 per cent of the population will live in cities, as defined by their dwelling and employment activities.

Primary Energy Production

Brunei's primary energy production is overwhelmingly dominated by the oil and gas sector. In 2005, Brunei produced 10.1 million tons (mt)

[6] United Nations. *World Urbanization Prospects: The 2003 Revision*. New York: Population Division, Department of Economic and Social Affairs, UN, 2004; UN, *World Population Prospects: The 2004 Revision*. New York: Population Division, Department of Economic and Social Affairs, UN, 2005.

[7] APERC, *APEC Energy Demand and Supply Outlook 2006*. Tokyo: Asian Pacific Energy Research Centre, 2006.

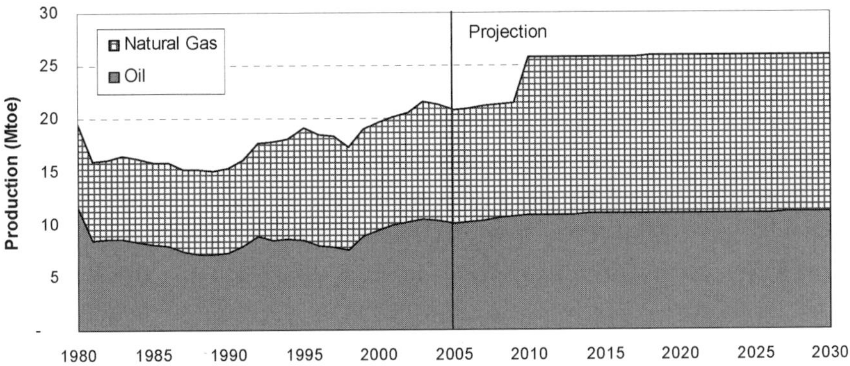

Figure 2: Petroleum Production in Brunei, 1980–2030

Source: British Petroleum, *BP Statistical Review of World Energy 2006*. London: BP, 2006; Datuk Paduka Haji Idris Haji Belaman, "Energy Situation and Emergency Response Policy: Brunei Darussalam", IEA/ASEAN Workshop on "Oil Security and Emergency Preparedness" and Associated Site Visits, Paris, IEA, 2003.

of oil and 12.0 billion m^3 (cubic metres) of natural gas.[8] The Brunei LNG Sdn Bhd's (BLNG) LNG plant was commissioned in 1973 and is located on the coast near Lumut, about 80 km from the capital Bandar Seri Begawan. There are five liquefaction trains in the BLNG plant, each capable of processing 5.3 million m^3 of gas per day. This gives an annual plant capacity of 7.2 mt of LNG. A plan has been drawn to expand the production capacity with the proposal for the addition of a new Train 6 with a processing capacity of 5.52 billion m^3 per year. The planned shipment is scheduled for around 2010, depending on the provision of new gas reserves.[9] The plan also includes further modernisation of the existing plant, similar to the rejuvenation programme carried out by Shell Global Solutions in 1994. As a result of the 1994

[8] British Petroleum, *BP Statistical Review of World Energy*. London: British Petroleum, 2006.

[9] Energy Information Administration (EIA), *The Global Liquefied Natural Gas Market: Status and Outlook*. Washington DC: EIA, 2003; BLNG Company Website at http://www.blng.com.bn, Brunei LNG Sdn Bhd [Jan. 2007].

programme, the Lumut plant now operates at 140 per cent of its design capacity.[10]

Taking into consideration the 6th LNG train addition, the natural gas production in Brunei is projected to reach an annual output level of 14.9 mtoe in 2010 and will stabilise around the 2010 production levels thereafter. To extend Brunei's oil reserves, the Brunei Oil Conservation Policy was introduced in 1980.[11] Given that the production level was originally set at 150,000 barrels per day and the Oil Conservation Policy was revised only in the mid-1990s to allow more flexible production, the total oil supply in Brunei is projected to increase by only 10 per cent during the forecast period, reaching 11.1 mtoe in 2030.

Total Primary Energy Supply (TPES)

In 1981, Brunei's TPES dropped a significant 28 per cent compared to 1980 and has been fluctuating ever since. Nevertheless, the TPES in 2004 rebounded to a level similar to that of 1980, reaching 2.7 mtoe, with oil and gas accounting for 27 per cent and 73 per cent of the TPES in 2004, respectively.[12]

Brunei's TPES is projected to grow 1.3 per cent annually during the outlook period, reaching 3.1 mtoe in 2030. Oil demand will be driven primarily by population growth and transport fuel consumption, and oil demand is forecast to increase at an annual rate of 2 per cent to 2030. In comparison, natural gas demand is driven mainly by the electricity sector. However, natural gas consumption will grow more slowly than power generation due to efficiency improvements in the electricity industry. Natural gas consumption is forecast to grow at an average annual rate of one per cent during the outlook period. As a result, oil and gas will account for 32 per cent and 67 per cent, respectively, of TPES in 2030.

---

[10] EIA, *Country Analysis Briefs: Brunei.* Washington, DC: EIA, 2006.

[11] Datuk Paduka Haji Idris Haji Belaman, "Energy Situation and Emergency Response Policy: Brunei Darussalam". IEA/ASEAN Workshop on "Oil Security and Emergency Preparedness" and Associated Site Visits, Paris, IEA, 2003.

[12] IEA, *IEA Energy Statistics*. IEA, 2006.

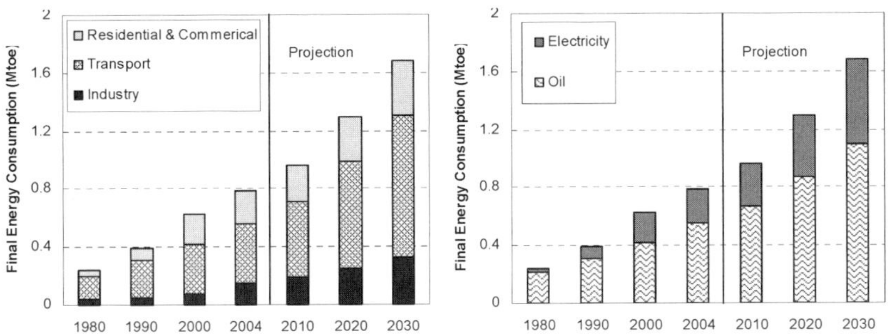

Figure 3: Brunei's Final Energy Consumption by Sector and Fuel Type, 1980–2030

Source: The 1980 and 1990 figures are based on APERC, *APEC Energy Demand and Supply Outlook 2006*. Tokyo: Asian Pacific Energy Research Centre, 2006.
The 2000 and 2004 figures are based on IEA, *Energy Balances of Non-OECD Countries*. Paris: IEA, various years.

Final Energy Consumption

In 2004, total final energy consumption in Brunei was 0.8 mtoe, and the shares of industry, transport, and residential and commercial sectors were 18, 52 and 29 per cent, respectively.[13] During the forecast period, final energy consumption is expected to grow at 3 per cent annually, and the share of transportation will increase to 58 per cent, mainly at the cost of the residential and commercial sectors, whose share will drop to 22 per cent in 2030.

Final energy demand is comprised mainly of oil and electricity. Electricity consumption is expected to grow at an average annual rate of 3.6 per cent to 2030, compared with the 10 per cent growth rate between 1980 and 2004. Oil consumption will grow at a slower annual average rate of 2.7 per cent. As a result, the share of electricity in final energy consumption is projected to increase from 30 per cent in 2004 to 35 per cent in 2030.

Electricity Generation

The Department of Electrical Services (DES) of the Ministry of Development is the Government agency responsible for the generation,

[13] *Ibid.*

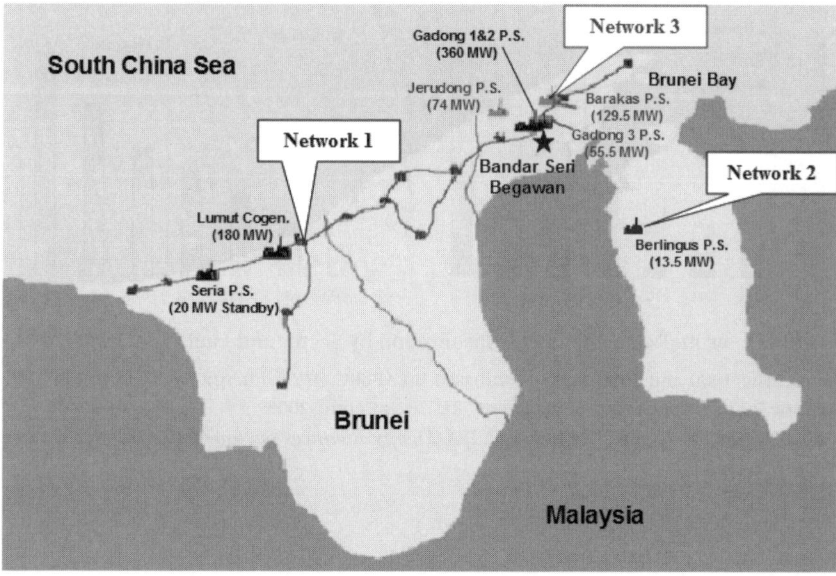

Figure 4: Electric Grid in Brunei

Sources: GENI, National Energy Grid, Global Energy Network Institute, 2006, at http://www.geni.org; Hj Abd Shawal Yaman, Brunei Darussalam. Workshop on the Promotion on Energy Efficiency and Conservation (PROMEEC) in Major Industry Building and Energy Management: SOME-METI Programme 2004–2005, Bandung, Indonesia, 26–27 Jan. ASEAN Centre for Energy, 2006. ; Marubeni Corporation, "Marubeni Awarded Brunei's First Combined Cycle Power Station", 2005, at http://www.marubeni.com/news/2005/050428e.html [5 Dec. 2006].

transmission and distribution of electricity in Brunei. There is also an independent power utility named Berakas Power Company Sdn Bhd (BPC). Metering and cable supply were privatised to make them more competitive and efficient. As illustrated in Figure 4, the country's electrical power system has three independent networks as below:

- Network 1 supplies power to Brunei Muara, Tutong and Belait District;
- Network 2 provides power in the Temburong District; and
- Network 3 in selected load centres in the Brunei Muara District.

DES is the operator of Networks 1 and 2, which have four generating plants with total installed capacity of 574 MW. BPC is the operator of Network 3 with three generating power plants. The total installed capacity of Network 3 is 259 MW and is inter-connected by a triangular network which is separated from the DES grid.[14]

Between 1980 and 2005, Brunei's electricity generation increased at an astonishing rate of 9 per cent annually, and natural gas-fired plants dominated the generation mixture. Of the 2.91 TWh electricity produced in 2005, natural gas power plants accounted for more than 96 per cent of the total, and the remainder was provided by diesel-fuelled engines. Electricity generation in Brunei was expected to increase annually at 2.5 per cent between 2005 and 2030. As a result, per capita electricity generation is forecast to increase from 7.8 MWh in 2005 to 9.6 MWh in 2030. Over the outlook period, natural gas is expected to maintain the dominant share in the electricity generation mix.

ASSESSMENT OF THE ENERGY CONSERVATION PROGRAMMES IN BRUNEI

National Assessment

Although one of Brunei's National Energy Policies is to "instil prudent use of energy in all its forms", energy efficiency improvement and conservation are actually not one of the main policy concerns for the Government. In the past, Brunei focused primarily on its energy conservation efforts in the areas of information dissemination, awareness campaigns and education and training.[15] For example,

---

[14] APERC, *APEC Energy Overview 2005*. Tokyo: Asian Pacific Energy Research Centre, 2005; Marubeni Corporation. "Marubeni Awarded Brunei's First Combined Cycle Power Station", 2005, at http://www.marubeni.com/news/2005/050428e.html [5 Dec. 2006]; ACE, "Member Countries-Brunei Darussalam-Electricity", 2006, at http://www.aseanenergy.org/energy_sector/electricity/brunei/introduction.htm [5 Dec. 2006].

[15] Hj Abd Shawal Yaman, "Energy Security for Prosperity", Stability and Sustainability Inception Workshop Promotion of Energy Efficiency and Conservation (PROMEEC) in Major Industry, Building and Energy Management — SOME-METI Work Programme 2006–2007, Kota Kinabalu, Sabah, Malaysia, 6–7 July 2006. ASEAN Centre for Energy, 2006.

the Brunei Energy Association was formed in 2002 to play a major role in the development of the energy industry and also in information dissemination, giving emphasis to energy conservation and efficiency. However, a survey conducted by the Energy Conservation Centre of Japan in 2005 revealed that Brunei lagged behind most ASEAN countries in terms of established measures for energy efficiency and conservation. Though the Government implemented energy conservation awareness campaigns and participated in international cooperation on energy auditing through the Promotion of Energy Efficiency and Conservation (PROMEEC) programme,[16] it has not worked on other proven measures such as promotion of energy service companies, establishment of energy management systems, incentives and subsidies for energy efficient equipment, energy labelling and demand side management.[17] However, the Government plans to do more to induce efficient energy use in the future as below:[18]

- Benchmark for lighting and air-conditioning;
- Development of energy codes of practices;
- Development of energy audit checklists; and
- Introduction of new technologies (building automation, direct digital controllers, etc.).

To assess the historical energy efficiency improvements in Brunei, and evaluate the effectiveness of existing energy conservation

[16] The PROMEEC project is further sub-divided into three projects, namely: (1) PROMEEC-Building; (2) PROMEEC-Major Industry; and (3) PROMEEC-Energy Management. The PROMEEC projects are a joint undertaking of the ASEAN Centre for Energy (ACE), Energy Conservation Centre, Japan and Energy Efficiency and Conservation Sub-sector Network (EE&C-SSN).

[17] Kazuhiko Yoshida and Takashi Sato, "Summary of Survey Results: Asean 10 Countries", PROMEEC. (Energy Management) for 2004–2005 Summary Workshop ASEAN Centre for Energy, 2005.

[18] Jean-Marc Alexandre, *Overview of ASEAN Experience*. Consultation Workshop, Establishment of the Energy Efficiency and Conservation Office of Vietnam, Dissemination of the Experience to ASEAN. Hanoi, 2006.

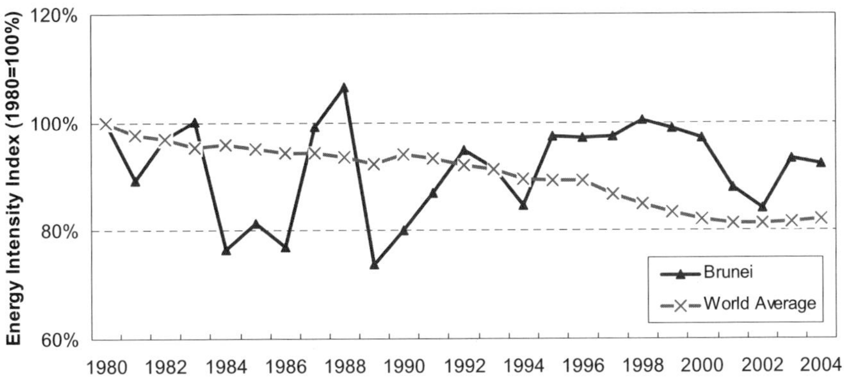

Figure 5: Indexes of Energy Intensity Based on GDP, 1980–2004

Source: British Petroleum, *BP Statistical Review of World Energy 2006*. London: BP, 2006; IEA, *IEA, Energy Statistics*. IEA, 2006; UN, *National Accounts Statistics Database*. Economic Statistics Branch of the UN Statistics Division, 2007.

efforts, the energy intensity of the economy as a whole was first calculated. In this study, energy intensity is defined as the ratio of energy consumption to output. Dividing primary energy consumption by GDP for both Brunei and the world and indexing the results to 1980 generated two economic energy intensity measures (see Figure 5). The energy intensity index shows that globally, the energy demand per unit of GDP declined a moderate 18 per cent between 1980 and 2004. In comparison, the similar index for Brunei fluctuated wildly during the same period, but generally shows a less impressive improvement (8 per cent) over the same period. The fluctuations in Brunei's energy intensity index not only indicate the significant impact of the 1997–1998 Asia Financial Crisis and the recent oil price spike on the economy, but also suggest that the quality of Brunei's historical energy and GDP statistics is relatively poor.

Energy intensity indicators are normally based on economic output. However, despite their popular use, indicators based on economic output are poorer measures of technological or process innovations than those based on physical output because monetary units are affected by many factors not associated directly with or highly correlated to energy consumption, such as costs of labour or the selling price of the final product.[19] In the following section, when the effectiveness of energy conservation measures in individual sectors is assessed, physical output-based energy intensity indicators are utilised whenever possible.

Electricity Sector

All of Brunei's installed power plants are natural gas-fired except for the Belingus Power Station in Network 2. Brunei's gas-fired power plants all operate single natural gas turbines, with the exception of the Lumut cogeneration facilities. As illustrated in Figure 6, the thermal efficiency of the Belingus Power Station is similar to the average levels of oil-fired power plants in Asia. In comparison, the overall efficiency of Brunei's natural gas power plants is extremely low, resulting in excessive waste of natural gas.

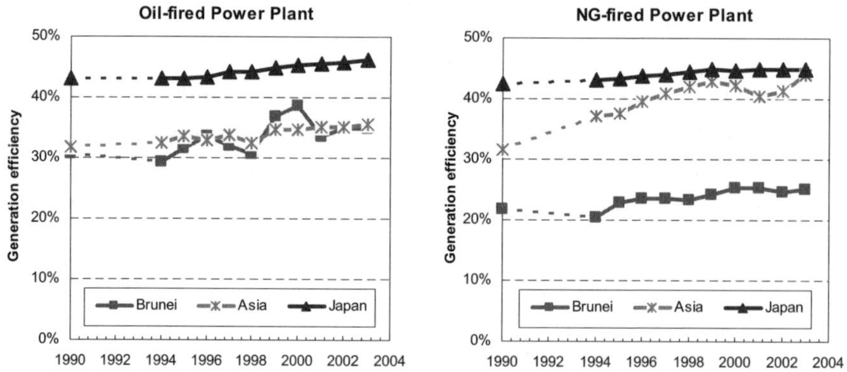

Figure 6: Electricity Generation Efficiency in Brunei

Source: IEA, CO_2 Emissions from Fuel Combustion 1971–2003. Paris: IEA, 2005.

---

[19] John Nyboer, Chris Joseph and Paulus Mau, *Development of Energy Intensity Indicators for Canadian Industry, 1990 to 2004.* Burnaby: Canadian Industrial Energy End-use Data and Analysis Centre, 2006.

The existing energy conservation measures in Brunei's electricity sector are these:[20]

1. Maintain natural gas as the major fuel for power generation.
2. Allow more efficient/larger generation stations to meet "base load", and the smaller/less efficient stations to meet "peak load".
3. Reduce the carbon dioxide content of fuel gas to produce higher electrical power output.
4. Construct greenfield power stations with designed thermal efficiency of greater than 45 per cent.
5. Prevent unauthorised connection by the provision of the Electricity Act (Amendment) Order in 2002.

The energy conservation measures 1, 2 and 3 led to a moderate efficiency improvement of the NG-fired power plants between 1994 and 2003. To comply with the energy efficiency requirement of the Greenfield power plants, Marubeni Corporation was awarded a contract in 2005 for the Bukit Panggal Combined Cycle Power Station, Tutong, Phase 1, with a total output of 110 MW. This project is the first combined cycle power station in Brunei, and is expected to be completed in July 2007. The electricity capacity growth in the future will come mainly from phases 2, 3, and 4 of the Bukit Panggal Combined Cycle Power Station, each of which is expected to be a 200 MW class power station.[21] After the completion of the Tutong power plant, the overall efficiency of natural gas generation stations in Brunei is expected to improve.

Figure 7 shows that the transmission and distribution losses from Brunei's electric grid are very high given the small size of the country. The recent improvement might be partly explained by measure 5 above, which prevents unauthorised connection to the grid.

---

[20] Department of Electrical Service (DES), The Electricity Act (Amendment) Order. Bandar Seri Begawan, 2002; Hj Abd Shawal Yaman, Brunei Darussalam component of Workshop on the Promotion of Energy Efficiency and Conservation (PROMEEC) in Major Industry, Building and Energy Management: SOME-METI Work Programme, 2004–2005, Bandung, Indonesia, 26–27 Jan. ASEAN Centre for Energy, 2006.

[21] Marubeni Corporation, "Marubeni Awarded Brunei's First Combined Cycle Power Station", 2005, at http://www.marubeni.com/news/2005/050428e.html [5 Dec. 2006].

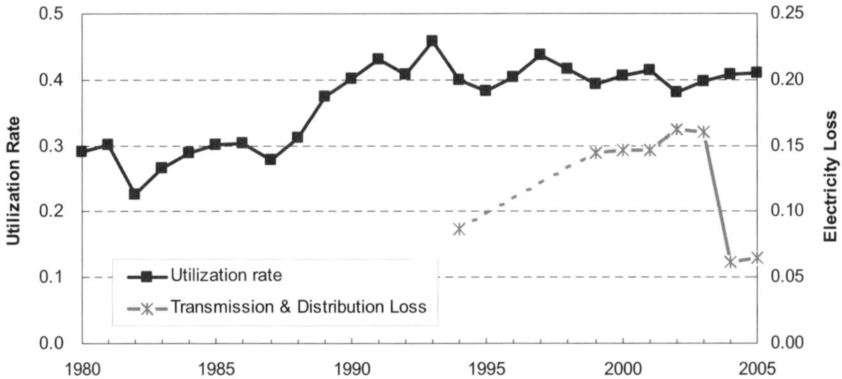

Figure 7: Electricity Plant Utilisation Rate and Electric Transmission Loss in Brunei

Source: Brunei Department of Statistics, *Brunei Sarussalam Key Indicators 2006.* Bandar Seri Begawan, Department of Statistics, Dept. of Economic Planning and Development, Prime Minister's Office, 2006; EDMC, *APEC Energy Database.* Tokyo: Energy Data and Modelling Centre, Institute of Energy Economics, 2007.

Cement Manufacturing Industry

In 1997, a clinker grinding plant with a capacity of 500,000 metric tons per year near the port city of Muara, was owned and operated by Butra Cement Sdn Bhd, which was a joint venture of Brunei, Indonesia and Taiwan. The German multinational cement company Heidelberger Zement AG acquired 50 per cent equity interest of Butra Djajanti Cement Sdn. Bhd in April 2000 and changed its name to Butra Heidelberger Zement Sdn Bhd (hereafter Brunei Cement). Brunei imported raw materials including clinker and gypsum from ASEAN neighbour countries for its cement manufacturing. All cement produced by the plant was consumed domestically.[22] However, Figure 8 shows that Brunei's cement demand plummeted after the Asian Financial Crisis in 1997 and 1998, and Brunei Cement has to operate at levels far below its design capacity especially after the 1998 collapse of the Amedeo Development Corporation. As a result, while the

[22] John C. Wu, *The Mineral Industry of Brunei.* Reston: United States Geological Survey, various years.

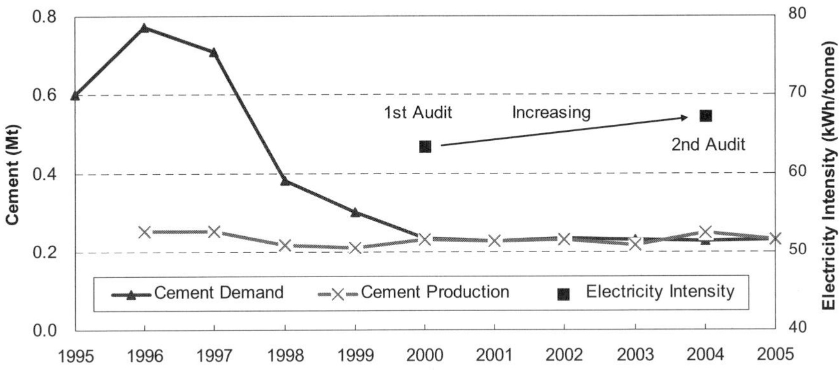

Figure 8: Cement Production and Demand in Brunei vs. Cement Production Electricity Intensity

Source: Brunei Darussalam — Brunei Cement. Seminar on the Promotion on Energy Efficiency and Conservation (PROMEEC) for Major Industry in Southeast Asia, 2005; USGS, Minerals Yearbook. Reston: US Geological Survey, various years; Hideyuki Tanaka, "Summary of Local Workshops and Follow-up Surveys in Major Industries at Brunei Darussalam, Cambodia, Indonesia and Philippines", Summary Workshop: Promotion of Energy Efficiency and Conservation SOME-METI Work Programme 2005–2006, 2006.

design specific power consumption of Brunei Cement is 45 kWh/ton of cement, the actual electricity intensity is at least 41 per cent higher than the design level between 2000 and 2005, leading to excessive waste of electricity.

The Government participated in the energy auditing activities of the PROMEEC project, and Brunei Cement was first selected to be audited in February 2001. And the proposed energy efficiency measures for Brunei Cement are listed in Table 1. Given the fact that the power consumption by Brunei Cement was 14.7 GWh in 2000, these measures could lead to more than 10 per cent of energy saving if implemented appropriately. However, Figure 7 shows that the electricity intensity in Brunei Cement actually increased 6 per cent, reaching 67 kWh/ton of cement in 2004, which was revealed during the follow-up energy audit on 14 December 2005. The unexpected energy intensity growth between 2000 and 2004 was caused by: (1) the frequent plant shut downs in 2004; and (2) the fact that Brunei Cement produces higher than necessary quality cement,

Table 1: Energy Efficiency Measures for Brunei Cement

Energy Efficiency Measures	Improvement
Reactivate capacitor bank	Increase power factor: 0.85–0.89. Expected power consumption reduced by 4.7 per cent
Use grinding aid	Reduce specific power consumption by 10–12 per cent or about 5.4 kWh/tonne cement
Shut down auxiliary equipment during weekly maintenance work	Potential power saving: 1,400 MWh/year
Reduce clincker unloading period from 7 to 4 days by transportation optimization	Potential power saving: 108 MWh/year
Install electric timer to turn off exterior lighting during daytime	Potential power saving: 14 MWh/year

Source: Adapted from Brunei Darussalam — Brunei Cement. Seminar on the Promotion on Energy Efficiency and Conservation (PROMEEC) for Major Industry in Southeast Asia, 2005.

which requires more clinker than otherwise. Moreover, Brunei started to import cement from China in July 2005, and as a result, Brunei Cement's utilisation rate is now constantly below 50 per cent, making Brunei Cement's ability to improve energy efficiency uncertain.[23]

Buildings

In 2004, the residential and commercial sectors represented nearly 90 per cent of final electricity consumption.[24] To curb the expected power demand surge in the future, it is important to improve the energy performance of buildings.

[23] Hideyuki Tanaka, "Summary of Local Workshops and Follow-up Surveys in Major Industries at Brunei Darussalam, Cambodia, Indonesia and Philippines", Summary Workshop: Promotion of Energy Efficiency and Conservation SOME-METI Work Programme 2005–2006 (2006).

[24] IEA, *Energy Balances of Non-OECD Countries.* Paris: IEA, various years.

The Orchid Garden Hotel in Bandar Seri Begawan was the first building selected for an energy audit. The first energy audit began on 6 November 2003 and ended on 10 November 2003. The total gross floor area of the hotel is 21,121 m^2, with energy and water intensity at 2.19 GJ/m^2 and 3.6 m^3/m^2 in 2002, respectively. In comparison, the benchmark energy intensity of the hotel category set by the PROMEEC project is 3.28 GJ/m^2.[25] The relative low energy intensity of the Orchid Garden Hotel can be explained by the following facts: (1) the hotel is only seven years; and (2) the hotel occupancy rate is extremely low for Brunei, with the hotel management even considering a 30–40 per cent occupancy rate during the Christmas and New Year holidays a boom.[26]

Brunei is located in a tropical climate region, and the average daily temperature is near 30°C.[27] As a result, air conditioning alone accounted for more than 70 per cent of the Orchid Garden Hotel's annual electricity consumption in 2002. Table 2 shows the suggested efficiency improvement measures for the hotel. Not surprisingly, increasing the indoor temperature setting alone could contribute more than 3 per cent of annual electricity savings, and the total measures listed in Table 2 could lead to 4.2 per cent of annual electricity savings if implemented appropriately.

The follow-up energy audit was conducted from 22 to 24 January 2004. However, the assessment results are kept confidential for public review, which probably suggests a negative efficiency trend between these two audits.[28] The third energy audit was conducted in 2006, and it is concluded that the Orchid Garden Hotel has achieved a 14.5 per cent improvement in its energy efficiency (about Brunei $3,000/month) after the last energy audit in 2003.[29]

---

[25] Akira Kobayashi, *Review 2003-2004 Energy Audit Orchid Garden Hotel.* Tokyo: The Energy Conservation Centre, 2006.

[26] Husin Ismail and Shafiyi Azahari, "Local Hotels Expecting A Holiday Boom", *Brunei Times*, 14 Dec. 2006; JPKE, *Brunei Economic Bulletin*.

[27] Economist Intelligent Unit (EIU), *Country Review 2006: Brunei.* London: EIU, 2006.

[28] Akira Kobayashi, *Review 2003-2004 Energy Audit Orchid Garden Hotel.*

[29] Lyna Mohamad, "Seminar On Energy Conservation", *Borneo Bulletin*, 10 Sept. 2006.

Table 2: Energy Efficiency Measures for the Orchid Garden Hotel

Suggested Improvement	Electricity Saving (kWh)	Costs (B$)	Per cent of Reduction
Repair of building automation systems			
Optimisation of AHU operating time	Depend on time		
Intermittent use of flow control units	Depend on time		
In-door setting of temperature	145,322	18,892	3.2%
Thermal insulation of hot water pipes	9,855	1,281	0.2%
Optimisation of the receiving transformer	2,310	300	0.1%
Adoption of efficient lamps	25,930	3,371	0.6%
Reduction of filter pump's operating time	6,424	835	0.1%
Total	189,841	24,679	4.2%

Source: Akira Kobayashi, *Review 2003-2004 Energy Audit Orchid Garden Hotel.* Tokyo: The Energy Conservation Centre, 2006.

Oil and Gas Sector

Brunei has proven crude oil reserves of 1.35 billion barrels, as estimated in January 2006. In 2005, Brunei produced 0.24 mbbl/d (million barrels per day), of which 0,22 mbbl/d was crude oil, and about 0.22 mbbl/d was LNG.[30] Oil production in Brunei peaked in 1979 at about 0.26 mbbl/d. However, to extend the life of mature fields and improve recovery rates, the Oil Conservation Policy introduced in 1980 set a production cap of 150,000 bpd,[31] and this policy was revised only in the mid-1990s to "at availability", which allows the oil companies to produce at available capacity without major investment.[32]

Brunei's oil industry is completely dominated by Brunei Shell Petroleum (BSP), which is a 50-50 joint venture between Royal Dutch/Shell and the Government of Brunei. BSP had been the only oil producer in the country and still operates the country's only oil refinery,

[30] EIA, *Country Analysis Briefs: Brunei.* Washington, DC: EIA, 2006.

[31] Barrels per day.

[32] Datuk Paduka Haji Idris Haji Belaman, "Energy Situation and Emergency Response Policy: Brunei Darussalam".

although the sector has opened to other players in recent years. The capacity of the refinery is about 8,600 bbl/d, and around 5,000–6,000 bbl/d of the refinery's output is used for local consumption.[33]

The Petroleum Unit reported that the total natural gas feedstock to the BLNG (part of BSP) was 9.19 mtoe (353.3 billion cubic feet). However, LNG exports in 2002 were only 8.81 Mtoe (352 million MMBtu). Considering that only a small amount of LPG is produced by Brunei LNG to supply domestic customers, the energy conversion efficiency of the Brunei LNG was about 96 per cent, which means that 4 per cent of the feedstock energy content was lost during the liquefaction and delivering process. In 2004, Brunei's petroleum refinery processed 672 kt crude oil, but only delivered 627 ktoe of refined petroleum products, and the implied conversion efficiency of 93 per cent is considered as inefficient according to international standards.[34] In summary, there is significant potential for energy efficiency improvement in Brunei's oil and gas sector.

BSP have committed to several voluntary initiatives in the past. The most recent example is the company's implementation of more aggressive environment standards. As a result, BSP was officially accredited with the Environmental Management System ISO 14001 certification in November 2005.[35] Nevertheless, the relatively low levels of energy efficiency performance of Brunei's oil and gas sector indicate that voluntary initiatives or policy that do not directly target energy performance might be insufficient to induce significant energy efficiency gains in the oil and gas sector.

RECOMMENDATIONS

In 2004, Brunei's energy intensity was about twice world average levels. The energy efficiency levels of its electricity industry, cement

[33] EIA. *Country Analysis Briefs: Brunei.*

[34] IEA. *Energy Balance of non-OECD Countries.*

[35] The ISO 14000 environmental management standards exist to ensure that products and services have the lowest possible environmental impact. See http://www.bsp.com.bn/main/hse/hse_info_iso.asp [17 Jan. 2007].

manufacturing industry and oil and gas sector are all relatively low; moreover, both Brunei's fuel consumption per capita and electricity use per capita are amongst the highest in Asia, which certainly indicates significant room for energy efficiency improvement and conservation. However, due to its enormous oil wealth, energy conservation has apparently not been the main focus of the Government in the past, which is evident as Brunei's past efforts are mainly comprised of information dissemination, awareness campaigns, education and training and irregular energy audits. Considering that the primary energy supply is solely oil and gas, and that the reserves/production ratio of oil was only 15 years in 2005, more aggressive measures should be considered by Brunei's decision makers to further tap the potential of energy conservation and wastage abatement in years to come.

Improve Energy and GDP Statistics Reporting

Reliable and consistent energy and GDP statistics are the basis for energy efficiency assessment. However, a comparison of Brunei's statistics over time and among different data collection agencies often shows unexpected changes and noticeable discrepancies. While the Department of Economic Planning and Development of Brunei has implemented an 'Action Plan to Improve National Accounts Statistics in Brunei Darussalam' by revising the national accounts series, especially estimates of GDP between 2000 and 2005, continuous efforts should be extended to cover national energy statistics and the period before year 2000.

Restructure National Development Plan towards a Softer Path

The total labour force (different from the working-age population) in Brunei reached about 160,500 in 2004, and is growing at 2.5 per cent annually. However, foreign workers made up 45 per cent of the total labour force in 2000, and the real unemployment rate seems to be higher than the official figure of 4.5 per cent, so Bruneians have become increasingly frustrated by the Government's failure to bring about promised economic reforms.[36] Moreover, realising its vulnerability towards the fluctuation of world oil prices, Brunei aims to diversify its economy away from the upstream oil and gas sector. Recent efforts have been focussed on attracting 3.52 billion USD foreign direct investment (FDI) in the Sungai Liang Industrial Park to build several energy-intensive plants such as (1) aluminium smelting; (2) elastomers; (3) methanol; (4) ammonia/urea; and (5) power generation.[37] However, as the jobs created directly from the aforementioned project are only about two thousand, of which the majority would be skilled foreign technical workers, the planned energy intensive industrial development might not fully meet the expectations of Bruneians. Considering that nearly one-half of Brunei's workforce is employed in the public sector, the Government should encourage development of the service industry to alleviate the employment challenges in years to come.

Set Minimum Energy Efficiency Standards for Greenfield Energy Intensive Plants

The Government signed a memorandum of understanding with American aluminum producer, Alcoa, on 25 September 2003 to study the feasibility of constructing a $1.5 billion gas-fired aluminium smelter in Seria.[38] The Brunei Economic Development Board has reportedly offered discounted gas to the plant.[39] To attract FDI in the petrochemical and other energy intensive industries, Brunei is likely to offer similar discounted energy deals in the future. However, to prevent the potential risks of low efficiency plant design, the Government should follow the example set for the electricity sector, and impose minimum energy performance requirements for any greenfield energy intensive plant.

---

[36] EIU, *Country Review 2006: Brunei.*

[37] John Perry. *Development of the Sg Liang Industrial Park.* The Brunei Economic Development Board, 2004.

[38] BEDB and Alcoa, "Joint Press Statement by the Brunei Economic Development Board and Alcoal Primary Development Group", 2003.

[39] EIA, *Country Analysis Briefs: Brunei.*

Reduce Energy Subsidies

Brunei's domestic gasoline and diesel prices are amongst the lowest levels in the ASEAN region, which partly explains the recent fuel consumption spike in the transportation sector.[40] Besides offering discounted electricity deals to the industry, the Department of Electrical Services also imposes a declining tariff system to its residential and commercial customers, which no doubt results in the over consumption of electricity in Brunei.[41] Moreover, in 2005, the DES revealed that electricity dues owed amounted to a whopping $250 million, of which $100 million are owed by the public and $150 million by the government and private sector.[42] To reduce excessive energy consumption, it is necessary for the Brunei government to increase domestic fuel prices and design a more effective electricity tariff system with more enforcement on electricity bill payment.

Implement Demand-side Management

Between 1990 and 2003, the peak load in Brunei increased 144 per cent, reaching 0.44 GW in 2003.[43] The significant increase of peak load could be partly explained by the declining electricity tariff the DES imposes on residential and commercial customers, and the discounted deals with the industry. However, if the peak load follows the historical momentum, it will become more and more difficult to meet demand surge and ensure energy efficiency improvement simultaneously. To tackle the peak demand challenge facing Brunei's electricity industry, a differentiated electricity tariff with peak load pricing should be considered.

---

[40] GTZ, *Fuel Prices in ASEAN Countries.* Eschborn: German Agency for Technical Cooperation, 2005.

[41] See Department of Electrical Services website at http://www.des.gov.bn/tariff.htm [25 Jan. 2007].

[42] Azlan Othman, "Brunei Begins to Audit Usage of Energy," *Borneo Bulletin*, 15 Dec. 2005.

[43] Hj Abd Shawal Yaman, Brunei Darussalam component of Workshop on the PROMEEC in Major Industry, Building and Energy Management.

Conduct Regular Energy Auditing for Major Energy Consumers

The Government participated in the energy auditing activities of the PROMEEC project, and the assessment results so far are mixed. While the Orchid Garden Hotel decreased its energy intensity by 14 per cent between the second and third audits, Brunei Cement's electricity consumption performance deteriorated. Nevertheless, if a national energy management framework could be established to allow industrial plants or buildings to be audited on a regular basis, the potential energy efficiency gains would be significant.

Promote Sustainable Building

The 'concept of sustainable building' reflects the idea of sustainable development and expands the concept of 'green building' to include buildings' impacts on society and the economy.[44] Although Brunei started to participate in building energy audit activities only recently, it should still consider the possibility of leapfrogging to the stage of sustainable building promotion. Compared with traditional building energy efficiency measures, the 'sustainable building' concept covers the full life cycle of buildings instead of a specific step (e.g., design, production, construction, operation, renovation and demolition). If a sustainable building concept could be initiated appropriately, it would be possible for Brunei to significantly lower the expected building power demand surge in the future.

Award Financial Incentives for Fuel Efficient Vehicles

In 2004, the transportation sector accounted for more than 71 per cent of final oil consumption in Brunei, and in recent years the motor vehicle imports tax has accounted for about one per cent of total Government

---

[44] Yoshitaka Ushio, "Promotion of Sustainable Building in Japan — EE&C and CASBEE", PROMEEC (Building) Projects for 2006-2007 Seminar and Workshop in Brunei Darussalam, ASEAN Centre for Energy, 2006.

revenue.[45] As the transportation sector is expected to grow strongly during the outlook period, it is necessary to curb the oil consumption spike in this sector. One possible solution might be lowering the imports tax rates for fuel efficient vehicles and increasing the tax rates for gas-guzzlers accordingly, while the impacts on the Government's income could be neutral, profound purchasing behaviour changes might be induced to improve the average energy efficiency performance of Brunei's transportation sector.

ACKNOWLEDGEMENTS

The views expressed herein are the author's own and do not necessarily represent the views of any organisation with which he is affiliated. The author would like to express his sincere appreciation to Dr. Elspeth Thomson for all her help.

[45] IEA, *IEA Energy Statistics.* JPKE, *Brunei Economic Bulletin*.

Chapter

5

Energy Conservation Policy Development in China

Fuqiu Zhou

CHINA' ENERGY CONSERVATION STRATEGY

Since the reform and opening programme began in 1978, the Chinese Government has put energy conservation work on the agenda, and made improving energy efficiency a major component of the national development strategy and a key component of national energy policy. In 1980, the Chinese Government finalised the energy guideline of "equal focus given to energy conservation and exploitation, giving priority to energy conservation in the near future", and positioned energy conservation as a long-term strategic task.

After the United Nations Environment Development Meeting in 1992, the Chinese Government began to study and stimulate sustainable development energy strategies. In March 1994, it issued *China's 21st Century Agenda,* devoted to improving energy efficiency and making energy conservation a key measure for sustainable development of energy. In March 1996, the 8th National People's Congress

Standing Committee passed the Guideline of the 9th Five-Year Plan and Prospective Targets to the Year 2010 for the National Economy and Society Development on the 4th Session. This proposed to change the economic system from a conventional planned economy to a market economy system and to implement the guiding ideology of following sustainable development and the strategy of "insisting on both resource development and resource economisation with the economisation as the first".

In November 2002, the 16th National Congress of the Chinese Communist Party (CCP) of China, proposed national long term development objectives of building a well-off society in an all-round way; to persist in using IT to propel industrialisation, which will, in turn, stimulate IT application. The hope was to blaze a new trail to industrialisation featuring high scientific and technological content, good economic returns, low resource consumption and low environmental pollution.

In March 2006, the 4th session of the 10th National People's Congress (NPC) passed the Outline of the 11th Five-Year Plan for National Economic and Social Development, to make resource conservation a basic state policy; develop a circular economy; protect the environment, accelerate the development of a resource-conserving, environment-friendly society; promote balanced development among the economy, population, resources and environment; promote the application of information technology in the economy and society, pursue development by taking a new road towards industrialisation, and adhere to resource saving, as well as clean and safe development in order to realise sustainable development.[1]

Energy efficiency and energy conservation have been considered by the Chinese Government for a long time because it realises that improving that these are a requirement for mitigating a medium- and long-term potential lack of energy and environmental damage and raising national economic competitiveness.

---

[1] *Outline of the 11th Five-Year Plan for National Economic and Social Development.* Beijing: The People's Press, 2006.

CHINA'S EFFORTS AND ACHIEVEMENTS IN ENERGY CONSERVATION

Energy Conservation Efforts

Over the past 20 years, the Chinese Government has been making arduous efforts to promote energy conservation and formulate a series of energy conservation policies and measures.[2]

Energy Conservation Planning

Since 1981, the Chinese Government listed energy conservation in the long-term national economic and social development plans as well as in the annual plans, specifying macro direction to local and sectoral work of energy conservation.

Energy Conservation Management and Service System

Since the 1980s, China has formed a relatively integrated energy management system. At the national level, the working meeting system for energy conservation under the State Council was established. It studies and makes decisions concerning major issues of energy conservation, inclusive of researching and examining guidelines, policies, regulations, plans and reform measures associated with energy conservation. It also arranges and coordinates energy conservation work. At the provincial level, the corresponding energy conservation managing organisations were established and energy management organisations with energy conservation management personnel were set at energy-intensive enterprises in charge of their energy conservation work.

Additionally, integrated energy conservation service systems were established to supply technical service and capital support. At present there are over 200 local and industrial energy conservation technical service centres which offer information and consultation services. The China Energy Conservation Investment Corporation provides

[2] Formulating Committee for China Medium- and-Long Term Energy Conservation Plan, *China Medium and Long Term Energy Conservation Plan*. Beijing: China Environmental Science Press, 2005.

capital support for energy-saving programmes. Besides this, many NGOs in China play important roles in the making of decisions for energy conservation, programme assessment, public propaganda, education and training, information service, international exchange, etc.

Energy Conservation Policies, Regulations and Standards

The Chinese Government has formulated, issued and implemented a series of energy conservation laws, regulations and standards to push and regulate the work of energy conservation. By the end of 2005, the central government issued and implemented approximately 20 energy conservation design norms for over 10 industries, over 120 energy conservation standards including general, management and methodology standards and approximately 20 product energy efficiency standards. These cover almost all energy conservation fields, establishing the foundational framework for energy conservation and the comprehensive energy standardisation system, and play an important role in improving the level of China's energy scientific management. They propel enterprises' energy conservation, improving economic and environmental returns, especially after the Chinese Energy Conservation Law was enacted in 1998, which marked the beginning of the use of the rule of law. In addition, governments employ various economic policies to promote energy conservation, including the setting up of a special energy conservation fund, offering preferential interest rates for loans devoted to energy saving projects, tax deductions and exemptions for new energy saving products and implementing peak and differential prices for electric power.

Energy Saving Technology/Product Promotion

The Outline of China's Energy Saving Technology Policy directs and pushes energy conservation technology advancements in key industries such as electric power, iron and steel, nonferrous metals, building materials and petrochemicals, for example adopting large capacity power generating units, improving middle and low voltage power generating units, developing cogeneration in the electric power industry; developing continuous casting technology and basic oxygen furnace (BOF) gas recovery technology in the iron and steel industry; steam self-supply of chemical fertiliser; residual heat recovery and re-use technology of all kinds for kilns; promoting circulating fluidise bed boilers and saving power by variable frequency speed adjustable technology for fans and pumps. In order to spread the timely replacement of normal electric motor products, the Government announced high energy-intensive products for elimination and energy saving products for popularisation, and devoted particular attention to the popularisation of new energy saving fans, pumps, transformers and steel for electric furnaces. By the end of 1998, 18 batches of 1,068 energy-saving electric motor products and 17 batches of 610 electric motor products for elimination were announced. Through the active promotion of energy-saving technologies and products, and limiting and phasing out of low efficient energy-intensive products, energy conservation work has advanced greatly.

Demonstration and Promotion of New Energy Conservation Mechanisms

Since the 1990s, along with the shift of China' economic system to a market economy, the Chinese Government has actively explored new energy conservation mechanisms including introducing international extensively applied integrated resources planning (IRP) methods and demand-side management (DSM) technologies, taking an active part in DSM demonstration and popularisation, introducing and demonstrating energy performance contracting management — a new market-oriented energy conservation mechanism and promoting it across the country; implementing energy conservation VA pilot at iron and steel industry enterprises, etc.

Energy Conservation Promotion, Education and Training

The Chinese Government promotes energy conservation and raises the public awareness of energy conservation through activities such as the annual Energy Conservation Promotion Week. In 1994, the Dalian China Education Centre was constructed with financial assistance and technology from the Japanese Government. This is currently the most modernised large-scale energy conservation training centre, undertaking training for senior energy conservation management and technical manpower. Energy conservation training centres have been established in every province and city, and in the main energy consumption sectors. Regional energy conservation training centres were constructed in Beijing, Tianjin, Shanghai, Chongqing, Harbin, Nanjing, Hangzhou and Xi'an, undertaking the training for enterprises' energy conservation management and technical personnel.

International Exchange and Cooperation in Energy Conservation

International exchange and cooperation in energy conservation includes introduction of international advanced energy-saving equipment and technology, information exchange, inviting international experts to diagnose domestic enterprises, pushing the application of energy saving technology and enterprise energy conservation by using capital and technical assistance from such international organisations as the World Bank, Global Environment Facility (GEF), Energy Foundation (EF), etc. These intensify the expansion of energy conservation, accelerate the shift in energy conservation mechanisms, induce policy studies of energy conservation, etc. At present, the ongoing international cooperative projects associated with energy conservation are the World Bank/China Energy Conservation Promotion Project and UNDP/China End Use Energy Efficiency Project.

Energy Conservation Achievements

With the co-efforts of the Chinese Government and Chinese society, energy conservation work to date has attained notable achievements. Energy efficiency has improved with energy consumption per unit of output value and energy consumption per product decreasing remarkably. Smaller increases in energy consumption support the continuing, rapid economic growth and help protect the ecological environment. The specific energy conservation achievements are outlined below.

Steady Increase in Energy Efficiency Level

The improvement in China's energy efficiency from 1980 to 2002 is given in Table 1. China's energy efficiency in 2002 was 33.4 per cent, an increase of 7.5 per cent from 1980. Of this, the increase resulting from end use energy efficiency was notable, reaching 49.6 per cent in 2002 from 34.4 per cent in 1980. This is attributed mainly to technical retrofitting in energy intensive industries, the development of the use of internal combustion locomotives and electric power on the railways, as well as the significant increase in the proportion of high quality energy use (gas and electricity) in the residential and commercial sectors.

Table 1: China's Physical Energy Efficiency (per cent)

	1980	1989	1997	2000	2002
1. Medial segment efficiency	74.0	72.4	68.8	67.8	67.3
Interim efficiency					
2. End use efficiency					
Agricultural sector	27.7	28.0	30.5	32.0	32.0
Industrial sector	38.7	40.5	46.3	49.6	49.0
Transport sector	21.2	25.4	28.9	28.1	28.0
Residential and commercial sector	29.1	42.5	54.8	66.2	68.1
Total	34.4	38.7	45.3	49.2	49.6
3. Energy efficiency (1 × 2)	25.9	28.0	31.2	33.4	33.4

Note: Medial segment refers to energy processing, transformation and storage. The industrial sector includes the building industry. The residential and commercial sectors include miscellaneous sectors.

Major Decrease of Energy Consumption per Unit of GDP

In terms of constant 1990 prices, energy consumption per 10,000 yuan of GDP dropped from 7.89 tons coal equivalent (tce) in 1980 to 2.83 tce in 2005, a decrease of 64.1 per cent. From 1981 to 2005, China's average annual GDP growth rate was 9.8 per cent, while primary energy registered 5.4 per cent and the elasticity of average annual energy consumption was 0.55. This indicates that since conducting active energy conservation work in 1980, China successfully demonstrated that a relatively low elasticity of energy consumption could support a relatively high economic growth rate, i.e., that energy conservation plays a very important role in supporting China's economic sustainable, rapid and sound development. During this period, China's average annual energy conservation rate was up to 4 per cent, an accumulated 1.09 billion tce, saving an equivalent of 770 million tons of carbon emissions. Thus China's energy conservation efforts yielded not only energy conservation and economic returns, but also made an important contribution to global GHG emission deduction (see Table 2).

Decrease in Energy Consumption per Product

Over the past 20 years, the per unit energy consumption of the main energy-intensive products dropped remarkably (see Table 3).

CHINA'S ENERGY CONSERVATION TARGETS FOR 2010 AND NEW POLICY MEASURES OF ENERGY CONSERVATION

Energy Conservation Targets for 2010

Despite making the above achievements in energy conservation, compared with developed countries, energy efficiency in China is still rather low.[3] Currently, China's energy consumption per unit of GDP is approximately 3, 4 and 8 times as much as that of the average world level, OECD countries and Japan, respectively. Moreover, energy consumption per unit of major industrial products was generally

[3] Wang Qingyi, *Comparison between China's Energy Efficiency, Energy Conservation and Environment with Foreign Countries*, 2005(6).

Table 2: China's Economic Growth, Energy Consumption and Energy Conservation Status, 1980–2005

	GDP (Billion RMB)	GDP Growth Rate (%)	Total Energy Consumption (Mtce)	Energy Consumption Growth Rate (%)	Energy Consumption Per unit of GDP (tce/10^4 yuan)	Elasticity of Energy Consumption	Energy Conservation Rate (%)
1980	763.8	7.8	602.8	2.9	7.89	0.37	—
1981	803.9	5.3	594.5	−1.4	7.39	−0.26	6.30
1982	876.4	9.0	620.7	4.4	7.08	0.49	4.22
1983	971.8	10.9	660.4	6.4	6.80	0.59	4.05
1984	1119.3	15.2	709.0	7.4	6.33	0.49	6.78
1985	1270.1	13.5	766.8	8.1	6.04	0.60	4.69
1986	1382.7	8.9	808.5	5.4	5.85	0.61	3.15
1987	1542.7	11.6	866.3	7.2	5.62	0.62	3.96
1988	1716.5	11.3	930.0	7.3	5.42	0.65	3.52
1989	1786.3	4.1	969.3	4.2	5.43	1.04	−0.16
1990	1854.8	3.8	987.0	1.8	5.32	0.48	1.93
1991	2025.3	9.2	1037.8	5.1	5.12	0.56	3.71
1992	2313.7	14.2	1091.7	5.2	4.72	0.36	7.92
1993	2636.3	13.9	1159.9	6.2	4.40	0.45	6.75

(Continued)

Table 2: (*Continued*)

	GDP (Billion RMB)	GDP Growth Rate (%)	Total Energy Consumption (Mtce)	Energy Consumption Growth Rate (%)	Energy Consumption Per unit of GDP (tce/10^4 yuan)	Elasticity of Energy Consumption	Energy Conservation Rate (%)
1994	2981.4	13.1	1227.4	5.8	4.12	0.44	6.43
1995	3307.3	10.9	1311.8	6.9	3.97	0.63	3.66
1996	3638.5	10.0	1389.5	5.9	3.82	0.59	3.72
1997	3976.2	9.3	1378.0	−0.8	3.47	−0.09	9.25
1998	4287.7	7.8	1322.1	−4.1	3.08	−0.52	11.02
1999	4614.9	7.6	1338.3	1.2	2.90	0.16	5.95
2000	5003.4	8.4	1385.5	3.5	2.77	0.42	4.51
2001	5418.9	8.3	1432.0	3.4	2.64	0.40	4.57
2002	5911.4	9.1	1518.0	6.0	2.57	0.66	2.83
2003	6503.9	10.0	1749.9	15.3	2.69	1.52	−4.78
2004	7159.7	10.1	2032.3	16.1	2.84	1.60	−5.50
2005	7868.6	9.9	2224.7	9.5	2.83	0.96	0.39

Sources: *China Statistical Yearbook 2005*; *China Energy Statistics Yearbook 2004*.
Note: GDP is calculated in constant 1990 prices.

Table 3: Energy Consumption per Unit for the Main Energy-Intensive Products

	1980	1990	2000	2004
Coal consumption of power supply /gce/kWh	448	427	392	379
Comparable energy consumption per ton steel /kgce/t (medium and large scale enterprises)	1201	997	784	705
Comprehensive energy consumption of cement /kgce/t	218.8	201.1	181.0	157.0
Alternating current consumption of electric aluminium /kWh/t	20342	16223	15480	15080*
Comprehensive energy consumption of crude oil processing /kgce/t	118.1	102.5	118.4	112.0
Comprehensive energy consumption of ethylene /kgce/t	2013	1580	1125	1004
Comprehensive energy consumption of synthetic ammonia /kgce/t (large-sized, natural gas)	1431	1343	1273	1220*
Comprehensive energy consumption of caustic soda /kgce/t (membrane process)	1870	1660	1563	1493
Comprehensive energy consumption of Soda ash /kgce/t				
ammonia soda process	571	560	467	455
non-ammonia soda process	386	387	313	325
Comprehensive energy consumption of calcium carbide /kgce/t	2570	2212	2190	2150
Comprehensive energy consumption of phosphorus /kgce/t	N.A	8583	7450	7340

Sources: State Grid Cooperation of China, China Steel and Iron Association, China Building Material Industry Association, China Chemical Energy Conservation Technical Association.

Note: * Refers to 2003; gce and kgce are the abbreviations for gram and kilogram of coal equivalent.

20–90 per cent higher than that of the same products abroad, averaging 40 per cent higher. This means China still has huge potential for energy conservation.

The Eleventh Five-Year Plan period will be a crucial time in building a well-off society in all respects. Constrained by various factors such as resources, environment, etc., it is very difficult to meet China's ever-increasing demand for energy solely by relying on exploration for more energy resources. China's economic and social sustainable development must take an energy saving-oriented path. Seriously pursuing the potential of energy conservation and continuing to improve the efficiency of energy use are strategic choices in support of realising the long-term national development target of building a well-off society in all respects, and the promotion of balanced economic development reconciling energy and environmental concerns. The Chinese Government has earnestly acknowledged this point, and in 2005 decided to build a resource-conserving society.

The Outline of the 11th Five-Year Plan for National Economic and Social Development, passed by the 4th session of the 10th National People's Congress in March 2006, set new energy conservation objectives for economic and social development: energy consumption per unit of GDP in 2010 will be reduced by 20 per cent compared to that at the end of the 10th Five-Year Plan period. This reflects the strategic plan of the Chinese Government to change economic and social development into an all-round, balanced, sustainable way and controlling the total amount of energy consumed during the 11th Five-Year Plan period and improving the economic returns of energy utilisation. It amounts to a good historical opportunity.

New Energy Conservation Policy Measures

Since 2005, the Chinese Government has made a series of arrangements for promoting energy conservation work. It has also initiated several major policy measures:

Revision of the Energy Conservation Law

Revision of the Energy Conservation Law was listed in the annual legislation plan of the National People's Congress and the amendments will be voted on in 2007. The revised laws will include more and effective economic incentive measures.

Commencement and Implementation of Ten Major Key Energy Conservation Projects

In 2005, the Chinese Government began implementing ten major key energy conservation projects. It is anticipated that 240 million tce (mtce) will be saved during the 11th Five-Year Plan period by implementing these ten key energy conservation projects and providing material support to fulfil the year 2010 energy conservation targets. These projects are:

- Coal-fired industrial boiler (kiln) retrofitting
- District cogeneration
- Residual heat and pressure utilisation
- Petroleum saving and substituting
- Motor system energy saving
- Energy system optimisation
- Building energy conservation
- Green lighting
- Government agency energy conservation
- Energy saving monitoring and testing, and technology service system building

Launching a One-Thousand-Enterprise Energy Conservation Activity

In April 2006 the NDRC launched a 1,000-enterprise energy conservation campaign. The one thousand enterprises refer to 1,008 enterprises whose annual energy consumption is above 180 thousand tce. Total energy consumption in 2004 of these one thousand enterprises was 670 million tce, accounting for 33 per cent of the total amount of national energy consumption and 47 per cent of industrial energy consumption. Thus, in order to realise the year 2010 energy conservation targets it is very important to focus on the energy conservation work of these one thousand enterprises. The campaign aims at dramatically improving the level of energy utilisation efficiency and the expected energy savings amount to about 100 million tce.

Establishment of GDP Energy Consumption Index

In December 2005, the NDRC issued a notice to implement the Communiqué on GDP Energy Consumption Index. Since 2006, the NDRC, National Energy Office and National Bureau of Statistics, will publicise at the end of June each year the indices of the previous year for each region, such as energy consumption per 10,000 yuan GDP, the decrease rate of energy consumption per 10,000 yuan GDP, energy consumption per 10,000 yuan value-added of industrial enterprises above designated size, and electricity consumption per 10,000 yuan GDP. The publicised data is based on figures verified by the National Bureau of Statistics. The implementation of the Communiqué on GDP Energy Consumption Index is a very important measure to expedite energy conservation.

SUGGESTIONS FOR CONFRONTING THE CHALLENGES FACED IN REALISING THE YEAR 2010 ENERGY CONSERVATION TARGETS AND CORRESPONDING POLICIES

Challenges in Realising the 2010 Targets

China has many positive conditions for energy conservation work. There have been tremendous economic gains over the past 20 years and the Chinese Government has considerable management experience in energy conservation. A standard system of regulations and laws, policy support system, supervision and management system has been established, as well as a technical service system for energy conservation. However, during the 11th Five-Year Plan period, China will undergo an acceleration in industrialisation and urbanisation, as well as a surge in civilian consumption. These factors will likely cause energy demand to soar during this period. The situation is still very tight for energy conservation work in the 11th Five-Year Plan period.

Lately, China's industrialisation has picked up speed. The energy intensive industries are mainly the petrochemical, iron and steel and building materials. China's steel output was up 24.6 per cent over the previous year; rolled steel output was up 24.1 per cent; aluminium oxide up 21.9 per cent; caustic soda up 21.4 per cent; and cement up by 10.0 per cent. The value-added of heavy industries accounted for 69.0 per cent of the total value-added of the industrial sector. In fact, the rapid increase in the production of energy-intensive products has driven the high increases in electricity and energy consumption in recent years.

Energy consumption per unit of GDP depends mainly on the structure of the industrial sector. The energy-intensive secondary industry is currently high in the industrial structure, 47.3 per cent in 2005, whereas, it is typically 30–40 per cent in other industrialised countries. Thus, a key approach to decreasing energy consumption per unit of GDP is to adjust the tertiary industry and optimise the secondary industry. Energy consumption per unit of output value in secondary industry is currently 2.3 times higher than that of the tertiary industry. The potential to reduce energy consumption per GDP through the adjustment and optimisation of industrial structure during the 11th Five-Year Plan period is very great.

However, in the undeveloped central and western regions of China, many provinces are still employing secondary industry to drive economic development during the 11th Five-Year Plan period. The greatest challenge for China in realising the year 2010 energy conservation targets is properly adjusting the structure of the three industries, especially optimising the internal structure of the secondary industry.

Relevant Policy Suggestions

(1) Study and stimulate policy measures to promote the development of service industries, and bring into play the effect of service industries on mobilising funds; adopt powerful measures to expedite the development of low energy consumption and high value-added tertiary industries, especially labour-intensive service and

modern service industries in order to shift the long-term backward status of the service industry and enhance its proportion in the national economy.
(2) Encourage development of new and high technology industries, giving priority to low energy consumption industries such as the information industry which can be a major driving force in economic growth.
(3) Encourage the transformation of traditional industries with new and advanced technologies; promote the optimisation and upgrading of the industrial structure.
(4) Stimulate the development planning policies of energy-intensive industries such as iron and steel, nonferrous and cement.

CONCLUSIONS

To sum up, here are the main conclusions:

(1) China's energy conservation strategy has all along been an important component of the national development strategy and one of the key components of the national energy policy. Improving energy efficiency and practicing energy conservation is required to substitute for the medium- and long-term potential lack of energy, alleviate environmental damage and raise national economic competitiveness.
(2) Over the past 20 years, the Chinese Government has been making arduous efforts to promote energy conservation by formulating a series of energy conservation policies and measures. With the co-efforts of the Government and whole society, China's energy conservation work has realised notable achievements, with energy efficiency continuously improving, as well as energy consumption per unit of output value and energy consumption per product. Continued progress in this area will support the ongoing rapid economic growth, and make a major contribution to the protection of the ecological environment.
(3) Compared with developed countries, China's energy efficiency level still is low. The new energy conservation objective is to

decrease energy consumption per unit of GDP in 2010 by 20 per cent compared to that at the end of the 10th Five-Year Plan period. In order to push the energy conservation work of the 11th Five-Year Plan and realise this target set for 2010, the Government has embarked on revising the Energy Conservation Law, implementing ten major key energy conservation projects, launching a 1,000-enterprise energy conservation campaign and establishing the Communiqué on GDP Energy Consumption Index.

(4) As considerable industrialisation of the central and western regions of the country is expected to take place over the 11th Five-Year Plan period, energy conservation work will be very difficult. The greatest challenge will be adjusting the industrial structure, especially optimising the internal structure of the secondary industry at a time when industrialisation is accelerating.

Chapter

6

Energy Conservation Policy Development in Indonesia

Bob Sugeng Hadiwinata

INTRODUCTION

Throughout the history of mankind, world civilisations have relied heavily on energy supplies. From ancient Egypt under the Pharaohs to the Industrial Revolution, human beings have always been dependent on energy to build shelter, cook, protect them from chilling winters, power their transport, light their homes, fulfil their thirst for consumer products and luxurious goods, etc. One important source of energy is fossil fuel. The quadrupling of oil prices in only four years (2001–2005) not only generated panic in the business and industrial sectors throughout the world, but also raised a new concern for global energy security. Overshadowed by a fear that oil and other fossil energy products will eventually run out, policy makers have become concerned about energy security.

In his book, *Energy and World Politics*, Mason Willrich maintained that energy security is often understood in two different contexts.[1]

[1] Mason Willrich, *Energy and World Politics*. New York: The Free Press, 1978, p. 67.

First, energy security may be viewed narrowly as the guarantee of sufficient energy supplies to permit the country to function during war. This interpretation was popular during World War II when Germany made drastic cutbacks in civilian energy consumption to support the war. Second, energy security may be viewed more broadly as the assurance of adequate energy supplies to maintain the national economy at 'normal' levels. The concept 'normal' economy is often associated with reasonable economic growth for ensuring societal welfare and allowing the national economy to function in a politically acceptable manner.

In order to maintain politically acceptable economic activities, state leaders try to devise policies that guarantee national energy security. In many cases, energy security policies include three major components. First, 'rationing' attempts to allocate available supplies and limit consumption. A country adopts this type of policy on the grounds that reductions in consumption will diminish the magnitude of many energy supply problems and extend the time for solving them. Second, 'stockpiling' aims at reducing an importing country's vulnerability to a supply interruption by providing a cushion against its effect. In this type of policy, a country may set aside sufficient national energy reserves to ensure security and hedge against abnormal price fluctuations. Third, 'diversification' attempts to ensure continuity of energy supplies by diversifying sources and suppliers. Developing alternative sources (coal, nuclear, solar power, hydropower, etc.) will reduce a country's dependence on a single energy source. Similarly, by opening contacts with other suppliers, a country will reduce its dependence on a single supplier of energy.[2]

Indonesia has many reasons to be concerned about its national energy security. Although this country is endowed with indigenous energy resources that make it a net energy exporter, it is increasingly facing a turning point and becoming a net energy importer. Indonesia is self sufficient in terms of energy supply except oil. It is the world's largest liquid natural gas (LNG) exporter and second largest coal

[2] *Ibid*, pp. 70–79.

exporter after Australia. In 2004, coal accounted for 46 per cent of total net exports, followed by natural gas (26 per cent), crude oil (20 per cent), and oil products (7 per cent). In 2004, exports of LNG accounted for about 41 per cent of total natural gas production.[3] The largest proportion of the LNG exports go to Japan (about 70 per cent), followed by South Korea (20 per cent), and the remainder to Taiwan, Singapore and Malaysia.[4]

While natural gas production has been increasing at a moderate rate of 1.1 per cent annually, growing from 71.1 million tons oil equivalent (mtoe) in 200 to 74.2 mtoe in 2004, oil production has been declining. Indonesia currently has proven oil reserves of 4.7 billion barrels, down 13 per cent since 1994. It has been a net oil importer since 2002. Oil production decreased from 70.6 mtoe in 2000 to 54.6 mtoe in 2004 due to depleting reserves and lack of investment for new exploration and development. Crude oil production also saw a slight decrease from an average of 1.10 million barrels per day (bpd) in 2002 to some 1.02 million bpd in 2003. A combination of political and economic crises, confusing rules and regulations pertaining to oil production and supply, and an unfavourable business environment prevented the Government from developing new exploration and investment in oil production.

While the economy is increasingly dependent on oil as an energy source, there is pressure on the Indonesian Government to reduce oil dependency by switching from oil to natural gas in the industrial sector, and from oil to coal and natural gas in the electricity sector. The Government is also promoting the use of renewable energy — especially bio-fuels — in the transportation sector.

This chapter examines Indonesia's attempts to pursue its national energy security policy in response to the growing energy demand and oil price increases over the past few years. It argues that the political transition from autocratic government to democracy in

[3] Asia Pacific Energy Research Centre (APERC), *APEC Energy Demand and Supply Outlook 2006*. Tokyo: Institute of Energy Economics, 2006, p. 33.

[4] US Department of Energy (USDOE), "Indonesia: Country Analysis Brief", 2006, at http://www.eia.doe.gov/emeu/cabs/indonesia.html [12 Dec. 2006].

1998 has allowed the Government to develop more market-oriented energy policies, especially in pursuing oil sector reform and alternative energy sources. The first section discusses the trends in energy production and consumption in Indonesia. This is followed by a discussion of the democratic transition in 1998 and the subsequent oil sector reform initiated by the Indonesian Government. The third section examines the Government's policies to ensure national energy security, especially during the oil price increase in 2005. Fearing that oil prices may continue to rise, the Government began to adopt a more rational pricing policy, energy saving regulations and renewable energy.

ENERGY CONSUMPTION AND PRODUCTION IN INDONESIA

Indonesia will eventually become a net importer of energy if the Government allows the current energy consumption pattern to continue. Having survived the Asian Financial Crisis of 1998, energy consumption has grown rapidly in recent years. Primary energy consumption increased from 137.4 mtoe in 2000 to 168.9 mtoe in 2004, growing at 5.2 per cent per year compared to 2.9 per cent between 1995 and 2000. Among the fossil fuels, coal increased at the fastest rate of 12.7 per cent per year, due to construction of coal-burning power plants in recent years. Natural gas grew at the second fastest rate of 8.4 per cent per year driven by the expansion of the industrial sector.

Meanwhile, Indonesia's final energy demand grew at 2.9 per cent annually. It is projected that by 2030 total final energy demand will reach 247 mtoe, more than double that of 2005 at 114 mtoe. The industrial sector has the largest share of energy consumption at 40 per cent, followed by residential (29 per cent), transport (28 per cent) and commercial (3 per cent) in 2005.[5] It is projected that Indonesia's primary energy demand will more than double from 171 mtoe in 2005 to 359 mtoe in 2030. By energy source, demand for coal is

[5] APERC, *APEC Energy Demand and Supply Outlook 2006,* p. 34.

projected to grow at 4.7 per cent per year, followed by oil and natural gas at 2.8 per cent, hydro at 2.6 per cent and renewable energies (bio-diesel, bio-mass, etc.) at 1.3 per cent.[6]

The increase in coal demand will be mainly driven by electricity generation as Indonesia promotes the construction of mine-mouth coal-powered electricity generation plants. Electricity demand is estimated to more than triple, growing annually at 4.6 per cent through 2030. In 2004, the electrification ratio was only 58 per cent. However, over the period of 2005 to 2030, it is planned to increase to 95 per cent in 2030. Consequently, total installed generation capacity is expected to increase almost four-fold, from 31 GW in 2005 to 108 GW in 2030. Of this total, coal will approximately account for 54 per cent of capacity requirements and natural gas, 40 per cent.[7] The increasing demand for electricity has much to do with the rapid expansion of industrial estates and residential areas. Table 1 reveals the growing pattern of energy consumption and electricity use in Indonesia.

Following the 1998 Asian Financial Crisis, Indonesia's economy began to recuperate. National GDP grew at 4.1 per cent in 2003, up from 3.7 per cent in 2002.[8] From 2005 to 2030, GDP is projected to grow at 4.6 per cent per year, from USD820 billion in 2005 to

Table 1: Electricity and Energy Use in Indonesia, 2000–2003

	2000	2001	2002	2003
Electric power consumption (kwh/per capita)	400	422	428	440
Energy consumption (kg of oil equivalent per capita)	695	728	745	753
Energy imports net (per cent of total energy use)	–59	–55	–54	–55

Source: International Energy Agency (IEA), "Energy Statistics and Balance of Non-OECD Countries", 2005, at http://devdata.worldbank.org/data-querry [10 Dec. 06].

[6] *Ibid.*, p. 35.

[7] *Ibid.*, p. 36.

[8] USDOE, "Indonesia: Country Analysis Brief".

USD2,795 billion in 2030. In 2003, the IMF-guided structural adjustment programme ended with success. In March 2003, the IMF disbursed its last financial assistance of USD469 million as part of its bailout package after reporting that Indonesia had made good progress in instituting reforms. The IMF review cited Indonesia's continued economic growth, decreasing inflation rates and strengthening banking sector as examples of progress made, while mentioning that more reform in other sectors (law enforcement, taxation, favourable business environment, etc.) are necessary.

The data in Table 1 indicates that Indonesia is currently a net exporter of energy, except for oil. It is estimated that the country's natural gas reserves are around 92.5 trillion cubic feet (tcf). Most are located near the Arun field in Lhoksemauwe, Aceh, around the Badak field in East Kalimantan, in smaller fields off-shore Java, West Papua and Natuna Island. Despite debate that the largest producer in the Arun field will soon run out, Indonesia's natural gas reserves are still promising. One project that holds great promise for Indonesia's LNG export is the Tangguh project in West Papua which is currently being explored by British Petroleum. It is estimated that production in the Wiriagar Blocks alone will produce at least 14 tcf ready for export. Completion of the 400-mile Natuna pipeline (one of the longest undersea gas pipelines in the world) has also brought natural gas directly from Natuna Island to Singapore. Operated jointly by Premier Oil, Conoco Phillips and Star Energy, the project will soon be expanded as new pipeline proposals connecting East Natuna with the Philippines are under consideration.

Indonesia also has 5.9 billion short tons of recoverable coal reserves. Almost two-thirds of the total coal reserves are located on Sumatra with the remainder in Kalimantan, West Java and Sulawesi. Indonesia has become the second largest coal exporter after Australia. Coal exports are sent to Japan, Taiwan, South Korea, the Philippines and Hong Kong. By 2010, Indonesia is planning to double its coal exports from 144 million short tons (mmst) in 2005 to 300 mmst. For this purpose, the Government invited private companies (both local and foreign) to develop private coal mines. For example, Clough Group of Australia was awarded a USD215 million contract to help improve the Indonesian firm GBP's Kutai mine in East Kalimantan. The development of new mines also involves another Australian firm with interests in coal mining, the Broken Hill Propriety.

Indonesia currently has oil reserves of 4.7 billion barrels, a decrease of 13 per cent since 1994. Much of the country's proven oil reserve base is located onshore, the largest of which is Duri and Minas Oil fields in Riau, Sumatra. Other oilfields can also be found offshore in Java Sea, East Kalimantan and the Natuna Sea. In 2002, a study by Indonesia's Directorate of Oil and Gas shows that new oil reserves in the Cepu block (Central Java) contain around 600 million barrels, about half of which is considered recoverable. Although the prospect for the development of the Cepu block will greatly boost Indonesia's oil production, especially when cooperation between ExxonMobil and Indonesia's oil company Pertamina comes online, Indonesia will continue to be a net oil importer. Between 2000 and 2005, oil production declined at an annual rate of 0.5 per cent. In 2003, crude oil production averaged 1.02 million bpd, down from the 2002 average of 1.10 million bpd.[9] The decline was due mainly to the natural fall-off or aging of oil fields, a lack of new exploration investment and Government will to respect business contracts with foreign firms, the omnipresence of regulatory hurdles that create disincentives to private firms, the dominant role of the state oil company (Pertamina) in determining national energy policies, etc.

With respect to energy prices, from the 1980s to 2005 Indonesia (and Iran) has gained attention for its highly subsidised oil and gas prices. In one of its working papers, the IEA indicates that energy-related subsidies in non-OECD countries tend to be higher than in OECD countries. Iran and Indonesia together subsidised oil more than the OECD energy subsidies in total. Iran subsidised oil product sales to the tune of USD11 billion in 2003, with a further USD3 billion going to electricity and some USD2.8 billion to natural gas. Energy subsidies in Iran were equal to 10 per cent of GDP — by far the highest share in the world.[10]

---

[9] *Ibid.*

[10] International Energy Agency (IEA), "Carrots and Sticks: Subsidising and Taxing Energy", Working Paper, Tokyo, 17 Jan. 2006, pp. 1–3.

Indonesia followed suit by subsidising oil at an average of USD6 billion per year between 2000 and 2005. A recent study by ADB, however, shows that the oil price increase in the world market in 2004–2005 raised the country's oil subsidies to over USD7 billion in 2004 and to more than USD12 billion in 2005 — equal to 5 per cent of GDP and almost 30 per cent of total spending.[11] For Indonesia, despite the financial burden that must be born by the Government, the oil subsidies which almost entirely go to consumers rather than producers are important. From the Government's point of view, oil subsidies are crucial in order to keep inflation low and allow the poor to get access to energy sources. While the legitimacy of the Government depends on its capability to satisfy popular energy demand at affordable prices, any policy that may trigger inflation generates public dissatisfaction and mass demonstrations which potentially disrupt regime continuity. Even a long standing autocratic leader such as Suharto could not stand such uproar when his New Order regime was brought down just a few weeks after he announced a dramatic increase in oil and electricity prices amid the Asian Financial Crisis in 1998.

Apart from being a financial burden to the Government, energy-related subsidies tend to negatively affect the country's energy sustainability. With the subsidised price, energy suppliers — including electricity generators and oil refineries — cannot earn a satisfactory rate of return. This low incentive leads to a lack of investment for developing infrastructure and replacing obsolete technologies in energy production and supply. This may explain why in the past few years several regions in Indonesia experienced frequent brownouts or blackouts due to inadequate electricity generation and transmission facilities. For example, on 18 August 2005, broad swaths of Java and Bali were powerless as grid failure forced several power plants to go temporarily offline. The Java-Bali grid experienced an electricity deficit of around 11,000 MW or almost 80 per cent of customary total supply of 14,655, for up to 12 hours in certain areas due to transmission problems.[12]

[11] *Ibid.*, p. 4

[12] US Embassy, "Indonesia: Energy Highlight August 2005", Jakarta, US Embassy Reports, 2005.

Between 22 and 24 November 2006, rolling blackouts recurred in several parts of Java and Bali including Surabaya (the capital of East Java province). The state electricity company, PLN, said that this time the Java-Bali grid suffered power supply shortages of 560 MW due to technical problems in several power plants.[13] Certainly, the lack of interest on the part of the domestic and foreign firms to make new seems to be the cause of the electricity generation deficits in Indonesia.

THE DEMOCRATIC TRANSITION AND OIL SECTOR REFORM

The year 1998 saw a dramatic political transition to democracy in Indonesia. Unable to contain the ongoing pro-democracy protests and demonstrations (widely supported by students, NGO activists, intellectuals, opposition leaders, workers and other elements of civil society), President Suharto was forced to resign in May 1998. It was unfortunate that attempts by the pro-democracy forces to install democracy in the country were tainted with anti-Chinese riots in major cities such as Jakarta, Solo and Medan killing dozens of people, many of whom were of Chinese origin. Under pressure from the pro-democracy activists, Suharto's successor, B. J. Habibie, began to liberalise the political system by introducing laws allowing new political parties to be formed and guaranteeing the political freedom of the citizen, conducting free and fair elections, gradually dismissing non-elected military representatives in the legislature, and removing state control on societal organisations.[14]

Following the democratic transition, the state's domination of most aspects of societal life began to draw to a close. Under the 'New

---

[13] *Ibid.*

[14] Bob S. Hadiwinata and Christoph Schuck, "Mapping Indonesia's Way Towards Democracy: in Search of Theoretical Frame", in *Democracy in Indonesia: The Challenge of Consolidation,* eds. Bob S. Hadiwinata and Christoph Schuck. Baden-Baden: Nomos, 2007, p. 16.

Order' regime, the government built its legitimacy on oil revenues.[15] However, the state-owned oil company, Pertamina performed rent-seeking. Monopolising drilling concessions and investment in oil production and supply, Pertamina served as the crucial financial machinery for the military-supported New Order regime, especially during the oil boom period in the early 1980s. In the words of Richard Robison:

> Pertamina constituted the channel through which the bulk of the state's revenue flowed, as well as the largest and most concentrated source of contracts for construction and supply. Pertamina therefore was the strategic focus of economic power and the crucial source of revenue. The autonomy and hegemony of the military was closely dependent upon its ability to maintain its control over this terminal and to prevent its absorption by any regularised state apparatus.[16]

Representing military interests, Pertamina carried with it some crucial functions including: (1) collection of oil revenues and coordination of production sharing with foreign oil companies; (2) catalyst and financier of major development projects favoured by the president and military leaders; (3) financier of the military; and (4) a source of wealth for loyal supporters of the New Order regime, especially within the military circles.[17] Under such circumstances, any attempt to make Pertamina accountable to the Ministry of Mines was always doomed to failure. For example, in 1967, the Minister of Mines, Slamet Bratanata, challenged Pertamina's director, Ibnu Sutowo over the allocation of drilling concessions and the process of tendering in oil exploration.

---

[15] The 'New Order' regime refers to the military-backed authoritarian government led by General Suharto from 1966 to 1998. The regime was a by-product of the political turmoil during 1965 and 1966 which led to the fall of President Sukarno and the massacre of nearly half a million members and affiliates of the Indonesian Communist Party (PKI) throughout the country.

[16] Richard Robison, *Indonesia: The Rise of Capital*. Sydney: Allen & Unwin, 1986, pp. 234–235.

[17] Richard Robison added another function category for Pertamina, that is, a core around which domestic capitalists could expand through special access to contracts for supply and construction as well as the provision of services which were based on personal connections. See Robison, *Indonesia: The Rise of Capital*, p. 234.

The minister was forced to withdraw his demand when President Suharto made it clear that Ibnu Sutowo would be directly accountable to him through the Cabinet Presidium and that Pertamina's operations became closed to public scrutiny and immune to public accountability.

With this unchallengeable role, Pertamina continued to establish itself as the dominant institution in Indonesia for more than two decades. The collapse of the New Order regime and the subsequent transition to democracy in 1998 allowed the Government to instigate economic reform. Under IMF's bailout programme Indonesia began to liberalise the national economy and privatise non-performing state enterprises. Military and other vested interests were removed from state bureaucracies and enterprises. As a result, from 2001 to 2004, the country's economy started to pick up. Inflation fell from over 12 per cent to 5 per cent. Interest rates dropped from 17 per cent to 7 per cent. Foreign exchange reserves rose from USD29 billion to USD36 billion. The ratio of government debt to GDP declined from more than 100 per cent to less than 70 per cent.[18] Based on the spirit of liberalisation, Indonesia carried out oil sector reforms. In October 2001, the Indonesian legislature passed the much-wanted Oil and Gas Law No. 22/2001 which limited Pertamina's monopoly on upstream oil development by the end of 2003. Also, Pertamina's regulatory and administrative functions were transferred to a newly formed body, the British Petroleum Migas (Oil and Gas Executing Body). This new body is accountable to the Ministry of Energy and Mineral Resources.

Under the new law, Pertamina no longer holds the retail and distribution monopoly for petroleum products. The law also stipulates that foreign firms are allowed to apply for licenses to retail petroleum products. In 2004, the Government awarded licenses to British Petroleum of the United Kingdom and Petronas of Malaysia to open petrol stations. In 2006, the French oil company, Total, was granted a license to open 100 petrol stations in Java. Foreign firms were also invited to invest in oil refineries. By 2006, there were nine oil refineries in Indonesia, the largest of which was located in Cilacap, Central Java, producing 348,000 barrels per day. New refineries — each of which

---

[18] Lex Rieffel, "Indonesia's Quiet Revolution", *Foreign Affairs* 83, no. 5 (2004): 106.

has a refining capacity of between 150,000–300,000 barrels per day — are still under construction in Pare-Pare, South Sulawesi, and Batam island in Central Sumatra. This project brought together local and foreign firms under the name of P.T. Kilang Minyak Nusantara. This USD6 billion-worth joint venture consists of the Al-Banader International Group of Saudi Arabia (40 per cent), China National Electrical Equipment Corporation (40 per cent), and P.T. Intanjaya Agromegah Abadi (20 per cent). The reform had attracted foreign interests to invest in oil exploration and production in Indonesia.

In August 2005, the Ministry of Energy and Mineral Resources announced nine tender winners for the July 2005 direct offer bidding round for oil and natural gas exploration and production. Among the winners were Conoco Phillips and Indonesian oil firms Energy Mega Persada and Star Energy. As indicated in Table 2, the Indonesian Government offered 13 new blocks for oil exploration and production. While nine blocks were awarded to different oil firms, four blocks

Table 2: Direct Offers Bidding Results for New Oil and Natural Gas Exploration, 2005

No.	Block Offered	Location	Contract Winners
1	Lhokseumawe	Onshore & Offshore Aceh	N.V. Zaratex
2	West Kampar	Onshore Central Sumatra	P.T. Sumatra Persada Energi
3	Bungamas	Onshore South Sumatra	P.T. Erry Guna
4	Bengkulu	Onshore/Offshore South Sumatra	P.T. Commissioning Service Indonesia
5	Citarum	Onshore West Java	P.T. Bumi Parahyangan Ranhill Energia
6	NE Madura V	Offshore Madura	Unsold
7	North Bali II	Offshore East Java	Unsold
8	East Kangean	Offshore East Java	P.T. Energi Timur Jauh
9	Taritip	Makassar Strait	Unsold
10	Sebatik	Onshore/Offshore East Kalimantan	P.T. Star Energy
11	Amborip VI	Offshore Papua	Conoco Phillips
12	Amborip V	Offshore Papua	Unsold
13	Wailawi	Onshore East Kalimantan	Benuo Taka

Source: US Embassy Jakarta, "Indonesia: Energy Highlights August 2005".

(North East Madura V, North Bali II, Taritip and Amborip V) remained unsold. With these new exploration deals, Indonesia expected to increase its oil and natural gas production to meet the growing domestic demands.

GOVERNMENT RESPONSE TO OIL PRICE INCREASE

When world oil prices increased dramatically to more than USD70 per barrel in August 2005, the Government had no choice but to reduce the oil subsidies which resulted in the dramatic rise of oil prices in the local market. In order to avoid social and political turmoil, the Government decided to increase the prices in four stages between January and August 2005 (see Table 3). This new price policy caused a significant reduction in domestic fuel consumption. About six months after the new price was introduced, the total consumption of fuel decreased significantly from around 191,000 kiloliters per day to 139,800, or about 27 per cent. Gasoline (mostly used for transportation) experienced the most substantial drop from 53,400 kiloliters per day to 33,700 (reduction of about 36.8 per cent). Diesel consumption (used by both transportation and industries) also dropped by 30.3 per cent from 77,000 kiloliters per day to 53,600.[19]

Table 3: Development of Industry Fuel Prices, 2005 (rupiah)

Fuel Type	January	March	July	August	Increase (%)*
Premium	2,100	2,870	4,060	4,640	14.3
High Speed Diesel	2,100	2,700	4,740	5,480	15.6
Diesel Oil	2,050	2,660	4,560	5,240	14.9
Kerosene	2,200	2,790	4,940	5,490	11.1
Fuel Oil	1,600	2,300	2,900	3,150	8.6

Source: Ministry of Energy and Mineral Resources, "Daftar Kenaikan Harga Minyak dan BBM 2005" (Oil and Fuel Price List of 2005), Jakarta, Directorate of Oil and Gas, 2005.
Note: * Increase from July prices.

[19] Koalisi anti Utang (Anti-Debt Coalition), "Solusi Krisis BBM: Rebut Kembali Keadulatan Migas" (Solution for Fuel Crisis: Regaining Oil and Gas Sovereignty), Press Release, Jakarta, Mar. 2006.

The substantial reduction in consumption is indicative of the serious impact of the soaring oil prices, especially in the transport sector. Soon after the new oil prices were introduced, the national association of public transportation businesses (ORGANDA) incited a series of strikes in major cities such as Jakarta, Bandung, Surabaya, and Medan causing major transportation problems in those cities. Some people changed their means of transportation from automobiles to motorcycles. Not only do motorcycles consume less fuel than cars, but riders can also avoid traffic congestion, especially during the rush hours. This shift contributed to a substantial reduction of fuel consumption in the transportation sector.

Energy conservation had already come to the Government's attention in the early 1980s. The Presidential Instruction (*Inpres*) No.9/1982 issued on 7 April 1982, for example, attempted to regulate energy consumption by focusing on energy efficiency by stipulating rules on lighting of public buildings, the use of air conditioning, electrical appliances and state-owned vehicles. Despite the detailed regulations to reduce energy consumption, the law was not enforced. Public buildings have not followed the terms and conditions set in the law; and nothing has been done in relation to the limited use of air conditioning, escalators, electrical appliances and vehicles in government offices.

About a decade later, as part of 'save the energy campaign', another law on energy conservation was introduced. This time the President issued a Presidential Decision (*Keppres*) No. 43/1991 on energy conservation. In this law, the President insisted on several action programmes which included: (1) a "save energy" campaign; (2) training sessions on energy conservation; (3) the invention and manufacturing of energy-saving tools; (4) initiating research and development in energy-saving technology; and (5) the application of an energy audit system in order to ensure efficiency. As with the previous regulations, the absence of rewards for the do-gooders and punishment for the wrongdoers, the *Keppres* was rendered ineffective and served more as government rhetoric than a regulating instrument. The save-energy campaign was not effectively conducted due to the lack of commitment on the part of the Government as well as insufficient funding.[20]

Amid the soaring oil prices between 2001 and 2005, the Government was desperate to reduce its increasing dependence on imported fuel. Thus, the President issued a Presidential Instruction (*Inpres*) No.10/2005 instructing all state ministries, provincial governors, and district heads to monitor and control energy consumption in their respective areas and/or territories. In order to ensure its effectiveness, the Ministry of Energy and Natural Resources issued a ministerial regulation (*peraturan menteri*) No.31/2005 specifying the Government's energy-saving policies which include: (1) the limited use of air conditioning, unnecessary lighting, escalators and other electrical appliances in government buildings, commercial buildings (shopping malls, entertainment centres, etc.), and private houses; (2) requiring cars with large engine capacities to use non-subsidised fuel; and (3) encouraging the use of renewable energy such as bio-diesel or bio-ethanol as substitutes to fossil fuel. In January 2006, the President issued a Presidential Regulation (*Penpres*) No.5/2006 which set the target of the national energy policy which includes: (1) a 20 per cent reduction in fossil fuel consumption by 2025; (2) a 30 per cent increase in the use of natural gas by 2025; (3) a minimum increase of 33 per cent in the use of coal by 2025; (4) a 5 per cent increase in the use of bio-fuel by 2025; and (5) a 5 per cent increase in the use of other renewable energy sources (bio-mass, nuclear power, solar power, water, wind, etc.) by 2025.[21]

Despite the detailed terms and conditions set in these new laws, Government attempts to set regulations and control national energy consumption were not successful. This failure can be linked to a number of factors. First, the Government lacks the capacity to enforce laws on energy conservation and efficiency. As mentioned earlier, it

[20] Prasetyo Roem, "Konservasi dan Penghematan Energy Menurut Peraturan" (Lawful Energy Saving and Conservation), unpublished paper, Department of Mining, Bandung Institute of Technology, Bandung, 2006.

[21] Presidential Regulation No.5/2006 on the "National Energy Policy" at http://www.presidensby.com [25 Jun. 2006].

failed to create incentives for those who were prepared to invest in the development of energy-saving technology as well as renewable non-fossil fuel. In the past few years, many private companies had promised to invest in the development of bio-diesel and bio-ethanol provided the Government help them with land preparation and transfers for growing *elaeis guineensis* (main ingredient of coconut oil), *aleurites fordii* (candle nut), *saccharum officinarum* (sugar cane), and *manihot esculenta* (cassava) to produce bio-fuel. At the same time, the Government was not able to punish those who continue to waste energy and use subsidised fossil fuels for their high-powered private cars. Indeed, petrol stations secured no authority to disallow high-powered cars to fill their fuel tanks with subsidised gasoline or diesel, although the law stipulated otherwise. Similarly, state officials had never carried out regular checks on commercial and state buildings with regard to their energy consumption.

Second, despite the rhetoric of conducting a serious energy conservation policy, the Government is lacking the political will to carry out research and development on energy-saving technologies and renewable non-fossil fuels. Unlike the American Government which formed a special task force to work on energy efficiency policy, the Indonesian Government still regards energy as a non-security issue, so the solution does not require action beyond the normal practice of public policy. For example, despite the awareness that biofuel is the most plausible way (given the availability of arable land to grow biofuel crops), the Indonesian Government did not find it necessary to set up a task force to work on this project more seriously. As far as biodiesel is concerned, the project was carried out by the BPPT (Centre for Science and Technology) which runs a factory producing about 1.5 tons of biodiesel per day. In July 2006 the BPPT launched another factory in Serpong, West Java, producing another 3 tons of biodiesel per day.[22] With no Government commitment to support the project in a more active way, the biofuel project may falter. If the

[22] Amalia Shintawaty, "Prospek Pengembangan Biodiesel dan Bioetanol sebagai Bahan Bakar Alternatif di Indonesia" (The Prospect for the Nurturing of Biodiesel and Bioethanol as Alternative Energy in Indonesia), *Economic Review* 203 (Mar. 2006): 6.

price of biofuel cannot compete with fossil fuels in the market and if the distribution lines of the biofuel are limited, consumers will continue to use fossil fuels and the renewable fuels project will collapse. Thus, before that happens, the Government must be more active in providing credit or tax incentives for investors in this sector, and begin to convince the public of the importance of replacing fossil fuels with renewable fuels.

Third, the failure to enforce the energy conservation laws can also be linked with weaknesses within the Government itself. For example, some regulations cannot be implemented because they are beyond common sense. For example, Ministerial Regulation No. 31/2005 of the Ministry of Energy and Natural Resources stipulates: "commercial as well as state buildings must follow the following regulation: room temperatures must be set at a minimum of 25°C, lighting must be kept at 15 watt per m^2, air conditioning and escalators must be turned on one hour before the business/office hours and turned off one hour before closing hour, and lifts must stop at only every two floors".[23] Many found these regulations ridiculous. Imagine, for example, a crowded commercial building (shopping mall) where the temperature is 25°C, lighting is 15 watt per m^2, the escalator is switched off one hour before the closing hour and lifts stop at every second floor. This would create chaos at closing time, and visitors would be bored by the dull environment due to the poor lighting and overly warm temperature. Thus, due to its own problems, the law has never been followed and has become the laughingstock of shop owners and civil servants.

CONCLUSION

In the light of its political transition to democracy, the Indonesian Government has managed to develop a more market-oriented energy policy to ensure its national energy security. In the oil sector, if in the past the Government pursued its rent-seeking behaviour by using Pertamina as the major source of state revenues during the oil boom

[23] Prasetyo Roem, "Konservasi dan Penghematan" (Lawful Energy Saving).

period, in the post-New Order era there has been growing pressure to reform the oil sector. While the country is still a net energy exporter (especially in natural gas and coal), many have predicted that in the next two decades or so, the growing domestic energy demand and depletion of fossil fuels will turn Indonesia into a net energy importer.

When the oil price reached USD78 per barrel in March 2006, the sense of panic increased. In a dialog on METRO-TV, one of the national television stations, a commentator suggested that the Government should impose a new policy in which cars with odd number plates may fill up only on certain days in the week, and cars with even plates fill up on other days. Another commentator suggested that high-powered private cars should be banned from the streets, while hybrid cars should be sold at subsidised prices.[24] These suggestions reflect the sense of desperation and panic among Indonesians with respect to the ambiguity of Government intentions and commitment regarding energy-saving policies.

Attempts to reduce dependence on oil, increase oil production, carry out energy-saving campaigns and develop renewable energies seem to indicate the Government's seriousness to guarantee national energy security. However, Indonesia will have to endure a long journey before it can finally guarantee sufficient energy supplies to carry out normal and politically acceptable economic activities. Rampant corruption among state officials, inconsistency in approaches towards production-sharing contracts and lack of law enforcement in the business sector has generated confusion over whether the country can truly make policies that guarantee national energy security.

[24] These suggestions were proposed by callers in the "Metro Hari Ini" programme run by METRO-TV, 20 Mar. 2006.

Chapter

7

Energy Conservation Policy Development in Malaysia

Mui Pong Goh

INTRODUCTION

The shift towards focusing on energy conservation in Malaysia has gained momentum in recent years particularly with the rapid rise of oil prices. Though an oil-exporter, it has not been immune to the growing realisation of the need to conserve its energy resources. However, its desire for rapid economic development has sometimes conflicted with these conservation considerations.

From Table 1 it can be seen that Malaysia's overall energy demand is expected to grow by 6.3 per cent annually between 2005 and 2010 due to anticipated higher GDP growth under the 9th Malaysian Plan (9MP), 2006–2010, compared to the 8th, 2001–2005. Under the 9MP, the Malaysian economy is expected to grow at 6.0 per cent per annum, compared to the 4.5 per cent GDP growth under the 8th Malaysian Plan (8MP).[1]

[1] Malaysia Economic Planning Unit, *Ninth Malaysian Plan 2006–2010*. Kuala Lumpur, Malaysia, 2006, pp. 15, 25, at http://www.epu.jpm.my/rm9/html/english.htm [15 Jan. 2007].

Table 1: Final Commercial Energy Demand by Sector, Malaysia 2000–2010

Sector	Petajoules			% of Total			Average Annual Growth Rate (%)	
	2000	2005	2010	2000	2005	2010	8MP	9MP
Industrial	477.6	630.7	859.9	38.4	38.6	38.8	5.7	6.4
Transport	505.5	661.3	911.7	40.6	40.5	41.1	5.5	6.6
Residential & Commercial	162.0	213.0	284.9	13.0	13.1	12.8	5.6	6.0
Non-Energy	94.2	118.7	144.7	7.6	7.3	6.5	4.7	4.0
Agriculture & Forestry	4.4	8.0	16.7	0.4	0.5	0.8	12.9	15.9
Total	1,243.7	1,631.7	2,217.9	100.0	100.0	100.0	5.6	6.3

Source: Malaysia Economic Planning Unit, 2006, p. 395.
Note: 'Industrial' includes manufacturing, construction and mining. 'Non-energy' includes natural gas, bitumen, asphalt, lubricants, industrial feedback and grease.

The largest consuming sector is the transport sector which accounted for 40.6 per cent of Malaysia's total energy consumption in 2000 and this percentage is expected to grow to 41.1 per cent in 2010. The next largest consuming sector is the industrial sector, at 38.6 per cent in 2005. The average annual growth rate of consumption by the industrial sector is estimated to be 6.4 per cent between 2006 and 2010 which is almost the same as the expected average annual energy growth rate for the whole country (6.3 per cent) for the same period.

The relative share of the commercial and residential sector has remained fairly constant at about 13 per cent in 2000 and 2005. However, the share is likely to fall slightly to 12.8 per cent in 2010.

Most of the energy consumed is in the form of petroleum products, accounting for 62.7 per cent of the total energy consumed in 2005, down from 65.9 per cent in 2000. This percentage is likely to fall even further to 61.9 per cent in 2010. Instead, there is an increase in the utilisation of natural gas. This gradual shift away from the heavy

Table 2: Final Commercial Energy Demand by Source, 2000–2010

Source	Petajoules			% of Total			Average Annual Growth Rate (%)	
	2000	2005	2010	2000	2005	2010	8MP	9MP
Petroleum products	820.0	1,023.1	1,372.9	65.9	62.7	61.9	4.5	6.1
Natural Gas	161.8	246.6	350.0	13.0	15.1	15.8	8.8	7.3
Electricity	220.4	310.0	420.0	17.7	19.0	18.9	7.1	6.3
Coal and Coke	41.5	52.0	75.0	3.4	3.2	3.4	4.6	7.6
Total	1,243.7	1,631.7	2,217.9	100.0	100.0	100.0	5.6	6.3
Per Capita Consumption (gigajoules)	52.9	62.2	76.5				3.3	4.2

Source: Malaysia Economic Planning Unit, 2006, p. 394.

Notes: Final commercial energy demand refers to the quantity of commercial energy delivered to final consumers but excludes gas, coal and fuel oil used in electricity generation.

The natural gas category includes natural gas used as fuel and feedstock consumed by the non-electricity sector.

reliance of petroleum is part of the government's plan to diversify its fuel supply.

This attempt to diversify the fuel supply is also evident in terms of the fuel mix used for the generation of electricity. Although natural gas is the main fuel used, its share has been on the decline, falling from 77.0 per cent in 2000 to 70.2 per cent in 2005. Its share is expected to fall further to 55.9 per cent in 2010. The contribution of oil as a fuel for electricity generation is very low, shrinking from 4.2 per cent in 2000 to 2.2 per cent in 2005. This share is likely to shrink to 0.2 per cent in 2010. Instead, there is a shift towards the use of more coal. The percentage of coal increased significantly from 8.8 per cent in 2000 to 21.8 in 2005 and is expected to increase to 36.5 per cent in 2010. Although Malaysia has an adequate supply of coal (approximately 1,712 million tons of proven reserves, 347 million tonnes of indicated reserves and 1,090 million tonnes of inferred

Table 3: Fuel Mix in Electricity Generation, 2000–2010

Year	% Total					
	Oil	Coal	Gas	Hydro	Others	Total (GWh)
2000	4.2	8.8	77.0	10.0	0.0	69,280
2005	2.2	21.8	70.2	5.5	0.3	94,299
2010	0.2	36.5	55.9	5.6	1.8	137,909

Source: Malaysia Economic Planning Unit, 2006, p. 399.

reserves), it imports coal from Australia, Indonesia, China and South Africa. This is due to the high cost of extracting the local coal.[2]

The contribution of other fuels such as hydropower or biomass is likely to see a small increase between 2005 and 2010: from 5.8 per cent in 2005 to 7.4 in 2010. A few months after the 9MP was released, the Malaysian Government announced that it would increase the proportion of hydro-power significantly over the next ten years. The Energy, Water, and Communications Minister Datuk Seri Dr Lim Keng Yaik announced that Malaysia would aim to generate 45 per cent of the required electricity from gas, 30 per cent from hydropower and 25 per cent from coal.

The increase in the use of hydropower is a result of the Bakun Dam Project located in Sarawak. This is a massive project with potential production of 4,000 MW and 5,000 MW of electricity.[3] A 670-kilometre network of undersea electrical cables to peninsular Malaysia is required. The cost of laying these cables alone is estimated to be RM9 billion.[4]

It is still too early to know if this dam project will yield this much energy. The Malaysian Government did not seem to have taken into account the possibility of transmission losses. Furthermore, this project has been heavily criticised by many non-governmental

[2] Abdul Rahman Mohamed and Keat Teong Lee, "Energy for Sustainable Development in Malaysia: Energy Policy and Alternative Energy", *Energy Policy* 34 (2006): 2390–91.

[3] "M'sia Goes for More Hydro-Electric, Less Gas-Powered Energy", *Bernama*, 12 Jan. 2007.

[4] "Bakun Cable Project Gets Green Light", *New Straits Times*, 13 Jan. 2007.

Table 4: Malaysia's Oil and Gas Reserves

	2002	2003	2004	2005	2006
Crude Oil Reserves (billions of barrels)	4.24	4.54	4.84	5.29	5.25
Lifespan	15	18	18	19	20
Natural Gas Reserves (trillion standard cubic feet)	87.5	89.0	87.0	85.2	88.0
Lifespan	34	35	34	33	34

Source: Petronas Annual Reports, 2003, 2004, 2005 and 2006.
Note: Petronas changed its classification and reporting of Malaysia's reserves to standardise with the industry's practice in 1 January 2006.

groups for the possible adverse environmental consequences.[5] Given all these difficulties and the fact that the project has been postponed several times in the past due to the huge financial outlay, it seems that the contribution of hydropower may not meet the government's plans.

Table 4 shows the current oil and gas reserves. Crude oil reserves increased from 4.24 billion barrels in 2002 to 5.25 billion barrels in 2006. However, there was a small decrease of 0.04 billion barrels of crude oil reserves from 2005 to 2006. At the current rate of extraction, Malaysia's oil reserves should last for another 20 years.

Malaysia has a larger quantity of natural gas reserves. These increased from 87.5 trillion standard cubic feet in 2002 to 88.0 trillion standard cubic feet in 2006. The Government has been careful not to over-exploit the lifespan of the gas reserves — which has been relatively constant at around 34 years.

---

[5] See for instance the International Development Studies Network website at http://www.idsnet.org/Resources/Dams/Bakun/BakunDam.html. See also Meenakshi Raman, "Environmental Struggles in Malaysia", *Development* 49, no. 3 (2006); 41 and Keong Choy Yee, "Energy Demand, Economic Growth, and Energy Efficiency — The Bakun Dam-Induced Sustainable Energy Policy Revisited", *Energy Policy* 33 (2005): 679–89.

Table 5: Key Energy Policies

Policy	Year Issued	Remarks
Petroleum Development Act	1974	
National Energy Policy	1979	
National Depletion Policy	1980	Revised in 1985
Four Fuel Diversification Policy	1981	
Fifth Fuel Plan	2000	This is part of the Eighth Malaysian Plan (2001-2005)

Source: Energy Information Bureau Website at http://eib.ptm.org.my/index.php?page=article&item=99,124 [5 Jan. 07].

MALAYSIA'S ENERGY POLICIES

Table 5 gives the key energy policies that the Malaysia Energy Centre has listed. It is evident that the Malaysian Government has generally been concerned about conserving its energy supplies. The first significant legislation pertaining to the energy supply made was the Petroleum Development Act in 1974 which gave PETRONAS the "exclusive right to explore and exploit Malaysia's onshore and offshore petroleum reserves".[6] This legislation was primarily aimed at regaining control of the petroleum industry as opposed to energy conservation.

The National Energy Policy was formulated in 1979. The three main objectives were supply, utilisation and environment. The supply objective was for Malaysia to have a reliable and cost-effective energy supply through the development of indigenous energy resources, both renewable and non-renewable. The utilisation objective was to promote the sustainable use of energy through the promotion of energy conservation. The environment objective was to minimise the negative environmental effects of the energy supply chain. Since then, there has not been a fundamental shift in the energy policy though there have been changes in terms of the focus on particular objectives.

[6] Bruce Gale, "Petronas: Malaysia's National Oil Corporation", *Asian Survey* 21, no. 11 (Nov. 1981): 1132.

The first major policy was the National Depletion Policy passed in 1980. This was the first major initiative to ensure the sustainability of Malaysia's oil reserves. (It was subsequently extended to cover natural gas as well.) The policy restricted production of oil from the major oil fields of over 400 million barrels of oil initially in place (OIIP) to 1.75 per cent. In 1985, the production limit was increased to 3 per cent of the OIIL. This effectively meant that oil production was restricted to 650,000 barrels of oil per day.

However, the volume of oil production has gradually increased over the years due to the discovery of new oil reserves. The average production of domestic crude oil and condensate has increased from 681,000 barrels per day (bpd) in 2000 to 727,000 bpd in 2005. At this rate of extraction, Malaysia's oil reserves could last for 19 years.[7]

The Four Fuel Diversification Policy was implemented after the series of oil crises in the 1970s when oil prices soared. Malaysia was then heavily reliant on petroleum as its main source of energy. As part of a move to reduce its dependence on any one particular fuel, the Government aimed to increase the usage of coal, natural gas and hydropower. This Four Fuel Diversification Policy was superseded by the 'Five Fuel Policy' which was part of the 8MP in 2001. The Government added renewable energy as another alternative source of energy to petroleum. Renewable energy, in this instance, referred to biomass, biogas, municipal waste, solar and mini-hydro.[8] As part of the 'Five Fuel Policy', the Government has also put out a National Biofuel Plan under which the Government would blend five per cent processed palm oil with diesel to form a new form of diesel (B5 diesel) to be used in the transport sector.[9]

INITIATIVES IN THE INDUSTRIAL SECTOR

Although there have been provisions and calls for greater efficiency in energy usage in Malaysia, it is only very recently that the

[7] Malaysia Economic Planning Unit, 2006, p. 396.

[8] Malaysia Economic Planning Unit, 2001, p. 324.

[9] Ministry of Plantation Industries and Commodities, *The National Biofuel*, Malaysia Ministry of Plantation Industries and Commodities, 21 Mar. 2006, at http://www.mpob.gov.my/html/pdf/Biofuel%20Policy.pdf [14 Jan. 2007].

Government shifted its focus towards increasing energy efficiency. These efforts have been previously handicapped by the lack of either a coordinating agency or a focal point to spearhead the energy conservation projects.

Hence, one of the key milestones was the establishment of the Malaysian Energy Centre (Pusat Tenaga Malaysia or PTM), established in 1998 as the national centre to increase awareness of energy conservation and utilisation of alternative energy sources to the public. It also conducts and coordinates research on energy.

The biggest project of PTM is the Malaysian Industrial Energy Efficiency Improvement Project (MIEEIP) which is a five-year (1999–2005) project co-funded by the United Nations Development Programme (UNDP), the Global Environmental Facility and the private sector.[10] It has since been subsequently extended for an additional two and half years to end on June 2007.[11] The focus on the industrial sector is due to the fact that the industrial sector is the second largest consumer of energy, after the transport sector.

The main objective of this project is to provide Malaysia with an "institutional and technical foundation for continued efforts to capture the energy efficiency potential within the industry sector and achieve reductions in greenhouse gas emissions."[12] A target to reduce greenhouse gas emissions by 10 per cent by 2004 was also set.[13] This target seems to have been optimistic since the subsequent UNDP (2006) report on MIEEIP merely indicated the possible reduction of gas emissions *if* the measures suggested were taken. It is also an indication that some of the targets might have been set without sufficient information.

---

[10] See "MoU Signed for Biggest Energy Efficiency Project in Malaysia", on United Nations Development Programme website at http://www.undp.org.my/uploads/pre260201.rtf [15 Jan. 2007].

[11] Information from UNDP website at http://www.undp.org.my/index.php?navi_id=54 [15 Jan. 2007].

[12] UNDP, *Achieving Industrial Energy Efficiency in Malaysia*. United Nations Development Programme, Malaysia 2006 at http://www.undp.org.my/uploads/Achieving_Industrial_Energy_Efficiency_2006.pdf, p. 13.

[13] From MIEEIP website at http://www.ptm.org.my/mieeip/indexabout.html [15 Jan. 07].

The MIEEIP initially focused its attention on eight energy-intensive manufacturing subsectors — wood, glass, rubber, pulp and paper, food, cement, ceramic, iron and steel. Three additional subsectors — olechemicals, plastics and textiles were subsequently added.[14] Table 6 provides an overview of the various components and objectives of MIEEIP.

One of the main energy technology demonstration projects is the High Efficiency Motors (HEMs). A study sponsored by the Danish International Agency found that 70 per cent of the electricity consumption in the industrial sector was by electric motors alone. About 50 per cent of the commercial sector's electricity consumption is by electric motors. However, only 2 per cent of these electric motors are HEM.[15] It is estimated that companies would be able to save 2–10 per cent of their running costs if they convert to the use of HEM.[16]

The Government launched a six-month long campaign beginning in May 2005 to encourage companies to use HEM. The campaign primarily involved the holding of seminars and publications of articles in the media. Sales of HEM in 2005 doubled to 2,800 compared to 1,400 units in 2004.[17]

INITIATIVES IN THE RESIDENTIAL AND COMMERCIAL SECTORS

While the MIEEIP was primarily targeted at the industrial sector, there were other initiatives introduced for the residential and commercial sectors. Many of the initiatives in these sectors were complementary to those introduced in the industrial sector. One of the key moves in promoting energy efficiency usage in the residential and commercial

[14] "Energy Project Extended to 3 Sectors", *Business Times* (Malaysia), 10 Oct. 2005.

[15] See "EC: HEM To Account for 20% of Motor Sales in Next Five Years", *The Edge Financial Daily*, 4 Apr. 2006.

[16] From the Energy Commission (Suruhanjaya Tenaga) website created to promote HEM usage at http://www.hem2profit.com.my/about.htm [10 Jan. 07].

[17] See "EC: HEM To Account for 20% of Motor Sales in Next Five Years", *The Edge Financial Daily*, 4 Apr. 2006.

Table 6: Components of MIEEIP

Components	Objective	Success Criteria	Remarks
Energy Use Benchmarking	To establish and develop energy use benchmarks for the various industrial sub-sectors	• Set up data collection and database systems for the energy benchmarking system • Establish industrial energy use benchmarks	
Energy Audit Programme	To achieve widespread use and practice of energy auditing to manage energy use	• Develop standardised energy audit procedures tools; • Conduct and evaluate energy audits in selected industrial sub-sectors	48 factories to be audited by 2005; 6 additional ones to be audited by 2007
Energy Rating Programme	To provide information on energy efficient equipment and energy rating programmes by establishing energy standards and labelling for key industrial equipment	• Disseminate information on energy efficient equipment and energy rating programmes • Develop a plan for a comparative energy rating programme for key equipment	

(Continued)

Table 6: (*Continued*)

Components	Objective	Success Criteria	Remarks
Energy Efficiency Promotion	To provide information on energy efficient practices in industries and technology applications	• Improve the information flow of information on energy efficiency • Establish a professional organisation of local energy specialists, consultants and technology developers and providers	
Energy Service Companies (ESCO) Programme	To enable PTM to determine the optimal structure for the development of ESCO industry which includes issues such as regulatory framework, financing and risk evaluation on projects	• Evaluate the capacity and capabilities of existing ESCOs • Develop and monitor the institutional and legal framework for the delivery of ESCO services	
Energy Technology Demonstration Projects	To demonstrate the applicability, technical and economic feasibility of energy efficiency technologies in various industrial sectors	• Implement potential energy saving technologies • Monitor and evaluate each project	

(*Continued*)

Table 6: *(Continued)*

Components	Objective	Success Criteria	Remarks
Local Energy Efficiency Equipment Support Programme	To replicate and promote energy efficient manufacturing improvements by local equipment manufacturers	• Identify potential improvements and new design for locally manufactured industrial equipment • Provide technical assistance and funding to eligible equipment design and manufacturing improvements projects	
Financial Institutional Programme	To establish an energy business fund and related framework	• Train local and financial institutions on funding energy efficiency projects in industry • Develop criteria for the finance companies for energy technology demonstration	

Source: UNDP, 2006.

sectors was the establishment of standards for equipment or appliances and having labels to provide relevant information.[18] Some care would need to be exercised to ensure that these standards are realistic and do not affect trade adversely (through the creation of non-tariff barriers). Having labels on the equipment or appliance is essential for end-users to compare the equipment.[19]

The Energy Efficiency Working Group had originally been tasked to 'develop standards' for refrigerators, air-conditioners and fans but the chair of the Working Group, Professor Faridah Mohd Taha admitted that the speed of development in developing standards and encouraging labels has been slow.[20] Malaysia is only slowly developing its capacity for designing and implementing the energy standards and labels.

From Table 7, we can see that only magnetic ballast (which is used in florescent lamps) has an established minimum energy standard.[21] There is a voluntary standard for motors while the establishment of minimum standards for clothes washers, iron and room conditioning units (both split and window types) is being considered.

It is likely that international or regional standards will be used for the equipment listed in Table 7. This is due to the fact that many of these products fall under a list of 20 priority products for which the ASEAN members have agreed in 1997 to adopt international

[18] Wiel and McMahon (2003) provide a helpful overview of the process of formulating a policy of standards and labelling. See Stephen Wiel and James McMahon, "Governments Should Implement Energy-Efficiency Standards and Labels- Cautiously", *Energy Policy* 31, no. 13 (2003): 1403–15.

[19] For more details on the labelling process on refrigerators and electrical fans in Malaysia, see also R. Saidur, H.H. Masjuki, M.Y. Jamuddin and T.M.I. Mahlia, "Labelling Design Effort for Household Refrigerator-Freezers in Malaysia", *Energy Policy* 33, no. 5 (2005): 611–18; and T.M.I. Mahlia, H.H. Masjuki, F.M. Taha, N.A. Rahim and R. Saidur, "Energy Labelling for Electric Fans in Malaysia", *Energy Policy* 33, no. 1 (2005): 63–68.

[20] David Cogan and Fanghei Tsau, eds., *Proceedings of the Seminar on Cooperation on Energy Labelling*. Seminar held in Kaohsiung, Chinese Taipei, 17–19 Nov. 2003. APEC Project EWG 03/2003. Singapore: APEC Secretariat, p. 6.

[21] The magnetic ballast is a devise used to regulate the voltage of the florescent lamp.

Table 7: Summary of Standards and Labelling Programmes by Equipment Type

Equipment Type	Minimum Standard	Labelling
Ballasts (Magnetic)	Ym	
Clothes Washers	U	
Fans		U
Freezers		Yv
Irons (Electric)	U	
Motors (3-Phase Induction)	Yv	U
Room Air-Conditioning (Split)	U	U
Room Air-Conditioning (Window)	U	U
Refrigerator		Yv
Refrigerator- Freezer		Yv

Source: APEC ESIS website at http://www.apec-esis.org/countrysummary.php?country=Malaysia [10 Jan. 07].

Note: Yv — Yes, Voluntary; Ym — Yes, Mandatory; U — Under consideration. The column 'Minimum Standard' refers to the minimum efficiencies that manufacturers must achieve for the particular appliance.

standards.[22] Despite the varying capacity of the ASEAN members (i.e., some members not having testing laboratories, and differences in voltages), the ASEAN members have agreed to use labels for ballasts (2004–2005), refrigerators (2005–2006) and air-conditioners, motors and fans in 2006–2009. Malaysia seems to be following, rather than being ahead of, this ASEAN trend of adopting labels for appliances.

There is no mandatory labelling of electrical equipment. Instead, only three items — freezers, refrigerators and refrigerator-freezers — are part of a voluntary labelling system.

A RM2.0 million campaign was launched for the promotion of energy efficiency labels for consumer products in 2005.[23] The Energy Commission (Suruhanjaya Tenaga) also set up a website (http://www.eefridge.com.my/index.htm), providing information encouraging the purchase of energy efficient refrigerators.

---

[22] See ASEAN website, particularly http://www.aseansec.org/15564.htm [10 Jan. 07].

[23] "Gov't Will Save RM11 BLN through Energy Commission's EE/ DSM Campaign", *Bernama*, 13 Sept. 2005.

ENERGY STANDARDS FOR BUILDINGS

On a larger level, the government has attempted to improve the level of energy efficiency in both residential and commercial buildings. In 1987, the Ministry of Energy, Telecommunications and Post initiated a programme to improve energy efficiency in buildings. The Ministry of Energy, Communications and Multimedia conducted energy audits of several government buildings in 2000.[24] However, it is unclear as to what the follow-up actions were or the basis upon which the audits were conducted.

The necessity for a standardised methodology for energy audit partly led to the creation of the Malaysian Association of Energy Service Companies (MAESCO) in 2001. By 2002, MAESCO had developed a guideline to standardise energy audit procedures in commercial buildings. With this guideline, a five-year series of energy audits of government buildings was initiated in 2002. The PTM Chief Executive, Dr Hassan Ibrahim pointed out that the next series of energy audits would be for commercial buildings.[25]

As per set out in the 9MP, the Malaysian Government is working to incorporate provisions to improve energy efficiency in the buildings. There is currently a voluntary Code of Practice on Energy Efficiency and Renewable Energy for Non-Residential Buildings (or MS1525). However, two ministries — the Housing and Local Government Ministry and the Energy, Water and Communications Ministry — are working to include energy saving requirements in Malaysia's Uniform Building By-Laws (UBBL).[26] There are three major components to the revision of the UBBL- thermal transfers of building walls and roofs, air-conditioning and ventilation, and energy-efficiency

[24] "Govt Will Save RM833 Million Annually Should Energy Efficiency Improve", *Malaysia General News*, 15 Aug. 2000. This Ministry was created on 1 November 1998 through a restructuring of the Ministry of Energy, Telecommunications and Post. This restructuring also resulted in the replacement of the Department of Energy and Gas Supply by the Energy Commission.

[25] "PTM Embarks on Five-Year Energy Audit of Govt Building", *Malaysia Economic News*, 22 Aug. 2002.

[26] "Low-Energy Office Buildings on Drawing Board", *New Straits Times*, 16 Nov. 2006.

lighting. Once these provisions are incorporated in the By-Laws, they will become mandatory.

The government is also considering introducing the "Efficient Management of Electrical Energy Regulation".[27] This is probably building on the Energy Commission's proposal that consumers using more than six million kilowatt-hours (kWh) of electricity a year would have to engage an energy manager to oversee energy efficiency measures for a building or industrial installation.[28] This proposal has been mooted since the formation of the Energy Commission in January 2002 and it is uncertain as to when it would finally be implemented.[29] Some of Malaysia's buildings have already won awards for energy efficiency. For instance, the Ministry of Energy, Water and Communications for Low Energy Office or the LEO Building won the 2006 ASEAN Energy Awards.

FINANCIAL INCENTIVES

Besides providing information on the level of energy efficiency, and legislating minimum standards for buildings, the Government has also offered a 'carrot' approach in the form of financial incentives to encourage companies to promote conservation. In 2002, tax incentives were provided to encourage companies that promote environmental protection and energy conservation.[30]

In addition to the above, in the 2001 Budget, the Government specifically granted biomass-based generating companies (that applied for incentives by 31 December 2002) exemption from income tax on 70 per cent of the statutory income for five years or a tax allowance of 60 per cent of qualifying capital expenditure incurred within five years. This allowance can then be used to offset up to 70 per cent of the statutory income.[31]

---

[27] Malaysia Economic Planning Unit, 2006, p. 409.

[28] "Ignorance Rules", *New Straits Times*, 12 Nov. 2005.

[29] "Energy Commission Assumes Responsibility", *New Straits Times*, 15 Jan. 2002.

[30] See Malaysian Industrial Development Authority, *Investment in the Manufacturing Sector*. Kuala Lumpur, 1 Mar. 2002.

[31] Malaysia Economic Planning Unit, 2001, p. 333.

Table 8: Available Incentives to Promote Energy Conservation and Efficiency

Objective	Target Beneficiary	Benefit for Applicants	Remarks
To encourage energy conservation	Companies providing energy conservation services	• Tax exemption of 70% of statutory income for 5 years or investment tax allowance of 60% • Import duty and sales tax exemption for equipment used in energy conservation	Applications must be received from 28 October 2000 to 31 December 2010. Projects have to be implemented within a year of approval of the incentive.
To encourage energy conservation	Companies which incur capital expenditure for conserving energy for own consumption	• Investment tax allowance of 60% on the qualifying capital expenditure incurred within 5 years with the allowance to be offset against 70% of statutory income for each year of assessment • Import duty and sales tax exemption for equipment used in energy conservation	Application must be made from 1 October 2005 until 31 December 2010

(Continued)

Table 8: (*Continued*)

Objective	Target Beneficiary	Benefit for Applicants	Remarks
To encourage use of renewable energy by companies	Companies which generate energy from renewable sources such as biomass, hydro power (not exceeding 10 MW) and solar power	• Tax exemption of statutory income for 10 years or investment tax allowance of 100% on qualifying capital expenditure incurred within 5 years with the allowance deducted each year of assessment to be off set against 100% statutory income for each year of assessment • Import duty and sales tax exemption for equipment used	Applications must be made between 28 October 2000 and 31 December 2010. Companies are required to implement their projects within 1 year of the approval of the incentive

Source: Malaysia Ministry of Energy, Water and Communications website at http://www.ktak.gov.my [10 Jan. 07].

ALTERNATIVE SOURCES OF ENERGY: RENEWABLE SOURCES OF ENERGY

Under the Eighth Malaysian Plan (2001–2005) or 8MP, the Malaysian Government planned to intensify the development of renewable energy as the fifth fuel. The target was for 5 per cent of the total power or 700 MW to be generated by renewable energy.[32] The mechanism to achieve this goal was via the Small Renewable Energy Power Programme (SREP). The programme was to allow small power generation plants which use renewable energy fuel to sell to the utility through the electricity distribution grid system. Individual power plants would be allowed to sell up to 10MW.

The SREP seemed popular with 68 SREP projects approved by 2005. However, the success of these individual SREP projects is unknown. The Energy, Water and Communications Minister Datuk Seri Dr Lim Keng Yaik pointed out that only two of these SREP projects had been implemented, producing 12 MW capacity.[33]

Under the 9MP the Government reduced the target to 300 MW of renewable energy to be generated.[34] Biofuel has emerged as one of the most attractive sources of renewable energy given Malaysia's large palm oil production. Furthermore, there is a large amount of waste materials associated with the palm oil industry such as palm kernel trunks and other residues that are not utilised and must be disposed of by incineration.

A five-year (2001–2005) project, 'Biomass-based Power Generation and Co-generation in Palm Oil Mills' was set up to use these palm oil residues to generate energy. In April 2005, the Renewable Energy Business Facility (REBF) was established to provide financial assistance to biomass-based power generation.[35] The facility was worth RM28 million. Part of the funds was to be used as a loan for the

[32] "5pc of Power to Come from Renewable Energy by 2005", *Business Times* (Malaysia), 25 May 2001.

[33] "Slow Progress in Sustainable Renewable Energy Source Development", *Bernama*, 15 Sept. 2005.

[34] Malaysia Economic Planning Unit, 2006, p. 408.

[35] "Renewable Business Facility to Offer Facility Help", *Bernama*, 11 Apr. 2005.

construction of a full-scale model of a power plant generated by biomass.

NUCLEAR ENERGY

The Government had previously toyed with the idea of using an alternative source of energy, namely, nuclear energy. This was in the immediate aftermath of a severe black-out in Malaysia in 1992.[36] However, this option was subsequently dismissed. There is a small nuclear energy department, the Malaysian Institute for Nuclear Technology Research (MINT) which monitors research on nuclear energy and periodically reviews the need for the nuclear option in the future. MINT also owns and operates a small one-MW thermal reactor, primarily for research and the production of radioisotopes for medical industry and agriculture.[37] While not eliminating the use of nuclear energy completely, the Malaysian Government has consistently affirmed that it does not have immediate plans to use nuclear energy due to the availability of other energy sources and the risk factors.[38]

In 2007, the Energy, Water and Communications Minister Datuk Seri Dr Lim Keng Yaik confirmed, "At the moment, we put the possibility of using nuclear as our source of energy at the back of our mind. Not during my time. Probably after 2030 when we should have exhausted our renewable energy."[39]

INITIATIVES IN THE TRANSPORT SECTOR

The transport sector seems to be the least successful in terms of the Government's moves to diversify fuel or conserve energy. It remains

---

[36] "Malaysian Groups Upset Over Energy Proposals", *South China Morning Post*, 1 Oct. 1992.

[37] "Very Unlikely for Nuclear Disaster to Occur Here", *New Straits Times*, 2 Oct. 1999.

[38] "No Nuclear Energy Plans", *New Straits Times*, 29 Nov. 2006; "Nuclear Energy Will be Govt's Last Choice", *New Straits Times*, 7 Mar. 2000.

[39] "M'sia Goes for More Hydro-Electric, Less Gas-Powered Energy", *Malaysia General News*, 12 Jan. 2007.

highly oil dependent. In 1983, the Government introduced a 'Go Gas' Initiative. However, it did not succeed despite the introduction of a sales tax initiative.[40]

The Government tried to increase the use of natural gas in the transport sector particularly in the taxi sector. In 1995, it increased the number of permits to the operator of taxi services at the Kuala Lumpur International Airport (KLIA) on the condition that the company would use more natural gas vehicles (NGV). However, even till 2003, the company, Airport Limo did not operate any NGV.[41] The reason given was the lack of infrastructure, particularly filling stations.

The Government has continued to encourage the utilisation of the NGV industry through several incentives. First, it stipulated that the retail pump price of NGV has to be at least 50 per cent less than that of premium grade petrol. Second, the import of conversion kits is exempted from import duty and sales tax. Third, the road tax for bi-fuel and dual fuel NGV has been reduced by 25 and 50 per cent respectively.[42]

However, these moves have met only with limited success. The number of NGV in Malaysia is still relatively small. The Minister of Transport, Datuk Sri Chan Kong Choy said in July 2005, that Malaysia had 14,565 NGV out of a vehicle population of about 7.4 million (excluding motorcycles). He pointed out that there were only 40 stations located in Malaysia, primarily located in Kuala Lumpur (including Klang, Seremban and KLIA), Johor Bahru and Prai, but that PETRONAS was planning to increase the number of NGV stations to 94 serving 54,000 NGV by 2010.[43]

---

[40] Information from Malaysia, Ministry of Energy, Water and Communications website at http://www.ktak.gov.my/template01.asp?contentid=154 [8 Jan. 07].

[41] "Still No Gas for KLIA Limo", *New Straits Times*, 26 Jan. 2003.

[42] Chan Kong Choy, Speech Given at the First Asia Pacific Natural Gas Vehicles Association Conference and Exhibition (ANGVA 2005) at Kuala Lumpur Convention Centre, 26 July 2005, at http://www.mot.gov.my/BM/Ucapan/260705%20-%20%20Asia%20Pacific%20ngv%20ass%20conference.pdf [9 Jan. 2007].

[43] *Ibid.*

The figures used by the Prime Minister in his budget speech later in the year seemed to be slightly different from those used by the Transport Minister. In October 2005, Prime Minister Badawi gave a more ambitious target in his 2006 budget speech when he said that the number of retail gas stations would double by 2007 from the then 51 stations in 2005.[44]

If the Malaysian Government is serious about reducing the reliance on petroleum in the transport sector, it has to tackle some of the 'infrastructural' issues of having even more gas refuelling stations as well as reducing the costs of conversion of vehicles to allow them to use natural gas or use both natural gas and petroleum. There also does not seem to be a comprehensive effort at addressing how vehicles might be converted from the use of petroleum to natural gas.[45]

While import duty and sales tax exemption have been proposed for the import of chassis with engines for monogas buses and trucks and conversion kits, the focus seems to be on encouraging the number of *new* vehicles utilising natural gas rather than a concerted plan to convert the existing vehicles to use natural gas.

The Government has not systematically tackled ways to reduce the car population either through financial disincentives or inducement to improve public transportation. The need for improvement of the public transportation network to reduce the number of passenger vehicles was also suggested by the Asia Pacific Energy Research Centre in the APEC Energy Demand and Supply Outlook 2006 publication.[46]

---

[44] "Biofuel Cuts Reliance on Petroleum", *Business Times* (Malaysia), 1 Oct. 2005.

[45] The conversion of vehicles from the use of petroleum to natural gas or a bi-fuel system does not necessarily mean that there is a positive environmental impact. In the case of Brazil, see Luz Dondero and Jose Goldemberg, "Environmental Implications of Converting Light Gas Vehicles: The Brazilian Experience", *Energy Policy* 33, no. 13 (2005): 1703–08.

[46] Asia Pacific Energy Research Centre, *APEC Energy Demand and Supply Outlook 2006: Projections to 2030 Economic Review*. Tokyo: Institute of Energy Economics, Japan, 2006, pp. 52–53.

THE MAIN OBSTACLE TO ENERGY CONSERVATION: FUEL SUBSIDIES

The efforts to encourage fuel diversification and conservation are unlikely to succeed unless the Government tackles the underlying reason for the relative lack of interest in diversifying and conserving energy, namely, fuel subsidies. It has twice tried to increase the price of petrol. The first was in July 2005 when it raised petrol prices by 10 sen per litre, diesel by 20 sen per litre and liquefied petroleum gas (LPG) by 5 sen per kilogramme.[47] In March 2006, the Government increased the price of petrol, diesel and LPG by 30 sen. This meant that the prices rose by 19, 23 and 21 per cent respectively.[48] Despite these price increases, petrol prices in Malaysia are still the cheapest in Southeast Asia after Brunei, a tiny oil-rich country. The Malaysian Government continues to subsidise fuel.

STRUCTURAL PROBLEMS IN THE ELECTRICITY SECTOR

The transport sector is not the only problematic sector in terms of energy conservation. The entire electricity sector requires urgent attention. It should not come as a surprise that the conservation of electricity has not always been a priority. For instance, the then Prime Minister Mahathir himself encouraged Malaysians to use more electricity in 1996.[49] Later in 2001, he encouraged all building owners in Kuala Lumpur to keep their lights on at night especially from 17 August to the middle of September "for tourism, the country and the celebration for National Day."[50] Mahathir urged, "If possible I would like to see Kuala Lumpur lighted just like the New York sky-

[47] "No Fuel Price Increase Currently, Says Nor Mohamed", *Malaysia Economic News*, 5 Jan. 2007.

[48] "KL Sets Up $1.9b Public Transport Fund to Soften Fuel Price Hikes", *Straits Times*, 17 Mar. 2006.

[49] "Compensation Must Be Justified: Dr M", *Business Times* (Malaysia), 22 Aug. 1996.

[50] "Keep Building Lights on For Merdeka Celebration", *Bernama*, 25 Jun. 2001.

[51] "Promote Malaysia as a Tourist Destination", *New Straits Times*, 26 Jun. 2001.

line... it would have the wonderful effects of the night silhouette."[51] Tenaga Nasional Bhd (TNB), Malaysia's main provider of electricity, was also supposedly committed to give a 50 per cent discount on electricity bills for buildings that use more electricity than they normally do.[52]

Part of the reason for the paradoxical encouragement to waste electricity rather than to conserve it is due to the way in which the electricity sector is set up and regulated. After the devastating 1992 power failure, Mahathir removed the monopoly of TNB in the generation of power and introduced private suppliers.[53] He thought that the introduction of these suppliers or Independent Power Producers (IPP) would prevent further power failure and ensure that there would be an adequate supply of electricity. A large number of IPPs was swiftly approved, resulting in a large supply of electricity coming onstream.

The increase in the supply of electricity coincided with the downturn in demand due to the onset of the Asian Financial Crisis (1997–1998). In 1999, peninsular Malaysia's generating capacity was 13,500 MW, though the demand was only 9,000 MW.[54] The reserve margin is very high relative to international levels. From Table 9 it can be seen that Malaysia's reserve margin was a staggering 50 per cent in 1995 though it fell to 34.1 per cent in 2000 before increasing again to 39.5 per cent in 2005. The reserve margin is expected to fall to 25.7 per cent in 2010. The penalty for having such a high reserve margin is high especially since TNG would

[52] "Getting Private Sector Commitment in the Push to Boost Tourism", *New Straits Times*, 13 Jan. 2002.

[53] The discussion of Malaysia's partial electricity sector is drawn from Jeff Rector, "The IPP Investment Experience in Malaysia", Programme on Energy and Sustainable Development, Stanford University, Working Paper Number 46, 17 Aug. 2005 and Thomas Smith, "Privatising Electric Power in Malaysia and Thailand: Politics and Infrastructure Development Policy", *Public Administration and Development* 23, no. 3 (2003): 273–83.

[54] Rector, "The IPP Investment Experience in Malaysia", pp. 10–11.

Table 9: Installed Capacity, Peak Demand and Reserve Margin, 2000–2010

Year	Accumulated Installed Capacity (MW)	Peak Demand (MW)	Reserve Margin (%)
1995	10,835	7,212	50.2
2000	14,291	10,657	34.1
2005	19,217	13,779	39.5
2010	25,258	20,087	25.7

Source: Malaysia Economic Planning Unit, 2006, p. 398; figures for 1995 taken from Malaysia Economic Planning Unit, 2001, p. 318.

Note: Peak demand refers to the maximum power demand registered by the system in a stated period of time. Reserve Margin refers to the amount of unutilised capacity of the power system at peak load as a percentage of total capacity. It is calculated by subtracting the peak demand from the accumulated installed capacity and then divided by 100.

be making a loss paying for electricity which it could not necessarily sell.

The entire manner in which the power generation was privatised does not stand up to scrutiny. There seemed to be little competition in the award of the IPP contracts.[55] Moreover, the variations in the Power Purchase Agreement (PPA) awarded to the IPP did not seem to be always economically justified or transparent.[56] The terms for these IPP were especially generous financially. TNB had to pay these IPP higher tariffs than its own cost of production.[57] TNB's cost of production is about 10 sen/ kilowatt-hour (kWh) but it has to pay at least one IPP 15.5 sen/ kWh. TNB also had to bear the risk of any possible increase in fuel prices. Since most of the new plants of the IPPs used natural gas which was supplied by PETRONAS, TNG had to bear

[55] Asian Development Bank, *Developing Best Practices for Promoting Private Sector Involvement in Infrastructure — Power*. Manila: Asian Development Bank, 2000, Appendix. p. 7 at http://www.adb.org/Documents/Books/Developing_Best_Practices/Power/ [8 Jan. 2007].

[56] Rector, "The IPP Investment Experience in Malaysia", fn33.

[57] Rector, "The IPP Investment Experience in Malaysia", p. 8; Asian Development Bank, *Developing Best Practices for Promoting Private Sector Involvement in Infrastructure — Power*, Appendix, p. 7; Smith, "Privatising Electric Power in Malaysia and Thailand", p. 278.

the cost of any increase in the price that these IPPs paid to PETRONAS.

Despite this, the Malaysian Government under Mahathir did not seriously consider renegotiating the IPP arrangement. It was also reluctant to allow any rise in electricity tariffs. Datuk Ahmad Tajuddin Ali, the then executive Chairman of TNB wanted to defer the number of new power generating projects until the demand for electricity increased and then renegotiate the PPA.[58] However, in 2000, Tajuddin was replaced.[59] Nonetheless, Rector noted that the privatisation of the electricity sector was a "qualified success" given that "timely expensive power is a far superior outcome than blackouts that discourage FDI and domestic investments, and stunt economic growth."[60] The Malaysian Government seems to share Rector's view in preferring excess supply of electricity. Despite acknowledging the reserve margin, the Energy Commission (Suruhanjaya Tenaga) noted that the reserve margin of 38 per cent in 2005 was "still tolerable for a developing economy like Malaysia in meeting the projected growth of electricity demand at 6–7 per cent per annum."[61] However, both Rector's and the Energy Commission's conclusions do not stand when reliability of the electricity supply is considered. Having a high reserve margin does not guarantee the reliability of the electricity supply. Other factors such as the performance of the grid also play a role.

Malaysia does not have a reliable power supply. Table 10 shows some of the major incidences of power failures. There is a high level of transmission loss. The acting Chairman/Chief Executive Officer of

---

[58] "Tenaga Proposes Putting Its Project on Hold", *Business Times* (Malaysia), 28 Aug. 1998; "Tenaga to Approach IPPs Over Unfavourable Terms", *New Straits Times*, 26 Aug. 1998.

[59] Smith, "Privatising Electric Power in Malaysia and Thailand", p. 278; "Tajuddin Snubs Tenaga's Chief Exe Post Offer", *The Straits Times,* 9 Sept. 2000. The Malaysian media was more positive about the change. See "Leave Political Hats Outside Boardrooms", *Business Times* (Malaysia), 6 Sept. 2000.

[60] Rector, "The IPP Investment Experience in Malaysia", p. 18.

[61] Suruhanjaya Tenaga, *Annual Report 2005*. Kuala Lumpur: Energy Commission, 2006, p. 14.

Table 10: Major Power Failures, 1992–2005

Date	Affected Areas	Cause	Remarks
29 September 1992	Whole of Peninsular Malaysia	Lightning which damaged the grid	Power failure lasted almost 48 hours[a]
3 August 1996	Whole of Peninsular Malaysia	Breakdown of switch gear outside the Sultan Ismail power station in Paka, Terengganu	Lasted for about 7 hours from about 5 pm[b]
September 2003	Perlis, Kedah, Penang, Perak and Kelantan	Tripping of power line between Bukit Beruntung in Rawang and Batu Gajah in Perak	Lasted for about 4 hours from 10 am.
13 January 2005	Malacca, Negri Sembilan, Kuala Lumpur and Selangor	Breakdown of switch gear at the Sultan Salahuddin Abdul Aziz Shah power station in Kapar, Selangor	Power failure started at about 12pm. Power restored at the last area about 5 hours later

Source: "Today's Power Disruption was Fourth Since 1992", *Bernama*, 13 Jan. 2006.
Notes: [a] Another source pointed out that it took about three days to restore power. See "Blackout in Malaysia Due to Failure of One Power Station", *Deutsche Presse-Agentur*, 4 Aug. 1996.
[b] Another source pointed said the blackout lasted for almost 24 hours. See "Blackout in Malaysia Due to Failure of One Power Station", *Deutsche Presse-Agentur*, 4 Aug. 1996.

the Energy Commission, Dato Engku Hashim Al-Edrus highlighted that about 15.3 per cent of the electricity generated was lost in the distribution and transmission grid in 2004.[62] In contrast, developed countries generally lose about 7–10 per cent.

A high reserve margin does not guarantee reliability of electrical supply. The former is only one component of the latter. Besides increasing the efficiency of the transmission of electricity, the

[62] Engku Hashim Al-Edrus, "Welcome Remarks made at the Seminar on Energy Efficiency in Buildings- How to Achieve Immediate Savings", Crowne Plaza Mutiara Hotel, Kuala Lumpur, 24 Jan. 2006, at http://www.st.gov.my/downloads/Speech-CEO-ST-Buildings-Jan06.pdf [15 Jan. 2007].

Government could also reduce gradually its reserve margin by interconnecting its electricity grid with neighbours such as Thailand and Singapore. This interconnection has been also agreed as part of the wider Trans-Association of Southeast Asian Nations (ASEAN) Power Grid (TAPG) agreement. Through participation in the TAPG, the TNB could reduce the reserve margin from 30 to 15 per cent.[63]

The high reserve margin also means that large subsidies provided for the generation of electricity are wasted. The state owned enterprise, Petroliam Nasional Bhd (Petronas) sells natural gas to the national utility TNB at a subsidised price of RM6.40 per million British thermal units (mmBtu) which is 80 per cent lower than the international price of RM30 per mmBtu.[64] This means that Petronas has lost RM38 billion between 1997 and 2005 by selling gas at the subsidised price.[65]

In 2005, ADB estimated that fuel subsidies cost the Government USD1.3 billion in 2004 and approximately USD1.7 billion in 2005. It also lost an additional USD2.1 billion from tax exemption on gasoline. In total, subsidies constitute 2.9 per cent of GDP in 2005.[66] There are some moves towards reducing the level of subsidies. The Badawi administration has been trying to reduce the budget deficit and has shown some fortitude in trying to reduce the subsidised energy. Prime Minister Badawi warned, "Higher crude oil prices will result in higher subsidises for fuel and also for other sources of energy, such as electricity... in due course, electricity tariffs will need to be adjusted...."

---

[63] Mohammad Zamzam Jaafar, Hwee Kheng Wong and Norhayati Kamaruddin, "Greener Energy Solutions for a Sustainable Future: Issues and Challenges for Malaysia", *Energy Policy* 31 no. 11 (2003), 1071.

[64] "Dr Lim Higher Gas Price Inevitable", *Business Times* (Malaysia), 11 Oct. 2006.

[65] In another interview, Petronas President and Chief Executive Hassan Marican said that the cumulative gas subsidy provided by PETRONAS from 1997 to 2005/6 was RM37.1 billion. See "Petronas Shows the Way to Go", *Business Times* (Singapore), 3 Jul. 2006.

[66] Figures in this paragraph are taken from Asian Development Bank, 2005.

[67] "TNB Rates Among Lowest in Region Despite Tariff Hike", *Bernama*, 24 May 2006; "Lim Announces a 12 PCT Average Increase in Electricity Tariffs", *Bernama*, 24 May 2006.

In June 2006, the government increased electricity tariffs by 12 per cent.[67] The last price revision was made 9 years ago in 1997. Yet the price of coal has increased by 69 per cent since then.[68] However, this was less than the 20 per cent requested by the Tenaga Nasional Bhd (TNB). The Government is likely to be cautious about further increases given the politically weak position of the Badawi administration.[69] Indeed, the Badawi administration did not agree to TNB's proposal of having the flexibility to set prices depending on the fuel price or simply the ability to review prices periodically.

He did, however, crucially force the IPPs to renegotiate their agreements with TNB.[70] By allowing TNB to renegotiate the contracts with the IPP, it should not only place TNB on a financially more sustainable platform but allow the prices of electricity to reflect more accurately the cost of the generation. The renegotiation would also enable the supply of natural gas from PETRONAS to the IPP to rise in price, resulting in greater awareness of the need for energy conservation.

The Government has also gradually paid more attention to electricity conservation in recent years. Demand Side Management (DSM) initiatives such as those mentioned earlier in this article have been gradually introduced. The chairman of the Energy Commission, Datuk Dr Mohammed Annas Mohammed Nor pointed out that Malaysia would be able to save more than RM110.0 billion in investment costs and the cost of energy not used in the next 15 years (till 2020) with the implementation of the DSM programmes.[71] Some 1,077 MW of electricity could also be saved up to 2015.

At the presentation of the 2006 Budget, Prime Minister Badawi directed all Government facilities to reduce their electricity usage by

---

68 "The Long Road to Tariff Hike", *Business Times* (Malaysia), 26 May 2006.

69 News of the price hikes of the fuel and electricity was greeted with public demonstration in Kuala Lumpur.

70 "Govt Hopes to Wrap Up Talks by Year End", *New Straits Times*, 22 Sept. 2006.

71 "M'sia Can Save RM11 BLN Through Energy Commission's EE/ DSM Campaign", *Bernama*, 13 Sept. 2005.

10 per cent in 2006. He also urged government buildings to maintain air-conditioning at 26 degrees Celsius to save electricity.

CONCLUSIONS

Unlike many oil exporting countries, Malaysia is not afflicted with the curse of poor natural resource endowment. Instead, its richness in resources has contributed greatly to its economic development. Despite being an oil exporter, it has been careful to diversify both its economy and its energy fuel needs.

Although the Government has in the past five years introduced a series of energy conservation initiatives in the industrial, commercial and residential sectors, its efforts are unlikely to succeed until it is able to tackle the root cause of energy wastage, namely, fuel subsidies. These are a drain on the country's revenues. With prices of energy kept artificially low, there is little incentive for the public and private sectors to conserve energy. In fact, the low electricity tariffs have been cited as one of the factors contributing to the lack of energy efficiency measures in Malaysia.[72] The low energy prices also mean that renewable sources of energy which are already generally more expensive than traditional fossil fuels become even less attractive as an alterative source.

Fortunately, the Government is gradually becoming more aware that its small reserves of oil and natural gas will be unable to supply the energy needs of its own economy in the medium term. However, time is running out.

---

[72] See for instance, "Ignorance Rules", *New Straits Times*, 12 Nov. 2005 and "In Search of Clean, Renewable Energy", *New Straits Times*, 9 Nov. 2003.

Chapter

8

Energy Conservation Policy Development in Korea

Jae-Seung Lee

ENERGY SUPPLY AND DEMAND IN KOREA

From the time that Korea launched its rapid industrialisation, energy consumption has increased sharply.[1] To sustain fast economic growth, a stable supply of energy became a primary task of the Government. Korea is currently the world's 10th largest energy consumer, surpassing its economic size ranking of 12th in the world.[2] By 2002, total final energy consumption had more than doubled since 1990 and was growing at an annual growth rate of 7.5 per cent, exceeding the economic growth rate during the same period.[3] Figure 1 shows the economic growth and energy demand trends for Korea. Per capita primary energy consumption grew from 2.2 tons of oil equivalent (toe) in 1990 to

[1] In this paper, Korea refers to the Republic of Korea (ROK).

[2] ROK, Ministry of Foreign Affairs and Trade, report submitted to the seminar "Energy Hegemony", National Assembly, 15 Nov. 2006.

[3] International Energy Agency (IEA), *Energy Policies of IEA Countries — The Republic of Korea*, 2004, p. 342 at www.iea.org/textbase/nppdf/free/2004/korea-conp04.pdf [17 Dec. 2006].

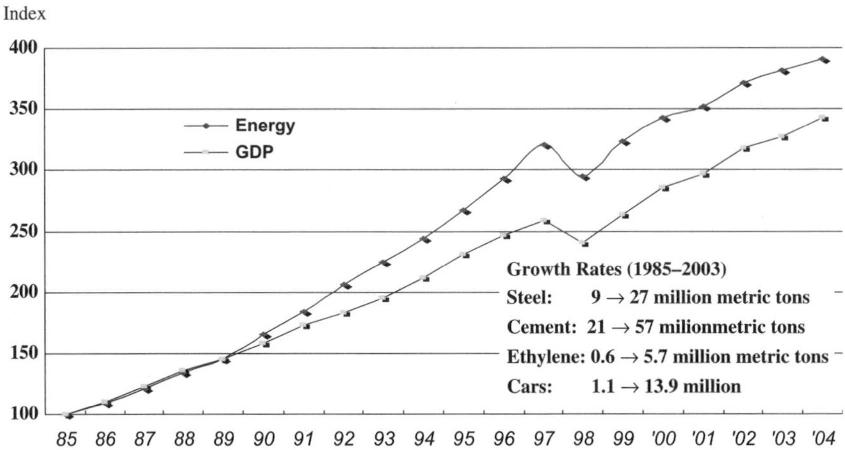

Figure 1: Economic Growth and Energy Demand in Korea (1985–2004)

Source: Ji-Chul Ryu, KEEI, report submitted to the seminar "Energy Hegemony", National Assembly, 15 Nov. 2006.

4.3 toe in 2002.[4] In 2005, Korea's total primary energy supply (TPES) reached 229 million tons of oil equivalent (mtoe). Table 1 presents the major energy indicators for Korea. Tables 2 and 3 provide total energy demand and sectoral predictions, respectively. Energy demand is expected to grow considerably in the forthcoming years.

Almost entirely dependent on overseas energy sources, Korea has been vulnerable to the external supply environment.[5] Securing a stable supply of energy has received utmost attention in Korea's energy policies, particularly following the two oil crises in the 1970s. Figure 2 shows the increasing volume of energy imports. Korea is currently the world's fourth largest oil importer (worth USD42.5 billion in 2005) and it relies heavily on the Middle East. Korea is also the world's second largest LNG importer. Energy imports account for 25.5 per cent (USD 66.7 billion in 2005) of total imports. Of the total

[4] The share of oil in TPES was 49 per cent, followed by coal (22), nuclear (16) and natural gas (10). (IEA, 2004, p. 342).

[5] Dependency on imported energy sources was 96.5 per cent in 2005. Self-supply of oil was only 4.1 per cent, much lower than that of Japan's 15 per cent.

Table 1: Major Energy Indicators for Korea, 1981–2005

	1981	1990	1995	2003	2005	Annual Increase (%)		
						'81–'90	'81–'05	'90–'05
Aggregate Energy Demand (million toe)	45.7	93.2	150.4	215.1	229.3	8.2	7.0	6.2
Energy Consumption per capita (toe)	1.18	2.17	3.34	4.50	4.75	7.0	6.0	5.4
Energy Intensity (toe/million won)	0.31	0.29	0.32	0.32	0.32	–0.7	0.1	0.7
CO_2 Emission (million t-CO_2)	135.9	239.0	366.9	474.4	—	6.5	('81–'03) 5.8	('90–'03) 5.4

Sources: Ji-Chul Ryu, Korea Energy Economics Institute (KEEI), report submitted to the seminar "Energy Hegemony", National Assembly, 15 Nov. 2006.

Table 2: Primary Energy Demand Forecast, 2005–2010

	2005	2006	2007	2008	2009	2010
Coal	84.7	87.6	92.5	100.9	104.9	104.9
(million tons)	(3.1)	(3.4)	(5.6)	(9.1)	(4.0)	(0.0)
Oil	764.2	769.7	778.8	787.3	794.7	800.4
(million bbl)	(1.6)	(0.7)	(1.2)	(1.1)	(0.9)	(0.7)
LNG	22.9	25.4	27.2	27.5	29.3	30.0
(million tons)	(4.8)	(11.0)	(7.3)	(1.0)	(6.3)	(2.4)
Hydropower	4.9	5.1	5.3	5.4	5.6	5.7
(TWh)	(-15.9)	(4.3)	(3.1)	(2.5)	(2.3)	(1.9)
Nuclear	147.5	149.8	150.2	150.4	150.6	166.8
(TWh)	(12.8)	(1.6)	(0.3)	(0.1)	(0.1)	(10.8)
Others	5.0	5.8	6.3	6.7	7.0	7.3
(million toe)	(25.9)	(16.2)	(8.5)	(6.0)	(4.7)	(4.0)
Total	229.6	236.8	244.2	251.6	257.7	263.7
(million toe)	(4.2)	(3.2) 54	(3.1)	(3.0)	(2.4)	(2.3)

Source: KEEI, *KEEI Mid-term Energy Outlook, 2005–2010*, Jan. 2006, at http://www.keei.re.kr/web_keei/en_outlook.nsf/BymainV/A4FD7D9C36C0415E492571100028CDFB?OpenDocument [22 Jan. 2007].

Note: Numbers in parentheses are percentage growth rates year-on-year.

Table 3: Sectoral Energy Demand Forecast, 2005–2010

	2005	2006	2007	2008	2009	2010
Industry	94.9	97.7	100.1	102.4	104.6	106.8
(million toe)	(2.1)	(2.9)	(2.5)	(2.3)	(2.2)	(2.1)
Transport	35.5	35.8	36.4	36.9	37.1	37.2
(million toe)	(2.4)	(0.9)	(1.9)	(1.3)	(0.7)	(0.0)
Residential/ Commercial/ Public	41.6	42.3	43.7	45.0	46.2	47.3
(million toe)	(8.4)	(1.6)	(3.4)	(3.0)	(2.7)	(2.3)
Final Total	172.0	175.7	180.2	184.3	188.0	191.3
(million toe)	(3.6)	(2.2)	(2.6)	(2.3)	(2.0)	(1.7)
Oil	729.7	734.1	742.3	750.3	757.2	762.7
(million bbl)	(1.5)	(0.6)	(1.1)	(1.1)	(0.9)	(0.7)
Anthracite	6.4	6.8	7.0	7.0	6.8	6.6
(million ton)	(11.0)	(5.1)	(3.1)	(0.0)	(-1.9)	(-3.1)
Bituminous	28.2	28.0	28.3	28.5	28.8	29.0
(million ton)	(-1.1)	(-0.8)	(1.4)	(0.8)	(0.8)	(0.7)
Electricity	332.4	352.8	371.5	389.1	405.0	419.4
(TWh)	(6.5)	(6.1)	(5.3)	(4.7)	(4.1)	(3.6)
Citygas	17.1	17.7	18.6	19.6	20.6	21.6
(billion m^3)	(10.8)	(3.5)	(5.3)	(5.2)	(5.1)	(4.9)

Source: *KEEI Mid-term Energy Outlook, 2005–2010*, Jan. 2006, at http://www.keei.re.kr/web_keei/en_outlook.nsf/BymainV/A4FD7D9C36C0415E492571100028CDFB?OpenDocument [22 Jan. 2007].

Note: Numbers in parentheses are percentage growth rates year-on-year.

energy imports, the portion of crude oil is the highest (63.7 per cent), followed by natural gas (13 per cent) and coal (7.8 per cent).[6]

Korea's energy consumption pattern has been another vulnerable aspect of its energy security. During the process of industrial modernisation, the Government emphasised energy-intensive heavy industries such as petrochemicals, cement, iron, steel, machinery and automobiles. As a result, heavy energy consuming sectors — petrochemical, steel and cement — consume 79 per cent of total industrial energy consumption.

[6] ROK, Ministry of Foreign Affairs and Trade, 2006, *op. cit.*

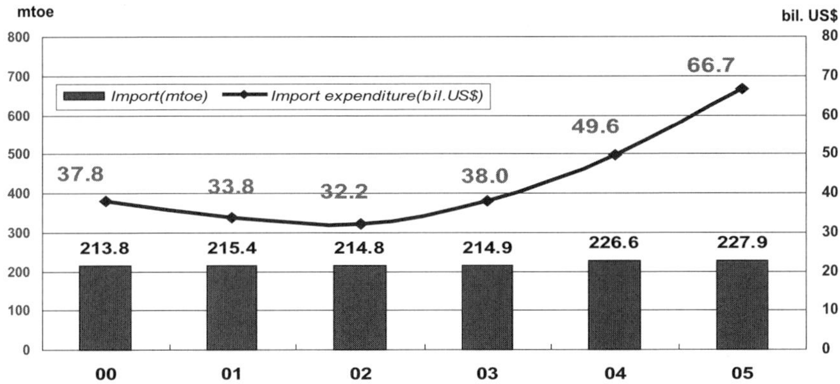

Figure 2: Korea's Energy Imports

Source: Ji-Chul Ryu, KEEI, report submitted to the seminar "Energy Hegemony", National Assembly, 15 Nov. 2006.

The steady rise in the number of automobiles has greatly increased energy consumption in the transport sector. Energy consumption in the residential and commercial sectors has also grown as income levels continue to rise.[7] The 'structural' problem in energy consumption made the improvement in energy mix difficult. The dependency on oil (45.7 per cent) is still high and the share of natural gas (12.9 per cent) is rapidly growing. Imported fossil fuels still comprise most of the energy consumed in Korea.

THE ENERGY SECURITY AND ENERGY CONSERVATION CHALLENGES

Since the oil crises of the 1970s, energy supply security has been the core of Korea's energy policies. Recent price hikes in the energy market have also posed a serious threat to Korea's energy security. The price of Dubai crude, the benchmark used in Korea, increased from USD33.6 in 2004 to USD49.4 per barrel in 2005 and averaged USD64.2 per barrel in April 2006.[8] The high oil price is expected to last into the near future.

[7] IEA, *Energy Policies of IEA Countries — The Republic of Korea*, 2004, p. 342.

[8] "Korea Promotes Energy Reforms, Conservation", *Korea Herald*, 15 Nov. 2004.

However, the challenge to Korea's energy security has stemmed not only from the supply side. Despite the high dependency on overseas energy sources, energy efficiency in Korea has not been high. The need to establish better energy efficiency policies is becoming more urgent. Thus, demand side management (DSM) has begun to receive new emphasis.

Environmental concerns, especially pertaining to climate change, are another challenge to Korea's energy security. Korea does not bear any obligation to reduce greenhouse gas (GHG) emissions under the Kyoto Protocol but is under heavy pressure to join in the regime given its economic size and the volume of its GHG emissions. Energy-related GHG emissions currently account for the largest component of total GHG emissions. Total GHG emissions increased at an annual average rate of 5.1 per cent from 1990 to 2002.[9] They presently rank ninth in the world, accounting for 1.9 per cent of the global total in 2002. The International Energy Agency (IEA) has warned that the growth of GHG emissions will continue unless considerable efforts are made to reduce emissions.[10]

In dealing with the challenges resulting from high energy intensity and external dependence, the Government has tried to curb energy consumption through various conservation measures. Energy conservation policies were improved following the oil shocks in the 1970s and the Gulf War. Energy conservation and efficiency policies are aimed at all components of the energy supply chain, from primary production to end-use.

Energy conservation contributes to demand-side management by reducing or slowing down societal demand for energy.[11] In the short run, the Government has implemented immediate measures to curb the growth of oil demand, especially in the transport sector, to deal

[9] CO_2 emissions in Korea increased from 226 million tons (mt) in 1990 to 451 mt in 2002.

[10] IEA, "Energy and Environment-Republic of Korea", 2004.

[11] Energy conservation is "the practice of decreasing the quantity of energy used while achieving a similar outcome of end use. Energy conservation is often the most economical solution to energy shortages in passive terms." This definition was drawn from http://en.wikipedia.org/wiki/Energy_conservation [12 Jan. 2007].

with sudden price hikes. In the medium and longer run, the Government has also tried to advance energy efficiency measures which have been implemented mainly by the Korea Energy Management Corporation (KEMCO).[12] The Government has also attempted to rationalise energy prices by using financial and tax measures.

The objectives of the energy efficiency programmes have been to encourage manufactures to produce more energy-efficient products by offering incentives, and to provide end-users more opportunities to purchase energy-efficient products.[13] Energy conservation has also been an important way to reduce climate change by curtailing GHG emissions. It has facilitated the introduction of renewable energy. Policies geared to efficient energy use have become intertwined with polices aimed at protecting the natural environment.

SURVEY OF ENERGY CONSERVATION IN KOREA

Legal Framework

Korea's energy conservation policies have been planned and implemented on the basis of the Rational Energy Utilisation Act (REUA) promulgated in December 1979. In the first place, the REUA attempted to ensure energy security in an emergency as well as to promote energy efficiency and conservation.[14] According to the REUA, the Basic National Energy Plan and the Basic Plan for Rational Use of Energy were to be drafted and implemented by the Ministry of Trade, Industry and Energy (MOTIE) (MOCIE in later years — see below) every five years[15] and the Action Plan for REUA was to be

---

[12] IEA, 2004, *op. cit.*

[13] IEA, "Energy Efficiency Programs in Korea" at http://dsm.iea.org/Files/Exco%20File%Library/Country%20Publications/programs.doc [8 Jan. 2007].

[14] Articles 6 and 7 of the REUA refer to emergency preparedness. Mi-Chung Ahn, "The Energy Conservation Program of the Republic of Korea", in United Nations, *Compendium on Energy Conservation Legislation in Countries of the Asia and Pacific Region* (1999).

[15] REUA, Articles 4 and 15.

drafted by the heads of the authorities concerned and local administrations.[16]

In 1997, the REUA incorporated a Pre-notification System of Energy Prices, which was designed to make consumers more responsive to energy policies. The revised REUA also prescribed the organisation and operation of the National Energy Conservation Promotion Committee responsible for reviewing and assessing the Basic Plan for Rational Use of Energy and other related matters.[17] In the same year, the Government established the 10-year National Plan for Energy Technology Development (1997–2006), which incorporates plans for renewable energy, clean energy and energy efficient technologies.[18]

The Government released The Second National Energy Plan in December 2002 following the publication of the Vision and Development Strategies for Korea's Energy Policy toward 2010.[19] The *2002* National Energy Plan projected the steady increase of total energy demand and emphasised the importance of energy conservation for rational use of energy and stable demand management.[20]

---

[16] REUA, Article 16.

[17] REUA, Article 15-2 (newly inserted in 1997). The Committee is headed by the prime minister and comprised of not more than 25 members including the prime minister. Ahn, *op. cit.*

[18] The main goals of the Plan are to meet 3 per cent of total energy supply with renewable energy by 2006; reduce 10 per cent of final energy consumption by 2006; and advance clean energy technologies, especially to reduce SO_x, NO_x, dust and CO_2. United Nations, National Energy Information–Korea at www.un.org/esa/agenda21/nat/info/country/repkorea/energy.pdf) [25 Dec. 2007].

[19] The objectives of the 2002 National Energy Plan are to: (1) promote a stable supply of energy; (2) promote energy conservation to enable a rational use of energy and stabilise energy demand; (3) minimise energy-related environmental damage and to promote the development of energy-related technologies; and (4) formulate directions and strategies for mid- and long-term national energy policies as well as the basic guidelines for all other energy plans by sector, source and region. (IEA, *Energy Policies of IEA Countries*, 2004, p. 341.)

[20] The Plan predicted that total energy demand would increase by an annual average of 3.1 per cent from 2001 to 2010 and by an annual average of 2.4 per cent from 2001 to 2020. Per capita energy demand, which stood at 4.1 toe in 2000, was projected to increase to 5.3 toe and 6.2 toe in 2010 and 2020, respectively. *Ibid.*

The Basic Energy Act was introduced in March 2006 and went into effect in September. The Basic Energy Act incorporated and replaced much of the REUA. On the basis of the Basic Energy Act, the Government set up a National Energy Committee led by the President. Together with the Basic Energy Act, 'Energy Vision 2030' was announced and provided the roadmap for future Korean energy policies including more active measures for energy conservation and efficiency improvement.

Institutions

The Government has long been actively involved in the energy sector.[21] Before 1978, the Ministry of Trade and Industry (MTI) was responsible for energy matters. In the wake of the second oil shock, Korea established the Ministry of Energy and Resources (MOER) to administer the planning and enforcement of national energy policies. Here, the MTI energy division formed the foundation of the new Ministry.[22] In 1993, MOER re-merged with MTI to form the Ministry of Trade, Industry and Energy (MOTIE). In March 1998, MOTIE was restructured and became the Ministry of Commerce, Industry and Energy (MOCIE). A new Office of Energy Efficiency and Conservation Policy was created in the MOCIE.

MOCIE is currently responsible for most energy-related activities including energy policy, supervision and reform of the energy industry, climate change issues and price control.[23] In addition, MOCIE plays an important role in sponsoring energy corporations and research institutes.[24] As Korea's energy sector is currently undergoing a transition, the Government is also changing the nature of its involvement.

---

[21] IEA, *Energy Policies of IEA Countries — The Republic of Korea*, 2002, p. 15 at www.iea.org/textbase/nppdf/free/2000/ korea-conp02.pdf.

[22] It included nuclear energy but the nuclear power safety programme remained under the Ministry of Science and Technology.

[23] UN, *op. cit.*

[24] For example, the Korea Electric Power Corporation (KEPCO), Korea Gas Corporations (KOGAS), Korea Institute of Energy Research (KIER) and Korea Energy Economics Institute (KEEI), etc.

The Korea Energy Management Corporation (KEMCO) is a non-profit government agency that was established in 1980 by MOCIE (then MOER) based on the REUA. KEMCO is responsible for implementing conservation policies. It functions as the national energy efficiency centre and the principal fund-allocating institution in energy conservation and efficiency matters.

ENERGY CONSERVATION PROJECTS

Policy Objectives

Energy conservation began to be ratified as one of the major pillars of energy policy. Based on the 1997 Basic Plan for Rational Use of Energy, Korea's current energy conservation policies attempt to:

- improve the trade deficit by reducing energy imports;
- strengthen industrial competitiveness by reducing production costs resulting from reduced energy use;
- contribute to global environmental protection by minimising CO2 emissions;
- enhance energy efficiency from production, distribution and consumption and develop an energy-efficient socio-economic structure;
- strengthen DSM in the power sector;
- develop best use market mechanisms to encourage energy efficiency investments;
- intensify regulations to an appropriate level in key areas (including energy efficiency standards for energy equipment and appliances);
- foster an energy- and resource-saving lifestyle by raising energy conservation awareness, properly adjusting energy price levels, etc.;
- strengthen international cooperation.[25]

[25] Ahn, *op.cit.*

Energy Conservation Policy Measures

Sectoral Categorisation

Based on the REUA and other policy measures, a series of energy conservation programmes have been promoted: energy savings and efficiency programmes including tax breaks, loan and subsidy programmes, energy conservation technologies, various pilot projects, energy exhibitions, and energy service company programmes, etc. These programmes can be categorised into cross sectional, industrial and residential/commercial.[26] Figure 3 summarises Korea's energy conservation policies.[27] Details of these are given below.

Policy Measures of Energy Savings and Conservation[28]

Management of Energy-Intensive Users

Extensive management programmes were introduced and aimed particularly at energy-intensive industries such as iron and steel, petrochemicals and machinery. The Government has the authority to announce Energy Management Guidelines to be adopted by heavy energy users. MOCIE can designate heavy energy users as Energy Management-Required Users (EMRUs) which "must report to the Government their annual production, energy facilities, equipment, annual energy use and corporate energy conservation plan along with the results of implementing the previous year's plan."[29] Approximately 200 companies are designated as special EMRUs

---

[26] UN, *op. cit.*, pp. 35–37.

[27] *Ibid.*

[28] Discussion in this section is based on Ahn, *op. cit.*; UN, *op. cit.*; and IEA, *Energy Policies of IEA Countries,* 2004. The author wants to note that some sentences from the above materials describe the general contents of the policy measures and may not carry a specific quotation.

[29] REUA, Articles 24 and 25.

Cross-Sectoral Policies

- A five-year conservation programme
- Tax and financial incentives for integrated energy suppliers
- Energy efficiency information, labelling and standards programmes
- Energy-saving labels
- Minimum energy efficiency performance standards
- Information labels for energy consumption
- Financial assistance for energy research and development
- Publicity campaigns and educational programmes
- Regional energy programmes with subsidies and assistance
- Local management programmes by energy supply companies
- The obligatory use of certified high energy efficiency equipment for designated types of buildings

Industrial Sector

- Energy audits
- Voluntary agreements

Residential and Commercial Sectors

- Insulating buildings to reduce energy use
- Monitoring the energy use of large energy-consuming buildings
- Five-year energy conservation plans for large energy-consuming buildings

Transport Sector

- Fuel efficiency labelling programme
- Tax incentives for the use of small cars
- Promotion of car sharing
- Investments in public transport and fuel efficiency targets

Public Sector

- Public procurement of certified high energy efficiency equipment
- Obligatory use of certified high energy efficiency equipment for public buildings

Figure 3: Categorisation of Korea's Energy Conversation Policy

and are required to set up and implement, through consultation with KEMCO, their own five-year Corporate Energy Conservation Plan within the framework of the Basic Plan for Rational Use of Energy.[30]

[30] These companies represent roughly 50 per cent of the total industrial energy use in Korea.

Energy Impacts Assessment on Energy-Intensive Projects

Large-scale, energy-intensive projects must be reported and monitored.[31] A governmental or public institution should set up an energy use plan and ask the MOCIE for consultation before carrying out a high energy-consuming project. These projects usually include large-scale urban development projects, energy resource development projects, industrial site or complex preparation projects, port and railroad construction projects, airport complex construction projects or tourist complex development projects.

Voluntary Agreements

A voluntary agreement (VA) is an agreement between an energy-intensive company and the Government. When a company sets a goal for GHG reduction and formulates a definitive action plan, the Government supports the company with various measures. Up to 2004, some 1,024 worksites participated in this programme.

Energy Saving in the Public Sector

The Energy Saving Performance Contracting Programme was launched to help the managers of public buildings reduce energy consumption. The central government agencies and local administrations are encouraged to use energy-efficient equipment and appliances. The Government monitors the effects of such energy-saving activities and provides related information through workshops and other public campaigns.

Waste Heat Recovery and Utilisation

According to the waste heat recovery and utilisation plan, energy users should try to recover and utilise waste heat produced in their workplace or help third companies utilise it.[32] The Government designates

---

[31] REUA, Articles 8–11.

[32] REUA, Article 35.

specific kinds of heat-using equipment and the installation or construction of this equipment must follow the rules and regulations set by the MOCIE and be inspected later by the city or province governor. Expansion of Combined Heat and Power provides mass energy consumers with heat and electricity through cogeneration including municipal waste incineration and industrial waste heat.[33]

Energy Service Companies

Energy Service Companies (ESCO) invest in energy utilising facilities that guarantee performance and later collect the invested capital and profit from the saved energy. The Government supports ESCO by providing them with relevant information on new energy efficiency technologies, and financial and taxation incentives.[34] To further promote ESCO activities, the Government began to provide preferential conditions for loans and taxation from 1997, such as the granting of tax incentives to a facility owner. As of 2005, 169 companies were registered as ESCO.[35]

Energy Auditing

Energy Auditing is an information transfer programme aimed at assisting energy consumers understand and employ technologies and practices that use energy more efficiently. Financial support can be provided depending on the performance.[36] Energy audits have been conducted mainly by KEMCO which has the human resources and latest audit facilities as required by governmental regulations and

---

[33] District Heating and Cooling and Industrial Complex Combined Heat and Power (CHP) are two major areas in this policy. Until 2004, District Heating has provided to 1,322,000 households in 21 districts, covering 10.6 per cent of total households. For the promotion of the programme, the Government enacted the Comprehensive Energy Supply Act in 1991.

[34] REUA, Article 22.

[35] Ahn, *op. cit.*

[36] REUA, Articles 30 and 31.

requirements regarding energy auditing.[37] Identified energy-saving measures are to be recommended together with technical assistance after auditing.

Demand-Side Management

DSM has been vigorously pursued due to growing difficulties in securing new power supply facilities to meet the rapidly rising energy demand. In particular, the dramatic increase in the use of air-conditioners since the beginning of the 1990s has been a severe drain on electricity reserves in the summers and resulted in power shortages on several occasions. To cope with this problem, the Government revised the REUA in July 1995 to make it mandatory for all the utilities to establish and implement a DSM investment plan on an annual basis and report the plan and its implementation results to the Government.[38]

Research and Development

The Government leads research and development (R&D) activities in collaboration with industry, universities and research institutes. Priority projects are financed by the Government budget and other energy-related funds. MOCIE supports R&D and dissemination of energy-efficient technologies, including renewable energy technologies and clean technologies. The Research and Development Centre for Energy and Resources (RACER) was founded as an affiliate of KEMCO in 1992 to take charge of managing the whole R&D process and financial assistance.[39] A Ten-Year Energy Technology Development Plan (1997 to 2006) was implemented in this perspective.

[37] It has six specialised audit teams for the chemical, metal, ceramic, textile, forest (pulp and paper) and food industries. Ahn, *op. cit.*

[38] REUA, Article 12.

[39] REUA, Articles 38–41.

Local and Regional Energy Plans

A series of measures have been developed to support local governments in rationalising energy use. According to the Basic National Energy Plan, all local governments are required to make and implement their own five-year regional energy plan.[40] MOCIE provides energy technology information to local governments and supports energy conservation projects that are suitable for their geographical and socio-economic needs and conditions. KEMCO also provides training and education to local governmental officials in charge of energy planning.[41]

Education and Training Programmes

Energy conservation education and training programmes have been fostered by both the Government and non-Government sectors[42] and include training and education for energy managers, and installing engineers or operators of heat-using equipment.[43] The Government also organises exhibitions and cultural events to publicise successful cases of energy conservation.[44] KEMCO is authorised to run an energy management training centre for this purpose and actively engages in public campaigns.[45]

Information and Statistics

The MOCIE supports energy conservation through collecting, analysing, processing and disseminating energy information. The management

[40] Article 5.

[41] Ahn, *op. cit.*

[42] The Ministry of Education and Human Resources designated 16 elementary and 16 middle schools as "Demonstration Energy Conservation Schools" in 2002; November is designated as the "Energy Conservation Month" and the first Friday of every month is designated as "Energy Conservation Day."

[43] REUA, Article 88.

[44] REUA, Article 20.

[45] KEMCO has organised an Energy Conservation Exhibition (ENCONEX) annually, and an Energy Conservation Convention, biennially.

and publication of energy statistics are implemented by the Government.[46] Energy-related institutions and major energy users may be requested to submit raw energy data to the Government.[47]

Energy Efficiency Policy Measures[48]

Energy Efficiency Standards

Energy efficiency standards are based on Minimum Energy Performance Standards (MEPS) and are aimed at stopping the manufacture and sale of products involving high energy inputs while Target Energy Performance Standards (TEPS) are designed to help manufacturers achieve higher efficiency. The MEPS aim to expel inefficient designs from the market, while the TEPS are designed to encourage manufacturers make more energy-efficient goods.[49] MEPS and TEPS are currently applied to six items: electric refrigerators, air-conditioners, fluorescent lamps, lamp ballasts, incandescent bulbs and passenger cars.

Energy Labelling

Energy labelling, started in 1992, aims to encourage manufacturers to consistently make products with high energy efficiency and to stimulate importers to introduce more energy-efficient products into the domestic market while helping consumers choose more energy-efficient goods.[50] The objective of this programme is to create a

---

[46] REUA, Article 13.

[47] KEMCO is designated to do this work for the Government. Raw energy data is currently reported to KEMCO from over 3,000 government-designated EMRUs. The processed data is also used to identify the energy conservation potential of those industries and is provided to other relevant organisations. Ahn, *op. cit.*

[48] Discussion in this part was also drawn from IEA, "Energy Efficiency Programs in Korea", *op. cit.*

[49] Ahn, *op. cit.*

[50] Seven targets of the programme are: electric refrigerators; electric air-conditioners; incandescent bulbs; fluorescent lamps; self-ballasted lamps; ballast for fluorescent lamps; and passenger cars.

'market push' by directing manufacturers to make products with higher efficiency. It is applied to both domestic and imported products. Energy labelling indicates the energy efficiency grade of the model from 1 to 5. MOCIE sets the provisions for this programme and KEMCO, authorised by the REUA, supervises its implementation and enforces the provisions.

Certification of High Efficiency Energy-Using Appliances

This programme certifies energy-using products whose energy efficiency is much higher than that of others.[51] Based on the REUA, the Government provides long term and low interest loans to certified companies to enhance the proportion of energy efficiency appliances in the market. Public organisations are encouraged to use certified products. The recommendations have been upgraded to obligatory regulations in lighting equipment and are planned to be gradually applied to all appliances.

Energy-Saving Office Equipment and Home Electronics

Launched in April 1999, this programme was based on MOCIE Announcement No. 1998-136, regarding the voluntary partnership between the Government through KEMCO and manufacturers. The objective of this programme is to induce manufacturers to voluntarily produce energy-saving products that meet the energy-efficiency guidelines proposed by KEMCO. Manufacturers producing high energy efficiency products are permitted to attach an energy-saving label of certification.[52]

---

[51] Subject appliances include: induction motors; 26mm32W fluorescent lamps; ballast for 26mm32W fluorescent lamps; self-ballasted lamps; reflectors for fluorescent lamps; sensor lighting equipment; heat recovery ventilators; windows; industrial gas boilers; domestic gas boilers; pumps; centrifugal water chillers; energy saving devices for monitors; and uninterruptible power systems.

[52] Subject appliances include: computers; monitors; all-in-one systems; printers; fax machines; combinations; printer/fax machines; copiers; TVs; VCRs; TV/VCRs; and combination units.

Financial and Taxation Assistance to Energy Efficiency Investments

The Government has provided long-term and low-interest loans from the Fund for the Rational Use of Energy along with tax incentives, for energy efficiency and conservation investments.[53] The Government provides tax incentives for energy efficiency investments with tax credits from 2005.[54] Korea's environment-based energy taxation system was also revised at the beginning of July 2000 to encourage the conservation of energy and promote protection of the environment.[55]

Environment and Energy Conservation

Reduction of Greenhouse Gases

Dealing with climate change has become a major environmental and energy concern. The action plan to address climate change has been pursued in several stages since 1999. The National Energy Plan, Basic Plan for Rational Energy Utilisation and Regional Energy Plan all specify environmental management measures through energy conservation. These plans to promote R&D on GHG reduction technologies included new and renewable energy, and medium- and large-scale technologies. They tried to expand GHG reduction programmes in all sectors — industry, transport, residential, waste management and agriculture — while establishing an integrated management system for energy conservation. These initiatives also facilitated the use of Kyoto mechanisms such as the Clean Development Mechanism (CDM) and Emissions Trading (ET). The revision of REUA in 2003

[53] The loans for installing energy-saving facilities or equipment offer favourable conditions. In most cases they have 3- to 5-year grace periods and 5-year repayment periods with interest rates of 2.75 per cent (floating rate), which are about half of the market or prime rates.

[54] UN, *op. cit.*

[55] This system aims to further improve the competitiveness of natural gas in the fuel market for industry or power generation. The tax levy on heavy oil was exempted from taxation while the tax rate on light oil and kerosene was raised.

emphasised the efficient use of energy and the reduction of GHGs. The revised act also broadened the scope of greenhouse gases, so that all six greenhouse gases prescribed in the Kyoto Protocol (carbon dioxide, methane, nitrous oxide, sulfur hexafluoride, HFCs and PFCs) are included.[56]

Alternative and Renewable Energy

In the case of new and renewable energy technologies, the Government introduced the Alternative Energy Development Promotion Act in 1987 under which a Basic Plan for the Development of New and Renewable Energy Technologies was established in the next year. According to the Basic Plan, new and renewable sources of energy (NRSE) were planned to contribute 2 per cent of total energy demand by the year 2006. Currently, eight energy sources (solar, bio, waste, small hydro, wind, hydrogen, ocean and geothermal) and two related technologies (fuel cell and coal utilisation technologies) are defined by the Alternative Energy Development Promotion Act as target technologies.[57]

Green Energy Family Movement

The Green Energy Family (GEF) Movement is a nationwide voluntary partnership movement among citizens, companies, NGOs and the press to enhance energy efficiency and reduce social costs. Launched in 1995, the GEF Movement attempts to enhance public awareness about benefits from energy-efficient equipment and facilities. GEF initiated the Green Lighting, Green Motor and Green Energy Design Programme.[58]

---

[56] IEA, *Energy Policies of IEA Countries,* 2004, p. 343.

[57] Ahn, *op. cit.*

[58] The Government will support ESCO with USD123 million worth of long-term low-interest rate loans and has triggered market development by pilot projects and procurements in the public sector.

PROSPECTS AND TASKS

Basic Energy Act and Energy Vision 2030

The Basic Energy Act was ratified in March 2006 and activated in September 2006. The Government also announced a long-term national energy policy plan, Energy Vision 2030, in November 2006.[59] The objectives of new energy policy were classified under: energy security, energy efficiency and environmental protection. This new energy vision came amid warnings that high crude prices would deliver a serious blow to the South Korean economy.

According to the Energy Vision 2030, the oil and natural gas self-development rate, which was 4.1 per cent at the end of 2006, is to be raised to 35 per cent by 2030. In addition, renewable energy, such as wind power, tidal energy and solar energy, which presently account for 2.2 per cent of the total, is to reach 9 per cent. The Government also plans to lower the country's petroleum dependency ratio from 44.3 per cent in 2005 to 35 per cent by 2030, through exploring foreign oil fields, as well as by restructuring the domestic energy industry so as to cut back on energy use across the nation.

In the MOCIE Policy Plan for 2006, the improvement of energy consumption structure, as well as medium- and long-term measures to increase energy efficiency were highlighted. Energy intensity in Korea has increased consistently as the economy has grown and was expected to grow quickly unless effective measures were put in place to curb demand and increase the efficiency energy use. The Government announced the goal to improve energy intensity from 0.299 (TOE/million won) in 2006 to 0.290 in 2008.[60]

Based on the ten-year National Energy and Resource Technology Development Plan (2006–2015), the R&D budget for energy and resources will increase from 409 billion won in 2006 to 588 billion won in 2008.[61] More quality-oriented measures are being implemented.

[59] "Government Announces 'Energy Vision 2030' Plan", *Maeil Business Newspaper*, 28 Nov. 2006.

[60] For comparison: Japan (0.106), Britain (0.152), France (0.200), US (0.221).

[61] IEA-CERT: Committee on Energy Research and Technology.

The introduction of ESCO certification and diversification of financial sources are also being pursued. Education and training programmes are currently being strengthened.

Policy Tasks

Energy conservation policies have received increased attention in Korea in recent years in an effort to reduce surging energy costs amid high oil and gas prices. The MOCIE and KEMCO have implemented a series of measures to save energy and improve energy efficiency.

However, current energy conservation measures are still insufficient to deal with the diverse aspects of energy security. The importance of DSM and efficiency improvement is expected to grow in the future. The IEA has made the following suggestions to the Korean Government:

- Make energy efficiency a high priority;
- Strengthen energy efficiency policies through additional measures;
- Facilitate the process of energy price reform so that fuel prices reflect costs;
- Ensure Korea's standards and energy efficiency norms comply with best international practice;
- Develop further energy efficiency policies as part of the effort to reduce greenhouse gas emissions.[62]

Improving energy efficiency is still key to future sustainable development in Korea. Many industrial equipment facilities became energy-efficient in recent years. The capacity to improve energy efficiency by introducing more modern technologies may not be as large as expected but these efforts are inevitable considering that a modest percentage gain can lead to huge energy savings.

A strategic approach to enhancing technological development of the energy and energy equipment sectors is also urgent. More progress needs to be made in the transport sector especially.

[62] IEA, 2004, *op. cit.*

The dramatic increases in the number of cars and in the average car size have sharply increased the use of energy. Since congestion and environmental problems have become acute, fuel prices need to be reviewed to fully internalise these externalities. In addition, active measures should be pursued to encourage the use of more energy-efficient cars and public transport.[63]

Cost-reflective pricing became essential in the industrial and commercial sectors to encourage the rational use of energy. Energy prices have been distorted in the past in ways that have led to inefficient energy use. Environmental and other costs should be internalised as much as possible. Better quality data collection is also required.

Building a more environment-friendly system and implementing market-friendly measures for energy conservation are other tasks. Of the total energy mix in 2030, the share of environment-friendly energy, including renewable energy, will be increased from 2.2 to 9 per cent.

Finally, more voluntary energy conservation is needed beyond the mandatory obligations. International cooperation in energy conservation technology development and transfer is becoming an external task of energy conservation policy in Korea.[64]

---

[63] In fact, state-run companies and government agencies will be required to designate one day of the week as a 'no-driving day'. The Government is considering expanding incentives to the owners of compact cars such as lower parking fees, taxes and insurance premium.

[64] K-GIN: Korea Global Innovation Network.

Chapter

9

Energy Conservation Policy Development in Japan

Yasuo Tanabe

INTRODUCTION

Japan is widely believed to be the world's most energy efficient economy. With the growing recognition of energy efficiency as a key to energy security and sustainability, attention has recently been increasing towards Japan's challenges and experience in this area. The lessons learned are worth reviewing, especially in the context of growing energy demand in the Asian region, which is causing concern about future energy security and environmental protection both at the regional and global levels. Until the early 1970s, Japan was a rapidly-growing, industrialising economy, similar to China and the ASEAN economies today. Following the two oil shocks of the 1970s, Japan has made vigorous efforts towards energy conservation and efficiency at all government, industry and public levels. As a result, it has achieved the highest energy efficiency levels in the world and continues to face challenges as it moves forward. How has energy efficiency been achieved in Japan? What are the features of Japan's performance so far? What remaining challenges lie ahead? Asia's emerging economies can and should learn from Japan's experience.

JAPAN'S ENERGY EFFICIENCY IN HISTORICAL PERSPECTIVE

Energy efficiency can be measured in various ways, but the most common measure at the global macro level is energy intensity, that is, energy consumption per unit of GDP. According to IEA statistics, Japan's energy intensity in 2004 was 0.11 toe/USD1,000 (based on year 2000 real prices). India was 0.84, Indonesia 0.88, China 0.91, Thailand 0.65, Singapore 0.25, and the US 0.22 (see Figure 1). In other words, Japan's energy efficiency is twice as strong as that of the US and Singapore, and eight times better than that of China and Indonesia. The average global and OECD energy intensities are 0.32 and 0.20 respectively, meaning Japan is three times better than the world average and twice as good as the OECD average. Japan's energy intensity level is the lowest in the world.

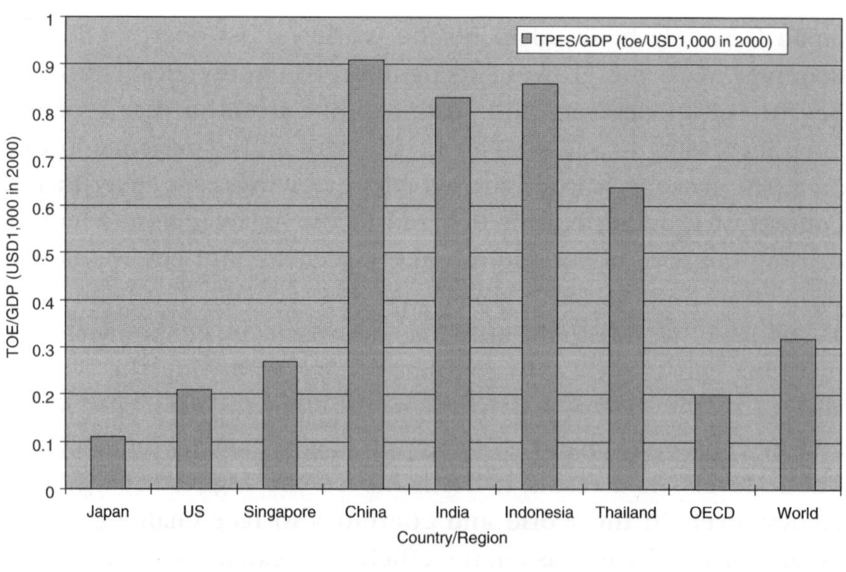

Figure 1: Energy Efficiency of Selected Countries/Regions (2005)

Japan's current energy efficiency level has greatly improved over the past 30 years. According to the Ministry of Economy, Trade and Industry (METI), Japan's energy intensity decreased by 37 per cent between 1973 and 2003.[1] The growth of energy consumption has been slower than economic growth in the last 30 years, thus elasticity has been below one. In contrast, from 1965 to 1973, Japan's average economic growth rate was 9.0 per cent while its average energy demand growth rate was 10.9 per cent resulting in an elasticity of 1.2.[2] Japan's performance in the 'pre-oil shock' period resembles the current Chinese situation, i.e., high economic growth with much higher energy consumption growth, but the pattern changed dramatically after the oil shocks of the 1970s.

Macro-level energy efficiency can be broken down into separate energy demand sectors: industry, transportation and others. The others are mainly residential and commercial and public services. A look at the sectoral energy demand progress in the post-oil shock era reveals that the industrial sector has shown higher performance than the transportation, residential and service sectors. In 1973, industry accounted for 63 per cent of total final energy consumption. The residential and service sectors accounted for 18 per cent and transport for 16 per cent. In 2004, the industrial sector accounted for 48 per cent, the residential and services sector for 28 per cent, and transport for 25 per cent (see Figure 2). In absolute terms, the increases in energy consumption between 1973 and 2004 were only 3.8 per cent for the industrial sector, while they were 112.8 per cent for the residential and services sector and 111.3 per cent for the transport sector.

In most industrial and industrialising economies, it is the industrial sector which has the heaviest demand. Within this category, heavy industries such as steel, chemicals, cement and paper and pulp are the most energy intensive, accounting for more than 70 per cent

---

[1] METI, "New National Energy Strategy", May 2006 at http://www.enecho.meti.go.jp/english/report/newnationalenergystrategy2006.pdf.

[2] EDMC, *Enerugii Keizai Tokei Yoran (Handbook of Energy and Economic Statistics in Japan)*. Tokyo: Energy Conservation Center of Japan, 2007, p. 15.

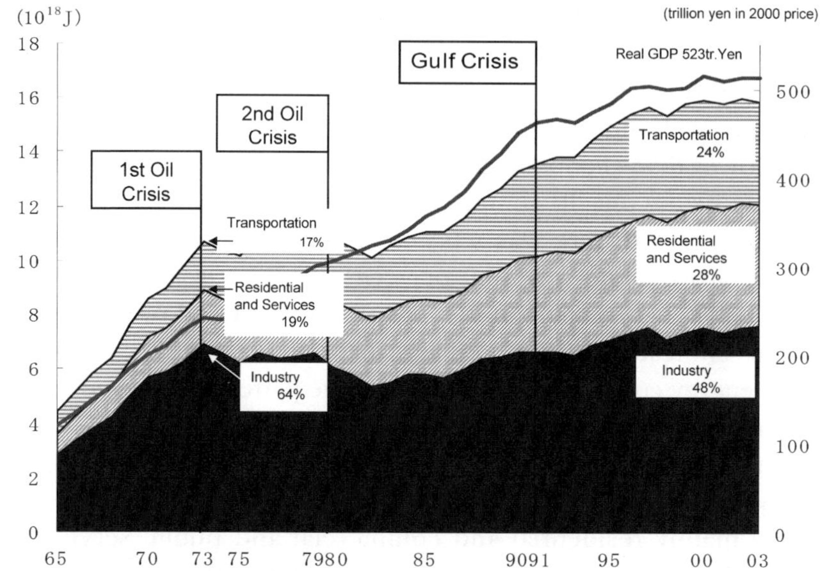

Figure 2: Energy Consumption by Sector in Japan

Source: METI

of total industrial consumption. Energy intensity measured by energy consumption per industrial production index is often used as an indicator for the industrial sector. Between 1973 and 2004, the energy intensity of Japan's manufacturing sector decreased 41 per cent, in the steel industry by 32 per cent, in chemicals by 47 per cent and in the paper and pulp industry by 46 per cent (see Figure 3). It is apparent that energy efficiency in the manufacturing sector, particularly among the heavy industries, has improved substantially, contributing to macro-level energy efficiency improvement.

The reductions in industrial energy are attributable to production, structural and efficiency factors. In general, industrial production grows in periods of booming business, leading to an increase in energy consumption, and contracts during recessionary periods, resulting in a decrease in energy consumption. However, when efficiency improves and/or structural change takes place, the growth rate of energy consumption is less than production growth. Therefore, care must be taken in viewing energy demand situations, distinguishing the factors causing energy consumption change. According to the

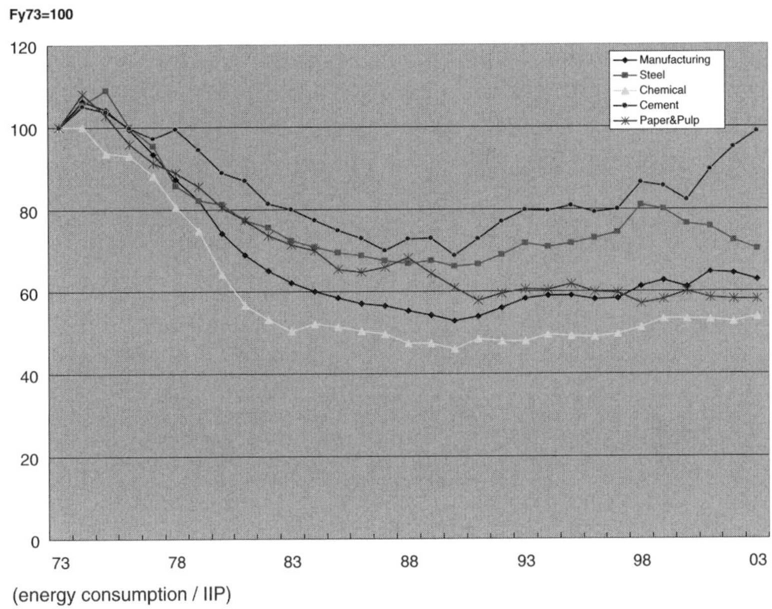

Figure 3: Energy Intensity by Industry in Japan

Source: METI

EDMC (Energy Data and Modelling Centre), the production growth factor in Japan in the 1970s and 1980s was offset by efficiency improvement factors of about 50 per cent and by structural factors of about 20 per cent. It may appear that Japan has transformed its industrial structure, but in real terms the share of the manufacturing sector as a proportion of total GDP has remained stable over the past 30 years at around 25–27 per cent. Within the manufacturing sector, the automobile and electric machinery shares have increased. This suggests that energy efficiency improvement in the industrial sector has been caused by efficiency improvement in each sub-sector as well as structural changes in manufacturing.

THE NEED FOR FURTHER IMPROVEMENT IN JAPAN'S ENERGY EFFICIENCY

Although Japan has thus far achieved the highest energy efficiencies in the world, the need for further improvement persists. The main reason is that Japan is committed to climate change mitigation. Japan's pledge under the Kyoto Protocol is to achieve a 6 per cent reduction in its greenhouse gases (GHG) emissions from the 1990 level by the first commitment period (2008–2012). In order to surpass this target, further strenuous efforts are necessary to improve energy efficiency. In 1998, the first General Principle to Promote Measures to Counter Global Warming was adopted at the Cabinet council. The General Principle sets a reduction goal for each greenhouse gas. Following the General Principles, the Energy Conservation Law was revised and the Keidanren adopted its Voluntary Action Plan (further details below). It was clear that energy efficiency improvements had slowed down throughout the 1990s compared with the 1980s. Energy consumption in the transportation, residential and service sectors had increased, while the energy efficiency levels in the industrial sector had been fairly stable.

The Government began stepping up its efforts following ratification of the Kyoto Protocol in the Diet in 2002. That year, the General Principle was revised to update the progress to date. Then, with Russia's ratification in 2005, the Kyoto Protocol finally went into effect and the Japanese Government also adopted the Kyoto Protocol Target Achievement Plan. This stipulates GHG reduction/absorption targets, consisting of energy-related CO_2 at +0.6 per cent, sinks including forests at −3.9 per cent and the Kyoto Mechanism at −1.6 per cent to clear the target of minus 6 per cent of total GHG emission compared with the 1990 level. Controlling the energy-related CO_2 emissions at that level has been found to be extremely challenging.

These efforts towards energy efficiency improvement are required not only in the short term but also in the longer term since climate change is a long term challenge for the entire planet. Recognising the importance of a long term commitment to improving energy efficiency, the Government announced an Energy Strategy Plan in 2006 which called on the country to improve its energy efficiency by 30 per cent in terms of energy consumption per GDP by 2030.

Another reason for Japan's need to further improve its energy efficiency is the recent growing concerns about energy security. There is consensus that energy efficiency is key to energy security.

In particular, it is very important for Japan to continue as the leader in energy efficiency in both the Asian and global contexts if the country wants to serve as a major soft power in the world.

Energy security concerns have grown since the turn of the century, coupled with the rising energy prices in international markets. Speculative transactions, triggered mainly by geopolitical events in the Middle East and Russia are considered the major causes of higher oil prices. However, it is far more important to recognise the fundamental supply-demand balance, particularly for countries such as China and India, and the supply bottlenecks in the world's oil and natural gas production regions. In other words, spare capacity has grown very thin. There were 6–7 million barrels per day (b/d) of spare oil production capacity throughout the 1990s. But since then, the demand for oil has grown substantially, while the increase in supply capacity in the Middle East has been limited and production in the OECD region has peaked. As a result, in the last two years the spare capacity for oil has been 1–2 million b/d which is a mere 1–2 per cent of global supply and demand.

The limited increase in global oil production capacity has led attention to the theory of an 'oil peak'. Debate has continued since the 1950s over whether the rate of oil production tends to follow a bell-shaped curve. There would be a significant impact on the supply-demand balance, and thereby on prices, if the demand for oil continues to grow when oil production is in the post-peak phase. Debate also continues over when oil production will peak, but there is no clear consensus. Thus, the answer is 'sooner or later'. However, a worldwide consensus has been formulated around the importance of the demand-side focus since the major source of the tightening market conditions is the unremitting growth in demand which is expected to continue for decades to come.

Substantial demand growth is expected not only in Asia, including China and India, but also in developed countries, led by the United States. Thus, a demand-side approach, or energy efficiency approach, has become an important policy goal for energy security objectives all over the world. China has adopted in its 11th Five Year Plan for 2006–2010 an energy conservation target that aims for a 20 per cent reduction in energy intensity measured in terms of energy consumption per unit of GDP. In its 2005 Green Paper the EU suggested it could achieve a reduction of energy consumption by 20 per cent compared to projections for 2020 on a cost-effective basis. The G8 Summit held in St. Petersburg in 2006, noted that "energy saved is energy produced and is often a more affordable and environmentally responsible option to meet the growing energy demand" and agreed to consider national goals for reducing the energy intensity of economic development by the end of 2006. Amid this international environment, policy-makers in Japan have committed to redoubling efforts to achieve further improvement despite already attaining the world's highest energy consumption efficiencies.

ENERGY EFFICIENCY AND CONSERVATION POLICIES IN JAPAN

Three factors have played significant roles in Japan. First, industries made substantial efforts to improve energy efficiencies in their production activities, based mainly on their own judgment of the need to rationalise energy consumption for the purpose of cost-cutting. Second, the general public also made significant contributions to energy conservation. This also stemmed mainly from the awareness of energy costs. Third, and most importantly, government policies had a positive, direct effect by way of categorical engagement, as well as an indirect effect by raising awareness.

The centrepiece of the Japanese Government's energy conservation policies has been the Law Concerning the Rational Use of Energy (hereinafter referred to as the Energy Conservation Law), which was enacted in 1979. The original law stipulated measures for three areas: factories, buildings and machinery and equipment. It stipulated specific guidelines for each area. For instance, factories that exceeded a certain energy consumption level have been required to appoint an energy manager and keep records of energy utilisation. Specific guidelines for energy efficiency were set for machinery and equipment, such as motor vehicles and air conditioners. The Law includes penalties such as the publicising of names of violating companies.

Generally speaking, the original Law was a soft regulation in both the substance of its obligations and severity of its penalties. However, it had a rather alarming effect as the first energy efficiency regulation in the form of a law. Industrialists and the general public responded very positively in abiding by the Law. Thus, the Law effected substantial improvement in energy efficiencies, especially in the industrial sector. The Law's requirements pertaining to factories were intended to force them to introduce energy management. Indeed, the major energy intensive industrial factories have competed among each other for better performance in their energy efficiencies based on their own energy management systems.

A major amendment to the Law was made in 1998. At that time, the Government recognised the strong need to further improve energy efficiency in all sectors of industry, transport and households in order to fulfil the obligations under the Kyoto Protocol. The Kyoto Protocol, agreed on by members of the United Nations Framework Convention on Climate Change in 1997, required Japan to reduce its greenhouse gas emissions in 2008–2012 by 6 per cent compared with the 1990 level. Stringent regulations were introduced in the Law after heated discussion and consultation, domestically as well as internationally, through the World Trade Organisation/Agreement on Technical Barriers to Trade process. The main changes to the Law were: (1) expansion of the range of factories required to conduct energy management; (2) the new obligation of submitting medium- to long-term plans on the rational use of energy for larger factories; and (3) the Top Runner Programme for machines and equipment. The Top Runner Programme requires manufacturers of certain designated products, such as motor vehicles, TVs and air conditioners to reach, within a certain time period, the prevailing and most energy efficient level commercially available for those products at the time of the regulation's introduction. For instance, manufacturers of air conditioners were required to reduce the energy intensity by an average of 63 per cent between 1997 and 2004 while the manufacturers of electric refrigerators were required to reduce the energy intensity by an average of 30.5 per cent between 1998 and 2004. Many stakeholders, both domestic and international, were nervous about these tough

Equipment	Base Year	Target Year	Initial Expected Energy Saving (%) compared to the Base Year at the Target Year	Actual Improvement Energy Saving (%) compared to the Base Year at the Target Year
Air Conditioners (below 4kW)	FY1997	FY2004	63.0	67.8
TV sets <Cathode Ray TV>	FY1997	FY2003	16.4	25.7
VCRs	FY1997	FY2003	58.7	73.6
Electric refrigerators	FY1998	FY2004	30.5	55.2
Electric freezers	FY1998	FY2004	22.9	29.6
Gasoline Passenger Vehicles	FY1998	FY2004	23.0 (*)	22.0

(*) 1995->2010

Figure 4: Actual Improvement of Energy Saving in Target Fiscal Year

Source: ECCJ

standards. However, the regulations in all categories of products were achieved by the original target year of 2004 (see Figure 4).

The Government has introduced various other policy measures in addition to the Law. The first was a financial scheme — a soft loan programme extended by the Development Bank of Japan (a public, policy-oriented bank wholly owned by the Government), starting in 1978. Under this programme, certain investments in plant and equipment for the purpose of energy conservation are financed at a preferential interest rate. The major beneficiaries have included steel and power generation companies. Another financial incentive was a tax reduction, which was also launched in 1978. The main feature of the tax scheme was a deduction of corporate or income tax, equivalent to 7 per cent of the acquisition cost of equipment for energy conservation used by small and medium-size enterprises (SMEs), and a special depreciation of up to 30 per cent on the acquisition cost of equipment in addition to ordinary depreciation.

The Government also extended a direct subsidy to business entities introducing certain advanced energy efficiency facilities by way of the NEDO (New Energy and Industrial Technology Development Organisation). This covers the industrial, housing and building, and residential and transportation sectors. On top of this, the Government developed its own research and development programme, the Moonlight Programme, which consists of large-scale projects such as magnetohydrodynamics (MHD) power generation, waste heat utilisation (vacuum heat pumps), high-efficiency cycle gas turbines (CCGT), fuel cell batteries, super conductivity electricity and ceramic gas turbines. Many technologies have been spun off and commercialised from the Moonlight projects. For instance, the latest CCGT has become highly energy efficient with a thermal efficiency over 50 per cent. Its high energy efficiency has driven the introduction of natural gas power plants in the electricity generation sector in Japan. The share of LNG thermal generation as a proportion of the total generation capacity has increased from 3 per cent in 1974 to 25 per cent in 2004. This has contributed substantially to diversification and efficiency of energy.

The Government also offered technical services to SMEs. The Energy Conservation Centre of Japan (ECCJ) conducted energy audits of SMEs in which specific lists of priority measures are prepared, including investment requirements and expected benefits. Based on these audits, SMEs have been able to make rational and reasonable investment decisions which have resulted in considerable improvement in energy efficiency.

The Government measures cover not only laws and regulations and financial programmes, but also include activities to provide proper information and to raise awareness of energy efficiency among the public. The first of the information/awareness-related measures was the Energy Conservation Award, started in 1990. Each year a grand prize is awarded by the Minister of Economy, Trade and Industry to the best performance appliance/system, and several others are awarded by the Director-General of the Agency of Natural Resources and Energy and the Chairman of the Energy Conservation Centre of Japan (ECCJ). The award is a tremendous motivation, especially to those in the industrial sector. Secondly, the Energy

Conservation Campaign has been effective, especially in times of high energy prices and/or emergency times such as the nuclear energy crisis in 2002–2003. Each year in summer and winter, the Government and ECCJ have been actively promoting energy conservation activities. In one instance, a famous actress was utilised in a campaign that gained great popularity. Thus, the consciousness of the general public has been raised.

Measures taken by industry, whether urged by the Government or based on the sector's own judgment, have also been an effective policy tool in Japan. One example is the Voluntary Environmental Action Plan of *Keidanren*, the economic association of Japanese industry. The *Keidanren* plan covers energy conservation measures for 35 industrial branches, accounting for 82 per cent of the total industrial sector's CO_2 emissions. The plan is to suppress year 2010's CO_2 emissions below the 1990 level. Industry commitments under the plan include reduction of absolute CO_2 emissions, reduction in overall energy consumption and improvements in energy intensity of output. *Keidanren* reviews the plan's implementation and publishes the results annually. The actual industry emissions under the *Keidanren* plan in FY2004 were 502 million t-CO_2, or 0.5 per cent below the FY1990 level, which demonstrates the effectiveness of the Keidanren's Voluntary Action Plan.

Another example of the industrial sector's voluntary measures is labelling. The Energy Star Programme, a US Government-led labelling initiative, has been applied to computers, computer displays, printers, fax and copying machines in Japan. The voluntary Energy Saving Labelling Programme was also launched in 2000 for air conditioning equipment, refrigerators, freezers, televisions and lighting. The label shows the energy efficiency of these products relative to their Top Runner targets mandated under the Energy Conservation Law (see Figure 5). The objective of the labelling programme is not only to inform consumers about the energy efficiency level in order to make wise purchases, but also to encourage manufacturers and importers to clear the Top Runner standards even if they are ahead of the target year.

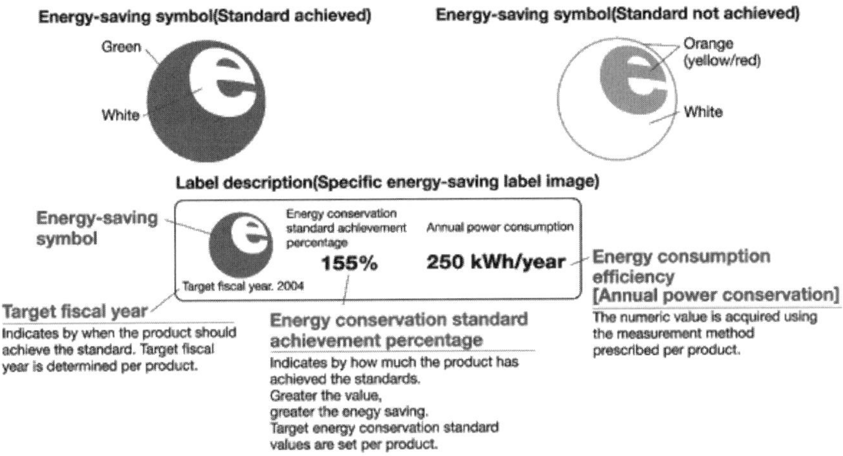

Figure 5: Energy Saving Labelling

EVALUATION OF THE VARIOUS MEASURES' SUCCESS

In general, the Japanese Government's policy measures for energy conservation and efficiency have been successful. It is important to note, however, that there is a fundamental precondition for the measures' effectiveness, namely market function and price signals. As Alan Greenspan pointed out, "the experience of the past 50 years — and indeed much longer than that — affirms that market forces play a key role in conserving scarce energy resources, directing those resources to their most highly valued uses." Economic theory tells us that demand and supply balance according to price signals in a proper market. When price goes up demand should go down, though the degree of the response to the price signal, or elasticity, differs depending on the markets. There are two ways to respond to price increases. One way, mainly short term, is simply to curtail the level of consumption. The other, mainly mid- to longer-term, is to introduce more efficient facilities and systems. In particular, when energy prices go up, the return on energy saving/efficiency investment in terms of cost-saving becomes larger. Energy price hikes create better opportunities for energy saving/efficiency investments.

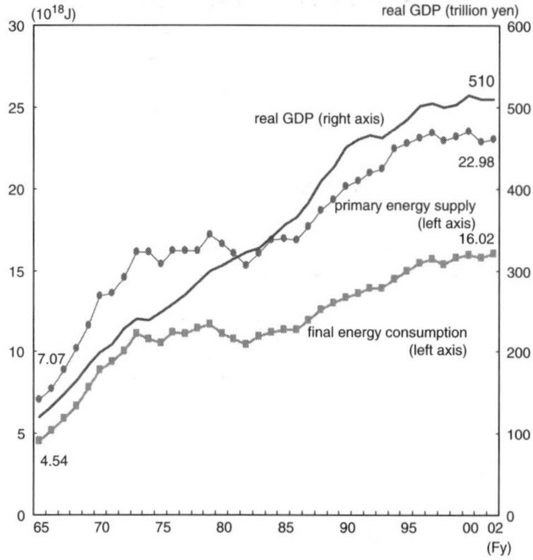

Elasticity

1965–'73	'73–'79	'79–'86	'86–'91	'91–2003
1.19	0.29	−0.11	0.85	0.71

Figure 6: GDP and Energy Consumption in Japan

Source: METI

In Japan, it is clearly apparent that energy efficiencies improved via the two channels in the late 1970s and early 1980s when oil prices were higher (see Figure 6). Japan had no energy price subsidies, thus international prices were carried over to domestic energy prices. The end users, both the industrial sector and the pubic, responded to the price signals. In other words, the market functioned properly in Japan. This is attributable partly to the conservation consciousness and cost-conscious nature of the Japanese people.

In particular, Japanese industry has been known for its total quality control (TQC), or *kaizen*, activities. TQC/*kaizen* is the act of improving total quality and productivity performance, including cost-saving, mainly in daily routine operations. It involves all members of a company — managers and workers — in their quest for increased productivity. Many factories in energy-intensive industries introduced or intensified their *kaizen* activities for energy-saving in response to the first oil shock in 1973–1974. In one example, steel factories introducing *kaizen* increased from 146 to 170. Energy-saving in the steel industry was critical, since about 20 per cent of operation costs were connected to energy. The management strategy of steel companies resulted in conservation of energy and reduction of costs. Thus, the first energy conservation phase in the steel industry after the oil shock in the 1970s happened more by operational improvement than investment in plants and equipment. These operational improvements in energy conservation were made possible by the shared interests of management and employees in cost reduction.

On the basis of the industrial sector's cost consciousness, Government policy measures also contributed to energy conservation and improvement in energy efficiency. The Energy Conservation Law, stipulated operational improvement activities in industry as an official guideline to be followed by industry factories. Companies and factories in the industrial sector competed with each other for better operation, following the requirements of the Law. The Law effectively disseminated operational know-how and promoted competition among companies, leading to better performance in energy efficiency in the industrial sector as a whole. In Japan, this interaction between industry and Government has served as a mechanism to cope with a variety of challenges. In the case of the Energy Conservation Law, the industries first introduced various activities for energy conservation, and then the Government passed a law with guidelines incorporating these industry practices.

The soft loan programme and tax incentives introduced in 1978 were intended to promote investment in energy conservation. The initial response by the industrial sector to the oil shock was mainly in the area of operational improvements and these had rapid effects on cost reduction. Also, management was hesitant to move towards larger energy conservation investments in view of the uncertainty of energy prices. However, many relevant parties in the late 1970s shared the belief that more ambitious longer-term investment in tandem with the latest technologies was needed for further improvement in energy efficiencies. In this regard, the soft loan programme and tax incentives matched industrial needs and had the effect on industry of triggering capital expenditure in the early 1980s. The industrial sector, led by the steel industry, made the decision to increase investment in energy conservation, motivated by the policy measures after the second oil shock in 1979–1980. The soft loan programme and tax incentives accelerated the return period on long-term investments. These are typical government intervention measures for rectifying market failure in which business decision-makers tend to be oriented towards the short term, avoiding long-term investment.

The Top Runner Programme, introduced in the amended Energy Conservation Law in 1998, was epoch-making. Its targeted efficiency levels are significant for most products, and have resulted in significant energy savings and CO_2 emission reductions. Follow-up monitoring shows that the Programme has had a positive impact on the efficiency of all products under its regulation. The most important policy impact is that it encourages manufacturers to develop more efficient technologies. Most researchers and engineers working in the manufacturing of regulated products are motivated by the Programme to develop more energy-efficient products as they are forced to compete in the market race for product development.

Government policy measures intended to raise the level of awareness and motivate industry and the public have also been successful. The Energy Conservation Award has been successful in motivating manufacturers to develop more energy efficient products. Each year the award ceremony attracts a large audience and media coverage. Manufacturers often trumpet their award-winning status in commercial advertisements. Researchers and engineers of the awarded products are often given special bonuses by their companies. The national campaign for energy conservation has been particularly effective when energy prices have been rising and/or crisis situations have been occurring. The Government urged people to refrain from driving automobiles and ordered an end to late-night TV broadcasting during the oil shock periods. These were crisis management measures. In 2003, the Tokyo metropolitan area was on the verge of a blackout due to suspended operations of nuclear power plants which were accounting for about 40 per cent of the Tokyo Electric Power Company's (TEPCO) operations. Publicity about malfunction and forfeited data forced the TEPCO to suspend nuclear power plant operations for inspection purposes. The TEPCO increased the operation of other power plants, such as those burning coal, oil and LNG, but this was not enough to meet the peak summer demand in 2003. The

TEPCO and the Government conducted a vigorous campaign to save energy. As a result, it was estimated that approximately one gigawatt of power was saved, averting a potential blackout. This example illustrates the fact that higher awareness on the part of consumers and users, inspired by campaigns, can change energy usage patterns and avoid energy shortages in crisis situations.

LESSONS FOR EMERGING ASIAN ECONOMIES: NEW "FLYING GEESE" MODEL

It is worth drawing lessons from the Japanese experience. Many Asian countries, such as China, Thailand and Indonesia, are currently in the midst of an industrialisation phase comparable to Japan's in the 1960s and 1970s. Their economies are expected to continue to grow at a relatively high rate and energy consumption will increase with this economic growth. These countries should aim for higher and more energy-efficient economic growth, or lower elasticity (ratio of energy consumption growth to economic growth). Between 1965 and 1973, when Japan was in its rapid industrialisation process, the elasticity was 1.19, meaning a higher rate of energy consumption growth than economic growth. Following the oil shocks, the elasticity fell to 0.29 in 1973–1979 and –0.11 in 1979–1986. Japan was forced into this by external supply shocks.

There are currently no comparable external supply shocks. It has mainly been an increase in demand which has caused hikes in energy prices. The emerging Asian economies should, of their own will, introduce energy conservation and energy efficiency strategies, based on lessons from the Japanese experience.

The first lesson which should be learned is to utilise the basic market function as much as possible. As described above, the Japanese experience shows that energy efficiencies improved when energy prices were higher and levelled when prices were lower. Demand and supply, to some extent, responded to the prices. If domestic prices are subsidised, market resilience will not fully work. Subsidised energy prices have been in place in much of Asia. Several Asian economies are in the process of removing them and moving towards international market prices. This trend should be supported.

Secondly, Governments must choose the right set of policy measures and launch these in the right sequence. Much depends on the level of energy efficiencies and the absorption capability of the general public and the industrial sector. The policy measure options are wide, from regulation (either hard or soft type), incentives (financial and/or tax), or government direct programmes (such as research and development, technical services, information dissemination and campaigns to raise awareness). In Japan, immediately after the 1970s oil shock, a strong move was initiated by industry towards energy conservation through operational changes rather than investment. Thereafter, industry shifted towards investment in new plant and energy efficiency systems. After certain business behaviour patterns were identified, the Law promoted similar activities as widely as possible. Also, the Japanese Government provided incentives to industry to motivate it toward energy conservation and efficiency.

Thirdly, in mobilising the industrial and general public, it is imperative that society as a whole becomes sustainability-oriented from the bottom up. This can be achieved through the use of both carrots and sticks. Cultures and lifestyles need to be changed. Japan has an image of being 'cool', a concept that has been discussed in terms of the soft power stemming from Japan's culture, such as anime, computer games, sushi and music such as J-pop. Most recently, however, 'cool Japan' has gained attention for its high performance in energy efficiency, rooted in the lifestyles of people and behavioural patterns of businesses. Japan should be ready to transfer its coolness to a 'cool Asia'. Needless to say, with respect to local cultures, there must be sincere sensitivity to local norms. However, no one doubts the value of energy efficiency. Therefore, there is no reason why Asian countries cannot start to strengthen themselves in becoming 'cooler' by the bottom-up power of the people rather than coercion from government authorities.

Asia should challenge the Environmental Kuznets Curve, a theory of development economics which describes the relationship between economic development (income per capita) and environmental

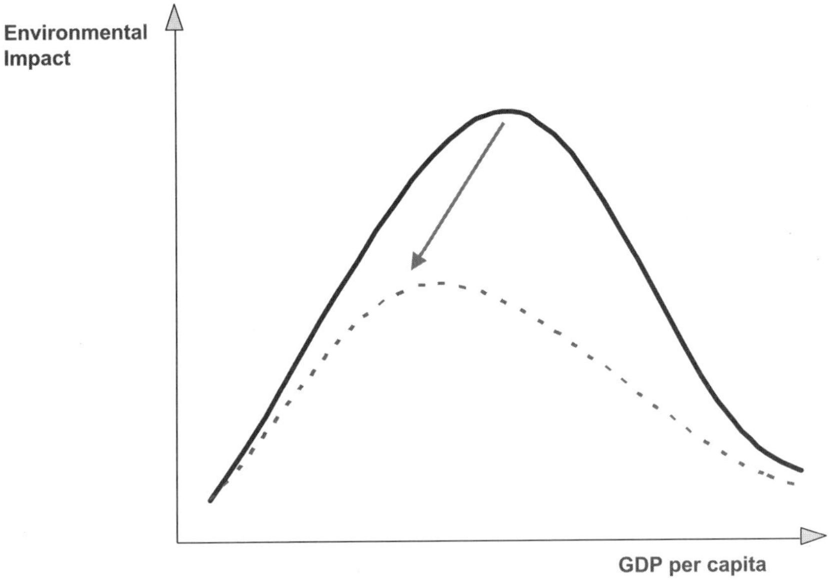

Figure 7: Environment Kuznetz Curve

pollution (energy intensity). As economies develop, environmental pollution (energy intensity) intensifies. At a certain point, the latter decreases while the former continues to develop. Thus, the relation can be drawn as an inverted U-shaped curve. However, it should be possible to accelerate the peak out phase and to lower the peak level phase (see Figure 7). Country A following country B (for instance China) can learn from and introduce experienced technology and know-how from country A (for instance Japan). That is the benefit of follower status. Japan has been a successful model in terms of energy efficiency. The emerging economies of Asia are fortunate to have a country such as Japan in this increasingly economically integrated region. Japan is ready to transfer its experience and the emerging Asian economies should learn from and utilise it. It is highly desirable that an integrated Asia can demonstrate the value of its economic development model, this new 'flying geese' model, to the world.

Chapter

10

Energy Conservation Policy Development in the Philippines

*Peter Lee U**

INTRODUCTION

As the millennium began, there was another global energy crisis. Oil prices skyrocketed to record highs, with adverse effects on the economy. Energy policy is once again a major concern, and in the Philippines, it figures prominently in the Five Point Reform agenda of the Arroyo administration. The stated energy policy, not surprisingly, devotes much attention to energy independence (See Table 1). The latest publicly available energy plan, the Philippine Energy Plan 2005 Update, targets an increase in energy self-sufficiency level from 56.6 per cent in 2005 to 60 per cent in 2010.[1]

* The author wishes to thank Mr. Jess Anunciacion, Ms. Helen Arias, Ms. Carmencita Bariso of the Department of Energy, Mr. Mauro Marcelo of the National Power Corporation, and Mr. Tony Arrobio of the Energy Management Association of the Philippines for their assistance in providing data and sharing their experiences in the field of energy conservation.

[1] The Philippines Department of Energy, "Philippine Energy Plan 2005 Update", vol. 1, p. 1.

Table 1: Philippine Energy Sector Framework

Five-Point Reform				
Economic Growth and Job Creation	Anti-Corruption Through Good Government	Energy Independence	Social Justice and Basic Needs	Education and Youth Opportunity

Energy Sector Agenda	Energy Sector Goals
Energy Independence and Savings • Increase reserves of indigenous fossil fuels • Aggressively develop renewable energy potential such as biomass, solar, wind, and ocean resources • Increase use of alternative fuels • Strengthen and enhance energy efficiency and conservation programme • Form strategic alliances with other countries	*60 per cent Self-Sufficiency Level by 2010* • Increase in oil and gas reserves by 20 per cent in ten years • Reduction in coal imports by 20 per cent in ten years • Increase in renewable energy capacity by 100 per cent in ten years • 100 per cent of Metro Manila buses running on CNG by 2010 • 5 per cent CME blend with diesel fuel for vehicles in 2010 • 5 per cent ethanol blend with gasoline for vehicles by 2007, to reach 10 per cent in 2010 • Convert retired and operating oil-based power plants to natural gas by 2010 • Forge energy agreements with existing and new energy partners • 19.8 MMBFOE (25.7 MTOE) average annual energy savings in ten years through the National Energy Efficiency and Conservation Program — 17.7 MMBFOE (2.9 MTOE) to come from energy efficiency and conservation program — 2.1 MMBFOE (2.6 MTOE) to come from alternative fuels for transport program
Power Sector Reforms • Create a transparent privatisation process • Create an investment climate attractive to investors	*Fair and Reasonable Energy Prices in a Competitive Environment* • Privatise 70 per cent of installed capacity of NPC's generating assets in Luzon and Visayas by 2007

(*Continued*)

Table 1: (*Continued*)

Energy Sector Agenda	Energy Sector Goals
	• Privatise TransCo • Continuously implement transmission system upgrade and expansion program • Implement WESM in Luzon and Visayas by 2006 • 100 per cent barangay electrification by 2008 • Opening of 14 "First Wave" NPC-SPUG areas to power sector participants • Single-digit national average systems loss of distribution utilities by 2010

Source: The Philippines Department of Energy.

The energy framework envisions increasing energy independence primarily by increasing indigenous energy sources, especially renewable energy sources and alternative fuels for transport. Meanwhile, there will be ongoing national energy efficiency and conservation campaigns. The framework also explicitly recognises the role of international cooperation in the pursuit of energy security. In her statement to the United Nations, President Macapagal-Arroyo echoed this vision, calling for collective development of alternative and indigenous energy sources, energy conservation, and regional stockpiling to address the energy crisis. As an example of international cooperation, she also cited the cooperative seismic operations the country undertook with China and Vietnam in the areas of the South China Sea that the countries competitively claim.[2]

The continuing reform of the power sector is the other policy imperative. While it may not appear to have a direct connection to energy independence, a more efficient and competitive power sector will also lead to energy savings.

Likc many countries, the Philippines was very much dependent on energy imports, specifically oil, and so when the first global oil

[2] Gloria Macapagal-Arroyo, "Statement at the High-Level Plenary Meeting of the General Assembly of the United Nations at its 60th Session", New York, 15 Sept. 2005 at http://www.un.org/webcast/summit2005/statements15/phi050915eng.pdf.

shock struck, the country was seriously affected. However, this may have been a blessing in disguise as it spurred the country to diversify its energy sources. From over 90 per cent reliance on oil in 1973 (which was almost entirely imported since the country had virtually no discovered oil resources of its own at the time) the Philippines has reduced that to about 45 per cent (see Figure 1).[3]

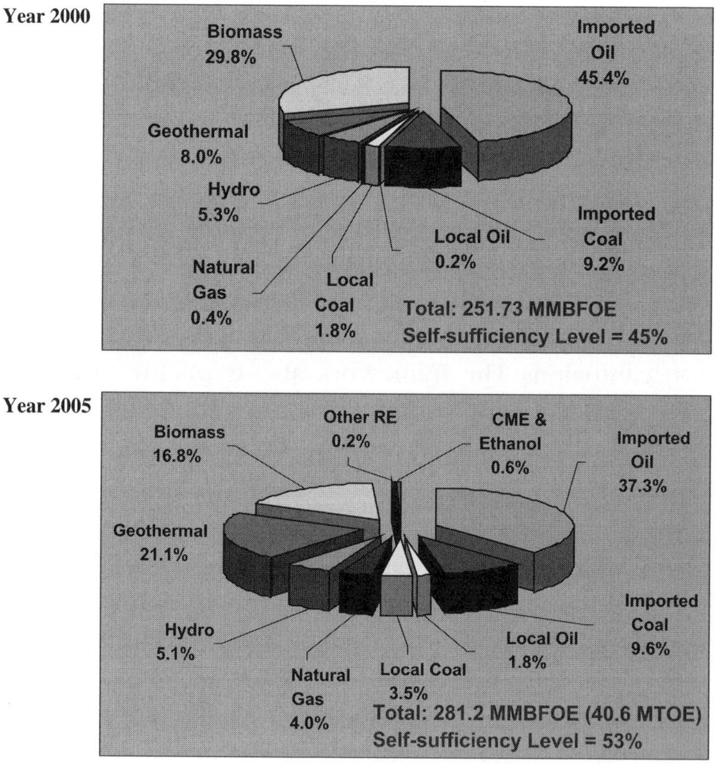

Figure 1: Primary Energy Supply Mix

Source: The Philippines Department of Energy.

[3] One may wonder why in Figure 1 there is a sizeable jump in the geothermal share of primary energy, while in Table 2 there is slight drop in the geothermal share of power generation. Apparently, there was a change in the assumed thermal efficiency of geothermal power plants used to compute total energy consumption from 35 per cent to 10 per cent in the two years.

Table 2: Power Generation Mix (per cent)

	1973	1998	2005
Oil	82.7	43.7	10.8
Hydro	17.2	12.2	14.8
Coal	0.1	22.6	27.0
Natural Gas	0.0	0.0	29.8
Geothermal	0.0	21.4	17.5
Total	100	100	100

Source: The Philippines Department of Energy.

In power generation, which has historically been a major user of oil, the country has also greatly reduced its reliance on oil (see Table 2). In the wake of the first oil crisis, geothermal sources were discovered and exploited. From zero, the country today derives close to 20 per cent of its power requirements from geothermal plants, making it second only to the United States in terms of geothermal power use.

In addition to the global energy crises which were beyond its control, the Philippines also suffered another energy crisis of its own making — the power shortages of 1992–1993. Thus, the country is certainly no stranger to energy crises.

A SHORT HISTORY OF ENERGY POLICY IN THE PHILIPPINES

When the first world oil crisis struck in 1973, and like most other countries, the Philippines had been consuming energy in a rather nonchalant manner. From an aggregate industry standpoint it was a supply shock. While additional quantities of oil could have been forthcoming, the market had artificially (from an economics sense, since the cause was geopolitical, having to do with the Arab-Israeli war) curtailed output (especially for the United States and its allies). See Figure 2(a).

By contrast, the current energy crisis today might be characterised as more of a demand shock. As Figure 2(b) depicts, the high oil

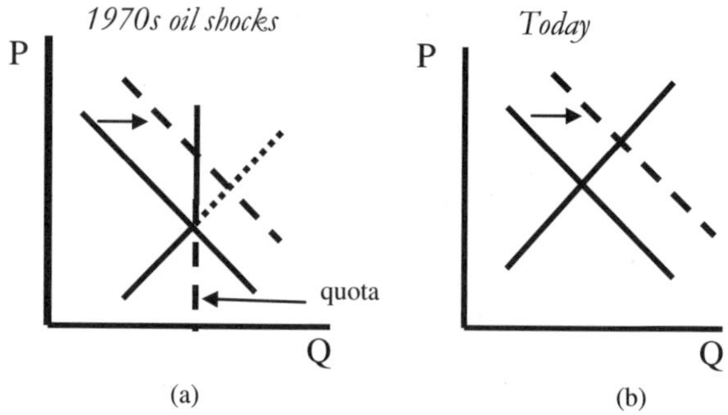

Figure 2: (a) OPEC Quota Limited Output Artificially. (b) Even with China's Insatiable Thirst, Supply Has Been Forthcoming, Albeit at Very High Prices.

prices today may be attributed to the strong global economic growth and consequent demand for energy and oil. Thus, in a sense there is no shortage of oil. More oil can be forthcoming from the market, albeit at higher marginal cost.

In the Philippines, the first global oil crisis also coincided with a period of political upheaval, with then President Ferdinand Marcos declaring martial law. This ushered in an era of greater government regulation and participation in the economy. In the energy sector, President Marcos then set up the Ministry of Energy and installed Geronimo Z. Velasco as the first minister of energy. Marcos had created the Philippine National Oil Company (PNOC) in November 1973 through Presidential Decree 334 with the objective of using it as a tool for energy security to ensure adequate energy supply.

Velasco had the full support of Marcos and ready access to him, which suggests that Marcos placed great importance on energy policy. Due to the authoritarian nature of his martial law regime, as Velasco himself would later admit in his memoirs, things got done faster:

> "Whether we admit it or not, the centralisation of decision making under the Marcos administration turned out to be conducive for building the energy infrastructure as quickly as possible. We could not have set up all those PNOC subsidiaries in quick succession, or made important

> investment decisions in a matter of days, if it were not for the fact that we answered only to the president."[4]

Oil was obviously at the top of the energy agenda of the day and Velasco proceeded to diversify the country's energy sources. The fruits of this have been demonstrated above. In the oil sector, Marcos directed Velasco to set out and obtain oil through government-to-government talks, in addition to relying on the global oil markets. Velasco was sent on missions to the Middle East, Indonesia, Malaysia and even China, to try to obtain oil through diplomacy. By 1981, the country had reduced its reliance on Middle Eastern oil to 68.6 per cent from 95.5 per cent in 1973 (see Table 3).

Marcos also issued Presidential Decree 87, the Oil Exploration and Development Act of 1972, mandating a more active government role in the search for indigenous oil incentives. It offered incentives to qualified investors in oil exploration. PD 87 also introduced a shift to the service contract system from the old concession system in farming out exploration activities. Caltex (Chevron and Texaco) signed the first service contract in 1973 to explore oil in Philippine waters.

The efforts at diversifying energy sources have long gestation periods and thus could not be expected to have an effect in time to avert the shortages. The oil crises of the 1970s also saw drastic measures taken such as gasoline rationing. Marcos issued General Order 41 on 10 December 1973 which effectively placed the oil industry under government control through PNOC, giving the latter supervision over the sale and distribution of all available stock of crude oil and oil products.

The Oil Industry Commission had been set up in 1971 even before the crisis and it was subsequently tasked to fix the pump prices of fuel. At this time, oil companies also had to obtain permission from the Commission to import petroleum products.

The oil industry had previously been a fairly free market. In the 1970s there were six oil majors operating in the country: Shell, Caltex, Esso, Mobil, Filoil (an affiliate of Gulf Oil), and Getty. There

[4] Geronimo Z. Velasco, *Trailblazing: The Quest for Energy Self-Reliance*. Manila: Anvil Publishing, 2006, p. 205.

Table 3: Crude Supply by Country of Origin

	1981		1973	
	MB	%	*MB*	%
Middle East	**41,663**	**68.6**	**65,581**	**95.5**
Saudi Arabia	31,312	51.5	37,787	55.0
Kuwait	7,122	11.7	18,897	27.5
Iran	—	—	5,955	8.7
Iraq	—	—	235	0.3
Abu Dhabi	—	—	1,604	2.3
Qatar	—	—	1,103	1.6
Dubai	—	—	—	—
Oman	—	—	—	—
Other Regions	**19,098**	**31.4**	**3,099**	**4.5**
Indonesia	5,202	8.6	—	—
Malaysia	3,515	5.8	3,099	4.5
Brunei	3,215	5.3	—	—
China (PROC)	4,527	7.4	—	—
Mexico	2,639	4.3	—	—
Total Crude Oil Imports	**60,761**	**100.0**	**68,680**	**100.0**

Source: Bureau of Energy Utilisation, *1981 Annual Report* (cited in Makasiar, 1984, see Footnote 11).

were four refineries: Shell, Caltex, Filoil and Bataan. As the business environment became more difficult and the Government started to regulate the industry, some started to leave the country. In 1973, Gulf Oil sold its shares in Filoil to the National Investment and Development Company, the investment arm of the Philippine National Bank, a Government-owned bank, and left the country. Also in 1973, Esso left as well, selling everything to the Philippine Government. Getty would later sell its Philippine holdings to Shell and when Mobil also left in 1983, only Shell, Caltex and the Government Petrophil/PNOC remained in the local market.

At the time of their departure, Esso and Filoil accounted for about a third of the oil market. Thus there was fear of a shortage in oil product supply. This was partly what precipitated Velasco's series of 'oil diplomacy' trips.

Velasco (see Footnote 4) recounts that during that period, the Government required the oil companies to maintain a minimum inventory of 60 days' supply of crude oil and oil products. They had a rule of thumb to implement rationing if the inventory dropped below 60 days but remained above 45 days. If stocks were to fall below 45 days but above 30 days, then the rationed amount would be halved. Finally, if stocks fell below 30 days, they would have declared an emergency and only hospitals, power generation systems and other critical areas would have been allowed to operate.[5]

Things did get bad enough at one point that the Government implemented rationing for two months. The system involved distributing coupons that were surrendered when purchasing gasoline. Private car owners were allotted coupons worth 200 litres per month but jeepney drivers were given more. And just as textbook economics predicts, with nonprice rationing systems, arbitrage and a black market arose. The jeepney drivers siphoned off fuel from their tanks after filling up and sold it to private car owners.[6]

On the demand side, restrictive measures were also undertaken. For example, certain activities such as motor sports (e.g., rallying, skydiving etc.) were banned; the importation, manufacture or assembly of motor cars with engine displacements exceeding 2,800 cc were prohibited; the use of air conditioning and the hours that businesses could use neon and other forms of lighting for advertising were regulated. Daylight saving time was even implemented to make the fullest use of the sun's light.

Under Velasco, the Ministry of Energy also set about diversifying the country's energy sources. While initially his mission was oil procurement, by 1976 attention shifted to the exploration, exploitation and development of indigenous energy resources. The Government

---

[5] *Ibid.*, p. 20.

[6] *Ibid.*, p. 20. Jeepneys are the colourful mini-buses converted from the WW II American military jeeps. They continue to be a mainstay of public transportation in the Philippines.

National Power Corporation (NPC) had been set up by the Americans even before WW II to exploit hydroelectric power. Velasco as Energy Minister was also director, and then chairman ex-officio of NPC, and further hydroelectric resources continued to be developed.

As mentioned above, geothermal energy may perhaps be the most significant indigenous energy success story. Actually, in August 1970, even before the crisis, Marcos had already issued Presidential Proclamation 739 mandating the development of geothermal steam for electricity. It tasked the NPC to develop the Tiwi and Makiling-Banahaw steam reservoirs.

Coal is the only fossil fuel known to exist in commercial quantities in the country. So it was natural that its development would also be pursued. Coal today accounts for about 10 per cent of the primary energy mix. Unfortunately, most of the coal is lignite with low heating value. Thus most of the coal used is imported. Nevertheless, the Government tried to develop local sources of coal, starting with Presidential Decree (PD) No. 972, the Coal Development Act of 1976. Two PNOC subsidiaries, Malangas Coal Corporation and the PNOC Coal Corporation were set up and commenced coal mining operations.

The next challenge was to create a market for the coal. To this end, the cement industry was targeted for conversion to coal. Probably because the coal was of lower quality, the Government had to subsidise the coal in order to make it attractive for the cement industry to switch. The Ministry of Energy had to guarantee that the price of coal from PNOC mines would not exceed 65 per cent of the industrial fuel oil price. The Ministry would shoulder any differential beyond that point. The cement plants were slow to convert, but by 1984 all 17 cement plants had done so.

The drive to diversify energy sources may perhaps have reached an initial zenith (even if just symbolically) with the construction of the ill-fated Bataan Nuclear Power Plant (BNPP). The NPC had contracted with Westinghouse Electric to construct a 620 MW capacity nuclear facility. It would have boasted the largest generation capacity of the time. However, because the original contract was perceived to be one-sided in favour of Westinghouse and also overpriced, Marcos allowed a panel to be constituted to renegotiate the contract. Critics later also charged that the project was not safe and that the plant site was subject to earthquakes. Together, these problems were its downfall. Then the Chernobyl and Three Mile Island incidents were the final nails in the coffin, and soon after Corazon Aquino came to power she mothballed the nuclear plant. By then, the plant was practically finished. Ironically, the country is still paying off the loans for the nuclear plant.

Thus, besides global energy crises, the Philippines also had an energy crisis of its own making: the power shortages of the early 90s. This was the result of mothballing the Bataan nuclear plant without providing for a replacement. The country paid dearly for this poor planning as it suffered daily power outages, lasting as long as 8 to 10 hours at their worst in 1993. Not surprisingly, economic growth languished in the early 1990s as a result of the power shortages.

Ironically, the power shortages of 1992–1993 briefly increased the dependence on oil for power again. Since oil fired or diesel plants were the fastest to install, significant oil/diesel capacity was created through the Independent Power Producers during those years as a short term solution to the crisis.

The discovery and development of natural gas has arguably been the second zenith of the Philippines' energy saga. Occidental Petroleum initially found gas at Camago in 1989. Shell Philippines Exploration subsequently partnered in the exploration and discovered the Malampaya gas field in 1992. The investment funds for developing the gas field poured in and constituted the single largest foreign investment in the country's history. Inaugurated in 2001, it provides natural gas via a 500 km underwater pipeline to three power plants (NPC-Kepco Ilijan, First Gas Power's Santa Rita and San Lorenzo plants) totalling 2700 MW in capacity. Production from Malampaya accounts for the great increase in the share of natural gas in the power generation mix between 2000 and 2005.

The use of alternative indigenous fuels for the transport sector were also explored. Programmes to promote the use of alcogas, or gasoline blended with alcohol, and coco-biodiesel had already been tried. Alcogas did not prosper for economic reasons. It cost more to produce one litre of the required water-free alcohol than one litre of gasoline. Velasco also attributed the failure of the alcogas programme to the decline of the sugar industry which precluded the programme from being able to source the quantity and quality of sugarcane needed to sustain the programme.[7]

Meanwhile, the early experiments with coco-biodiesel seem to have failed for technical reasons. Impurities were not sufficiently removed, causing the coconut oil to coagulate. As Velasco relates, transport operators who used the early coco-biodiesel had literally to use a toothbrush to clean the engine of the coagulated coconut oil.[8]

CURRENT POLICIES AND PROGRAMMES FOR ENERGY CONSERVATION

As oil prices climbed once more in 2005, the President issued Administrative Order 126 on 13 August 2005 directing the enhanced implementation of the government's energy conservation programme. It adopted measures to limit the use of petroleum products to essential activities with the target of reducing public sector fuel consumption by 10 per cent (of the average monthly consumption in the first semester of 2005). Similarly, electricity consumption is also to be reduced by 10 per cent of the monthly average consumption in the first semester of 2005.[9] For example, in the use of air conditioning, government offices were instructed to limit their use between the hours of 9 am and 4 am and to set the thermostat at 25°C. Government vehicles were also directed to use Department of Energy (DOE) certified biofuels and CNG where possible. A task force was set up including officials from the DOE that conducted random inspections of various government offices' compliance and rated their energy conservation practices.[10]

---

[7] Velasco, *Trailblazing: The Quest for Energy Self-Reliance*, p. 73.

[8] *Ibid.*, p. 75.

[9] Raphael P. M. Lotilla, "Energy Efficiency and Conservation Measures in Government", presentation slides, 19 Aug. 2005.

[10] *Ibid.*, p. 20.

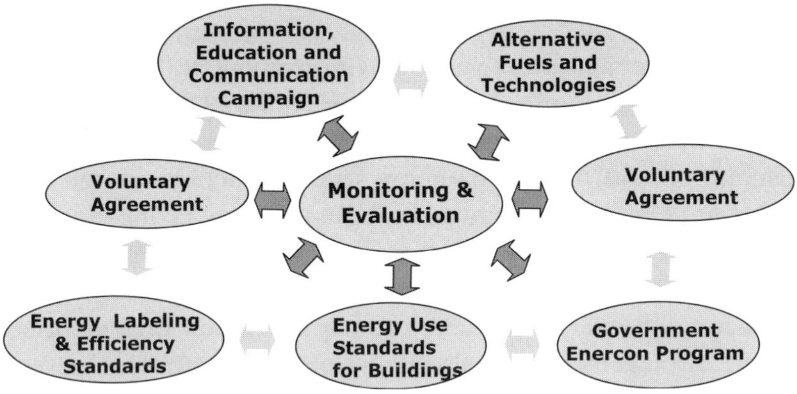

Figure 3: National Energy Efficiency and Conservation Programme

Source: The Philippines Department of Energy.

The DOE administers a National Energy Efficiency and Conservation Programme under the Energy Efficiency and Conservation Division of the Department's Energy Utilisation Management Bureau. The programme has two prongs: a fuel efficiency and conservation programme, and an electricity efficiency and conservation programme. Figure 3 portrays the programme's various components and strategy.

Information, Education and Communication Campaign

The DOE has been waging a public education programme to increase awareness of energy conservation and efficiency issues among the general public. Advertisements have been taken out in the mass media and seminars/workshops and in-house briefings have been conducted. The campaign seeks to encourage the public to car pool, have a car-less day and take public transport instead, and to correct bad habits such as idling while waiting. Information and tips on saving energy have been disseminated through brochures. The DOE has also been holding regular fuel economy runs where stock automobile models are tested under standard conditions. This can provide useful information to car shoppers on fuel consumption of automobile models they may be considering.

Voluntary Agreements

These are partnerships between the Government and private sector to facilitate energy efficiency and conservation. One track aims to improve the energy efficiency of firms through monitoring of energy consumption and providing energy services such as energy audits which can help firms identify potentials for energy savings, including the adoption of energy efficiency and conservation technologies. Other activities include recruiting leading fast food chains to use placemats with energy efficiency and conservation themes printed on them. Arrangements are also being pursued with malls, airports and other commercial institutions to rationalise vehicle use:

a. Park and Walk — Encourage mall patrons to park and walk around the commercial area rather than driving to each new shop;
b. Park'n Fly — Make available nearby parking for overnight and short duration travellers so that they can park, take their trip, return and drive home. This eliminates the added trip of someone else who would have to drop them off and pick them up again;
c. Park'n Ride — Encourage people to park and take public transport, especially the MRT or LRT (mass rail transit system in Metro Manila);
d. Park'n Pick — Encourage setting up of taxi queues so that taxis minimise roaming while looking for their next customer.

Energy Labelling and Efficiency Standards

This programme hopes that by labelling appliances' energy consumption ratings, consumers can make better informed choices and thus enhance consumer protection. This programme consists of: (1) energy labelling and standards for room air-conditioners (see Figure 4); (2) energy labelling for refrigerators and freezers; (3) energy labelling for compact fluorescent lamps; (4) ballast loss labelling and standards for fluorescent lamp ballasts; (5) energy labelling for luminaires; (6) energy labelling for linear fluorescent lamps; (7) energy labelling for household electric fans; (8) television stand-by power reduction;

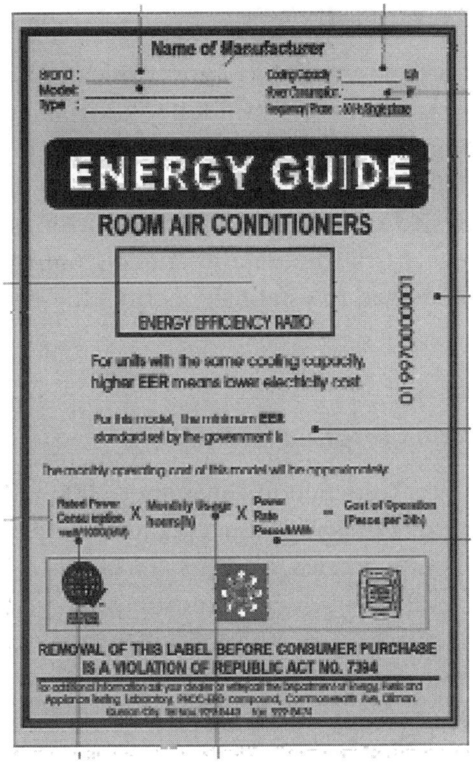

Figure 4: An Example of Energy Label for Room Air-Conditioners

(9) performance certification of fans and blowers; and (10) performance certification for electric motors. For room air conditioners, in addition to mandatory labelling, they are also required to comply with energy standards while labelling is pursued only in the case of the other products listed. The DOE has set up a laboratory for this purpose, the DOE Lighting and Appliance Testing Laboratory.

Besides appliances, the programme has expanded to include fuel efficiency guide labelling in motor vehicle dealer showrooms. Car manufacturers are encouraged to display in their dealers' showrooms the fuel economy rating and results of their entries in previous DOE fuel economy runs in order to promote consumer awareness of fuel economy in their choice of a new vehicle.

Energy Use Standards for Buildings

The DOE will coordinate with local government units in implementing energy use standards. Its Guidelines for Energy Conserving Design of Buildings and Utility Systems is already a referral code of the National Building Code. It plans also to enlist the United Architects of the Philippines to disseminate information on energy use standards for buildings. Lastly, the data obtained from the energy audits of buildings will be used to establish energy efficiency standards for buildings.

Government Enercon Programme (GEMP)

The programme was launched in December 2000 with the objective of integrating energy efficiency concepts into the procurement practices of government agencies, bureaus, and offices. It also includes monitoring and evaluating energy consumption reports of government agencies. Seminars are conducted for government building administrators and engineers who are encouraged to establish an energy management group within their respective organisations. One specific programme cited here is the replacement of all 40-watt fluorescent lamps with more energy efficient 36-watt fluorescent lamps.

Energy Management Programmes

The DOE's energy management programmes strive to improve energy utilisation of firms, including electric power generators and distribution utilities through the conduct of energy audits, financing of energy efficiency and conservation projects, heat rate improvement in power plants, system loss reduction in power distribution utilities, demand side management and recognition programmes.

The DOE has been providing energy audit and other services for a fee to private companies. This dates back to the days when it was the Ministry of Energy. The Ministry of Energy had set up an Energy Management Consultancy Service (EMCS) to provide energy services to the private sector. It had four divisions: consultancy and project engineering, testing and standards, planning and evaluation, and education. It was equipped with a Fuels and Appliance Testing Laboratory (FATL) for testing fuels and energy consumption of various appliances. The EMCS offered energy audit and project engineering services to industry and gathered energy consumption and production statistics from companies. It also provided training courses on various aspects of energy management.

The System Loss Reduction Programme is actually mandated by law, RA 7832 or the Anti-Pilferage of Electricity and Theft of Transmission Lines/and Materials Act of 1994. The law requires electric utilities to reduce systems losses, which would otherwise be passed on to consumers in the form of higher rates. Table 4 compares the targets with the actual historical average system losses of recent years. While there has been some improvement, it shows that utilities were still not reaching the mandated targets.

Demand Side Management (DSM) is enshrined in RA 9136 or EPIRA where it is the stated policy of the state to "encourage the efficient use of energy and other modalities of demand side management." DSM programmes are undertaken by utilities to encourage and influence end users' energy demand so that system resources are efficiently utilised. An example of a DSM programme is time of use or peak load pricing. Other examples are programmes to encourage end users to switch to efficient lighting such as CFLs, or to undertake measures to improve their power factor.

The Heat Rate Improvement Programme seeks to improve the operational capability of the NPC's thermal, coal, and diesel power plants and thereby improve their operating efficiency, reduce operating cost and extend plant life.

Table 4:

	RA 7832 Target (%)	2000 Actual (%)	2001 Actual (%)	2002 Actual (%)
Electric Cooperatives	14	16	15.4	15.4
Private Utilities	9.5	11.6	11.4	11.0

Source: The Philippines Department of Energy.

An obstacle to implementing energy efficiency projects and acquiring technologies, especially in small- and medium-scale enterprises, is the cost. Thus the DOE has initiated, with funding from the United States Agency for International Development (USAID), the Technology Transfer for Energy Management Demonstration Loan Fund (TTEM-DLF). This fund extends loans to qualified energy efficiency and conservation projects.

Lastly, the Philippines Efficient Lighting Market Transformation (PELMAT) Project, funded by the United Nations Development Programme-Global Environment Facility (UNDP-GEF) propagates the use of energy efficient lighting systems to reduce energy consumption and at the same time, greenhouse gas emissions. Incandescent lamps are still widely used in households and the project seeks to establish policies to encourage switching to energy efficient versions of linear fluorescent lamps, compact fluorescent lamps, ballasts (low-loss electromagnetic and electronic), and energy efficient luminaires.

Alternative Fuels and Technologies

At present, the DOE has been pushing for the passage of the biofuels bill which would mandate the use of ethanol and coco methyl ester (CME) as fuel blend for gasoline and diesel. Not surprisingly, the coconut and sugar industry lobbies are also behind this bill, since its passage would enlarge the market for the sugar and coconut sectors as they are the main sources contemplated at the moment.

It is worth noting though, that even without the benefit of such a law, some of the new entrants in the local oil industry have already started selling ethanol blended gasoline and coco-biodiesel (diesel blended with CME) in the local market. Even an oil major, Shell, has announced that it will start offering ethanol blended gasoline in some of its selected gasoline stations. The majors (Shell, Petron and Caltex) probably still have 85 per cent of the market and would be hard put to immediately supply the entire demand. Understandably, they have been slower than the smaller new players to introduce the alternative fuels. This underscores a potential short term problem: insufficient capacity to produce ethanol and CME. There may be insufficient capacity to produce the volume needed were it to be made mandatory. Thus some quarters argue that from the energy security perspective, it may simply resort to importing ethanol instead of importing oil. And of course, even at the highest contemplated mix in the bills (about 10 per cent for ethanol and 2 per cent for CME) the amount of oil product displaced will be relatively small.

Renewable Energy Bill

Another item on the legislative agenda is passage of the Renewable Energy Bill. The proposed law aims to develop renewable energy sources by mandating power generators and utilities to source a minimum proportion of their power from renewable sources. The bill proposes offering fiscal incentives for investment in renewable energy, such as tax holidays and tax- free importation of equipment and materials. The bill also seeks to open access to the energy grid for all renewable energy sources and give priority dispatch for wind power and other intermittent generators.

Getting Prices Right

While many of the above Government programmes have focused specifically on shaping consumers' preferences with respect to energy conservation and efficiency, there are other energy policies that will have an impact on the supply side of energy. These involve the restructuring of the oil and power sectors; the former by the Oil Deregulation Law and the latter, by the Electric Power Industry Reform Act (EPIRA). While these policies do not ostensibly have energy efficiency or conservation as their objective, they do aim to restructure the respective industries to make them more competitive, which in a market setting, often means improved efficiency. A precondition is that prices are transparent and allowed to seek market equilibrium.

The Philippines energy sector has been and continues to undergo radical restructuring, following the general trend of restructuring in the economy. The administration of President Aquino started the trend towards liberalising various sectors of the economy which accelerated during the administration of President Fidel Ramos. It was during his term that the downstream oil industry of the Philippines was liberalised through RA 8479, The Downstream Oil Industry Deregulation Act. The next major milestone came early during President Macapagal-Arroyo's term with the passing of RA9136 or the Electric Power Industry Restructuring Act (EPIRA).

Deregulating the Downstream Oil

Deregulating the downstream oil industry has not been easy. It illustrates the pitfalls awaiting regimes that formerly regulated and/or subsidised fuel prices if they attempt to reform and return to market determined prices. The original act deregulating the industry, RA 8180, was actually overturned by the Supreme Court, forcing lawmakers to replace it with the current RA 8479. The industry had first been deregulated in 1997. Unfortunately, the Asian Financial Crisis struck soon after, with the consequent depreciation of the local currency. Since the country imports virtually all its oil requirements, this meant a consequent increase also in the prices of petroleum products. Having been used to a regime of regulated prices where prices did not change very often, the frequent pump price changes after deregulation sparked public outcry, leading to a repeal of the original law after some legislators filed a case with the Supreme Court.

The Oil Deregulation Act ended oil price regulation and allowed firms to freely set their prices. Prior to that, pump prices had been set by the Energy Regulatory Board. Oil prices were set much like that for public utilities; companies were allowed to recover costs plus a return on the allowed rate base. An Oil Price Stabilisation Fund (OPSF) had also been set up as a buffer fund to stabilise prices. The principle was that when regulated prices were above costs, the differential would go into the fund. When regulated prices were below costs, the oil companies would draw from the fund.

When the fund ended with the onset of deregulation, the fund was in deficit, i.e., the Government owed the oil companies. This meant that prices had not been raised often enough. This was not surprising, since it is always politically easier to cut prices when costs go down than to raise prices when costs go up. It also implied that oil had been priced cheaper on average over the OPSF period than market forces would have dictated.

However, old habits die hard, and even in 2005, cries were still heard for a return to regulating the industry, price regulation and the OPSF. Thus the DOE formed a six-man independent committee to review the oil deregulation law. The author had the honour of serving as member of the said committee. It submitted its report in June 2005 and argued for the retention of oil deregulation and allowing the market to set prices. It cautioned strongly against subsidies and recommended the enhancement of the DOE's enforcement powers to allow it to chase erring industry players who made it an uneven playing field.

While oil prices in the Philippines are now generally set by market forces, there are still some exceptions. First there is an implicit subsidy built into pump prices because certain oil products such as diesel, kerosene and liquefied petroleum gas are taxed less (or even carry zero excise tax) than the other products (e.g., unleaded gasoline and aviation fuel). The rationale for this is that the poor and low income classes are more likely to consume the first group of products. While it is admirable to help the poor, indiscriminately keeping the price of diesel low for example, also benefits the wealthy who own diesel powered cars. In fact, it has sparked a shift in the composition of motor vehicle demand towards more diesel vehicles. And in the Philippines, the poor are not in a position to purchase cars.

Restructuring the Power Sector

Meanwhile, the restructuring of the electric power industry is proceeding, albeit behind the schedule as contemplated in EPIRA. The Act contemplates dividing the power industry into generation, transmission, distribution and supply. Generation and supply will be opened to market forces while transmission and distribution will remain regulated monopolies. A spot market for power, the Wholesale

Electricity Supply Market, where generators and buyers of power can bid has in fact started to operate.

The power sector is widely perceived to be inefficient and the Philippines has among the highest electricity rates in the region. This has been the bane of business for a long time and the major motivation for restructuring the sector. Part of the problem has been that prices have not been transparent. An intricate system of inter-regional/grid and inter-class subsidies developed. Thus for example, industrial and commercial consumers subsidised the residential consumers while the Luzon grid subsidised the Visayas and Mindanao grids. One achievement of the restructuring is that rates have been unbundled and much of the cross- subsidising has been or will soon be eliminated.

As in the case of oil prices, it was difficult to raise prices and many times in the past, the state-owned NPC had their rate hike petitions turned down or downscaled. This explains in great part the huge losses of the NPC in recent years, which ironically therefore pose a major strain on the Government budget as a result. That electricity rates are still among the highest in the region suggests how inefficient the sector must be.

SUMMARY AND CONCLUSION

There is nothing like a crisis to spur action. Makasiar made a very interesting observation: there was a lot more energy legislation from 1972 to 1981 (a total of 89 pieces) compared to the entire period 1911 to 1971 (12 pieces).[11] The danger of course in waiting for a crisis to happen before acting is that rash decisions may be made.

Looking back over Government responses to the energy crises, it seems that the response to the 1973 shock was more drastic and that the Government intervened with a heavier hand then compared to

[11] Gary S. Makasiar, "Structural Response to the Energy Crisis: The Philippine Case", in *Energy and Structural Change in the Asia-Pacific Region*, ed. Romeo M. Bautista and Seiji Naya. Manila: Philippine Institute for Development Studies and Asian Development Bank, 1984, p. 308.

the response to the high oil prices in recent years. In fairness, one might argue that oil was more critical to the economy at that time, accounting for over 90 per cent of primary energy needs. Today, oil is much less important and the measures taken have not needed to be as drastic. To be fair, to a great extent, this is due to the energy diversification policies undertaken following that first oil shock.

The Government conservation programme today seems to rely more on public education and information and is more voluntary in nature, in contrast to the reaction in the 70s, which included more restrictive measures, even on private sector activities. This time around, government regulation of activities has mainly been limited to the public sector; i.e., government offices and establishments.

Therefore, what is the proper course of action to take in an energy crisis? And what is the role of the government, if any? The economic theory of demand suggests that consumers will naturally seek to consume less of a good when its price increases due to substitution (alternatives become relatively cheaper) and income (the consumer is poorer) effects. The magnitude of the response will depend on the elasticity of demand. It is usually thought that the demand for essential goods would tend to be inelastic, or not that sensitive to price. Some would argue that energy falls into this category and some studies suggest that demand for oil and electricity in the Philippines is relatively inelastic.

Thus when the price of oil increases, one would expect that there should be a shift to other energy sources, i.e., there is a natural tendency to conserve oil. This is exactly what happened in recent Philippine history following the oil shocks and the 1990 Gulf War. Figure 5 suggests that when crude oil prices go up, the petroleum intensity of the Philippines goes down, i.e., fewer barrels of fuel oil equivalent (BFOE) are used per 10,000 pesos of gross domestic product.

This serves as an argument for the importance of getting prices right. Then consumers can weigh the costs and benefits of their energy consumption and decide whether to continue consuming or to conserve. For example, energy labelling can be effective only if consumers face the right prices for energy. Otherwise, if oil

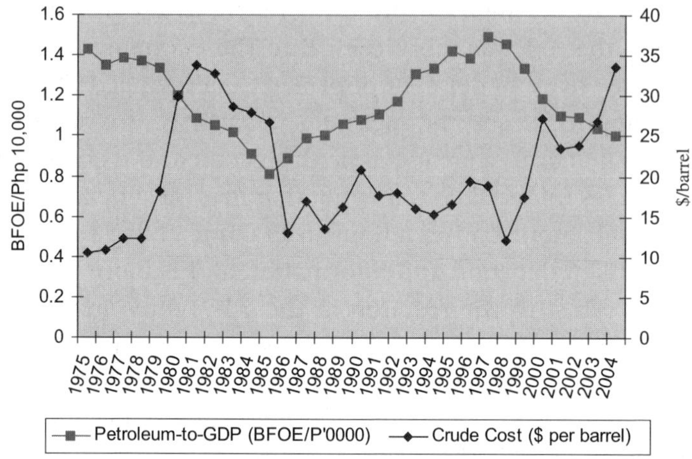

Figure 5: Petroleum Intensity of the Economy

prices and electricity rates are kept artificially low by subsidies, consumers may continue to purchase energy inefficient automobiles and appliances.

The same behaviour can be expected of firms. There is a natural incentive to be energy efficient. Being efficient reduces cost and therefore increases profits. When contemplating an energy saving investment or capital expenditure, the firm will naturally weigh the costs and benefits of such investment. If energy prices are artificially low, the incentive to make the investment is blunted. Note that energy efficiency may also be a competitive advantage. Thus one might not expect firms to share their energy conservation experiences or practices, especially if the savings are substantial.

Perhaps the other lesson is that government legislation and bills may not always be necessary. As pointed out, some new players in the petroleum sector have started to market ethanol blended gasoline and coco-biodiesel even without the benefit of a law being passed mandating it. Time and again, markets have demonstrated that as long as the incentives facing buyers and sellers justify it, products will be brought to market. This has proven true whether in markets for commodities or for fuel, or energy conservation and saving technologies.

Thus care must be taken in crafting laws and regulations so as not to distort incentives and stifle this initiative. The same is true for granting fiscal incentives.

When should the government intervene? According to economic theory, energy is generally a private good, i.e., rivalrous and excludable in consumption. Rivalrous means that one person's consumption of a unit of the good precludes others from consuming the same unit. It is excludable in that someone who does not pay the price of energy can be excluded from consuming it. If prices are 'correct' and property rights are well defined, the buyers and sellers can do their individual cost and benefit calculations and the market can be left to do its work in this case.

However, when there are externalities, the price mechanism will not work properly because costs and benefits may not be fully internalised. For example, if person A's consumption of energy emits pollution that inflicts harm on others then person A may not (and people generally do not) take this into account in his cost-benefit calculations. Thus, he may consume too much energy and emit too much pollution. In such cases, government may correct the externality by making the polluter internalise the added cost to others of his consumption by a tax or the threat of sanctions if the consumer does not take steps to mitigate or reduce the pollution to acceptable levels.

Waging a public information and education campaign is another possible role for the government. Markets work properly only if buyers and sellers have complete information, including information on the benefits of a product. Some information or knowledge may be public goods, i.e., there are benefits to the general public being aware of these, but it may not profit private persons or groups to disseminate such information or knowledge. This may be the case for example, of information or tips on energy saving practices. Energy labelling is another possible project for the government. Producers of energy efficient products have an incentive to proclaim their products are efficient while producers of inefficient products would not wish to reveal that fact. Thus a credible third party is needed to certify and disseminate this information to the public. However, note that there are also firms who have been able to provide such information to consumers commercially, e.g., Consumers' Digest, car magazines, etc.

Lastly, the government should be the first to practice energy conservation and efficiency. The government operates using resources owned by the people for the people (or should). As Milton Friedman once noted, people tend to take the least care in spending money when they are spending other people's money. Thus, it is commendable that the Philippine Government took the lead to implement energy conservation measures and report to the public those government offices that have excelled as well as those who have performed poorly in energy conservation. As another saying goes, practice what you preach!

Chapter

11

Energy Conservation Policy Development in Singapore

*Gavin Hearn Yuit Chua**

INTRODUCTION

In Southeast Asia and likewise for other parts of the world, 'green' is once again in vogue as ASEAN (the Association of Southeast Asian Nations) member states scramble to adopt more sophisticated energy strategies. Plans to adopt biofuels and nuclear energy are crowding the headlines. The rate of energy diversification is progressing at an unprecedented manner.

Promising developments are occurring at the regional level, with ASEAN member states signing an MOU on 27 July 2006 at the 24th ASEAN Ministers of Energy Meeting (AMEM) in Vientiane, Laos. Specifically, the ministers agreed to work towards the finalisation of the new ASEAN Petroleum Security Agreement and the Memorandum

* The views expressed herein are the personal views of the author. The author welcomes critical feedback that can be forwarded to hearnyuit@gmail.com.

of Understanding on the ASEAN Power Grid for possible signing at the 25th AMEM to be held in 2007 in Singapore.[1] The AMEM also witnessed more urgent calls for cooperation in renewable energy use. At the fifth meeting of the SOME (Senior Officials Meeting on Energy) + 3 Energy Policy Governing Group in Singapore in February 2007, improvements were also made in terms of oil stockpiling as well as cooperation in the areas of energy efficiency and conservation.[2]

But the general prescription remains for the region to develop a more robust energy policy emphasising conservation and efficiency. Singapore has joined the upper end of the energy security curve in this respect, alongside South Korea, Japan and Australia vis-à-vis other countries in the region.[3] A recent Morgan Stanley study attested that Singapore continues to be "an 'efficient oil user' (being a 'major global bunkering port'), with oil consumption as a percentage of gross domestic product estimated at 4.0–5.5 per cent this year (2005) — the same range as Hong Kong and Taiwan".[4] As discussed below, however, caution must be given to any complacency.

RECENT AND PROJECTED ENERGY CONSUMPTION PATTERNS

According to the Report on Energy Efficiency in Singapore released by the Government in 2000, Singapore's energy demand grew at 11.9 per cent from 1980 to 1995.[5] By comparison, GDP grew at 7.6 per cent over the same period. The Government report

[1] "ASEAN Strives for Regional Sustainable Energy Future", *Xinhua News Agency*, 28 Jul. 2006.

[2] "Summary Record of the 5th Meeting of the SOME + 3 Energy Policy Governing Group", ASEAN Centre for Energy at http://www.aseanenergy.org [1 Jan. 2007].

[3] "Asian Countries Spurred to Become More Energy Efficient", *Dow Jones International News*, 2 Mar. 2006.

[4] "Southeast Asia Faces Diverse Energy Challenges: Morgan Stanley", *Agence France Presse,* 19 Oct. 2005.

[5] Inter-agency Committee on Energy Efficiency (IACEE), *Energy Efficiency in Singapore Report,* 2000, at http://www.nccc.gov.sg/Newsroom/IACEE%20report.shtm [1 Jan 2007].

acknowledges that the increasing challenge for subsequent years is to minimise energy consumption whilst maintaining economic growth and global competitiveness.

According to the International Energy Agency (IEA), in 2004 Singapore imported a total of 100,703 thousand tons of crude oil, petroleum products and gas (oil equivalent — ktoe) (see Table 1).[6] In turn, the country — as the world's third largest oil product trading centre and largest fuel oil storage location and bunkering port — exported petroleum products amounting to 53,598 ktoe.

As the largest marine bunkering centre in the world, international marine bunkers also accounted for 22,737 ktoe in Singapore. Deducting the amount of exports and contribution from international marine bunkers yields a total primary energy supply (TPES) of 25,586 ktoe. Such computations are contentious when it comes to the country's international ranking in energy intensity, which refers to the "amount of energy consumed by a country for a given GDP".[7]

The World Economic Forum, United States Energy Information Administration (EIA), and British Petroleum (BP) give Singapore higher rankings in comparison to other developed economies.[8] The most recent case is from a report by investment firm Lehman Brothers that ranked Singapore's energy consumption per capita at about 11 tons per year, second among 21 countries on the list, behind Qatar (21 tons per capita) and twice that of some European countries such as France, Germany and Britain which posted five or just under five tons each.[9]

The National Environment Agency (NEA) and the Ministry of Trade and Industry (MTI) have disputed such rankings by pointing out that the calculations have been skewed by including marine bunkers and the

---

[6] Eighty per cent of crude oil imports and 60 per cent of condensate imports come from the Middle East.

[7] Singapore Ministry of Trade and Industry, *Economic Survey of Singapore*, Third Quarter, 20 Nov. 2006 at http://app.mti.gov.sg/default.asp?id=148&articleID=5901 [1 Jan. 2007].

[8] *Ibid.*

[9] "Singapore Disputes 'Energy Guzzler' Label", *Straits Times*, 3 Feb. 2007.

Table 1: Energy Balance for Singapore (ktoe on a net calorific value basis), 2004

	Coal	Crude oil	Petroleum products	Gas	Nuclear	Hydro	Geothermal, solar, etc.	Combustibles, renewables and waste	Electricity	Heat	Total
Production	0	0	0	0	0	0	0	0	0	0	0
Imports	7	45,079	50,332	5,286	0	0	0	0	0	0	100,703
Exports	0	−1	−53,597	0	0	0	0	0	0	0	−53,598
International Marine Bunkers**	0	0	−22,737	0	0	0	0	0	0	0	−22,737
Stock Changes	0	0	1,219	0	0	0	0	0	0	0	1,219
TPES	7	45,078	−24,784	5,286	0	0	0	0	0	0	25,586

*Totals may not add up due to rounding.

** International marine bunkers are not subtracted from the total primary energy supply for world totals.

Source: International Energy Agency (IEA) at http://www.iea.org/Textbase/stats/balancetable.asp?COUNTRY_CODE=SG&Submit=Su.

sale of jet fuel to international airlines in energy consumption. Given Singapore's heavy reliance on these two activities, as well as the manufacturing sector and the consumption of fuels as feedstock in the petrochemicals industry, they cause Singapore to rank unfavourably.[10] Also in response, NEA and MTI deferred to the IEA's computations that have removed the contribution of marine bunkers from Singapore's energy consumption per capita to render 230 mtoe (per million USD GDP at 2000 prices) as compared to EIA's 480 mtoe and BP's 409 mtoe.[11] Based on IEA's calculations, Singapore ranks competitively above the developed economies of Australia, New Zealand and Finland (see Figure 1).

Electricity generation accounts for the largest proportion of primary energy consumption at 48 per cent, as compared to manufacturing (33 per cent), transport (17 per cent), buildings (1 per cent) and consumers/households (1 per cent). For secondary consumption (use of electricity), manufacturing accounts for 44 per cent, transport 5 per cent, buildings 30 per cent, consumers/households 18 per cent and others (aviation) 3 per cent (see Table 2).

Electricity generation has witnessed rapid growth in terms of volume and capacity as well as in peak demand (5.6 per cent per annum compounded) corresponding with the country's economic growth pattern. In nine years alone, from 1995 to 2004, the total electricity generation volume grew by 67 per cent.[12] In 2005, Singapore's total electricity consumption stood at 34,761 GWh, having increased at a compounded annual growth rate of over 4.7 per cent since 1996. The 'domestic' share of total consumption has increased slightly while that of the 'manufacturing' sector and 'other industries' has decreased (see Table 3).[13] In the long term, Singapore's electricity demand is expected to grow at a rate of about 4 per cent per annum.[14] By 2018,

[10] Singapore Ministry of Trade and Industry, *Economic Survey of Singapore*.

[11] *Ibid*.

[12] *Ibid*.

[13] "Energy Conservation: Educate the Consumer", *Business Times*, 30 Oct. 2006.

[14] Speech by Mr S. Iswaran, Singapore Minister of State for Trade and Industry, delivered at the Financing Energy Projects in Asia Conference "Energy Security and Alternative Energy" held 18 Oct. 2006 at http://app.sprinter.gov.sg/data/pr/20061018996.htm [1 Jan. 2007].

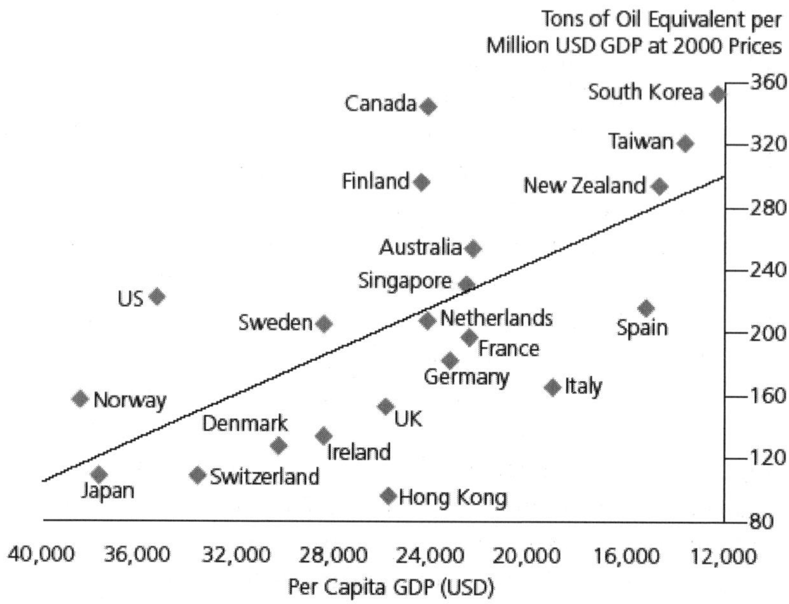

Figure 1: Energy Intensity and Per Capita GDP (USD), 2003

Source: IEA, reproduced by the Singapore Ministry of Trade and Industry, *Economic Survey of Singapore, Third Quarter 2006*. 20 Nov. 2006 at http://app.mti.gov.sg/default.asp?id=148&articleID=5901.

the Energy Market Authority (EMA) intends to have two-thirds of Singapore's gas supply conveyed by pipeline and the rest in the form liquefied natural gas (LNG).[15]

ENERGY RESOURCE GOVERNANCE POLICY

The Government responded more anxiously after acknowledging the country's poorer standing "behind most developed countries," especially in the World Economic Forum's *World Competitiveness Yearbook 2000*, where Singapore ranked 25th out of 45 countries in terms of energy intensity or the amount of commercial energy consumed

[15] Currently, 80 per cent of Singapore's electricity is generated from gas, of which 80 per cent is imported from Indonesia with the rest imported from Malaysia.

Table 2: Primary and Secondary Energy Consumption by Sector (per cent), 2004

	Electricity generation	Manufacturing	Transport	Buildings	Consumers/ Households	Others	Total
Primary Consumption	48	33	17	1	Approx. 1		100
Secondary Consumption		44	5	30	18	3	100

Source: Singapore Ministry of Environment and Water Resources (MEWR), "Singapore's National Climate Change Strategy — Consultation Paper" 2006, p. 12 at http://www.mewr.gov.sg/nccs/introduction.htm. These figures were used to compute key carbon contributors in 2004.

Table 3: Historical Electricity Consumption Breakdown, 1987–2005

Year	Domestic (%)	Manufacturing (%)	Other Industries (%)	Total (%)
1987	16.28	45.62	38.09	100
1989	15.34	47.71	36.95	100
1991	17.61	46.92	35.47	100
1993	18.17	45.32	36.51	100
1995	18.94	44.36	36.70	100
1997	19.23	44.45	36.32	100
1999	19.70	42.97	37.33	100
2001	20.22	41.35	38.43	100
2003	20.34	42.85	36.80	100
2005	19.42	43.17	37.41	100

Source: Singapore Department of Statistics, *Singapore Yearbook of Statistics*; and Energy Market Authority at http://www.ema.gov.sg/FILES/historical_electricity_consumption.pdf.

per dollar of GDP.[16] According to the *World Competitiveness Yearbook 2006*, Singapore rose by a mere five places in 2002, and trailed behind other Asian countries such as Hong Kong (1st) and Japan (5th).[17] Furthermore, the report also noted that Singapore ranked 33rd in terms of electricity costs for industrial clients in 2005, at 7.3 US cents per kWh vis-à-vis Taiwan's 5.5 US cents and Malaysia's 5.6 US cents.[18]

Key reasons provided by the Government for the unfavourably high energy intensity ranking (apart from skewed data computation by ranking agencies) are the country's being a 'tropical city-state' — and hence intensive usage of air-conditioning and associated lifestyle habits — and the fact that Singapore is an "urban city with no rural base".[19]

In recent years, sound energy resource governance has become a more urgent priority than ever for Singapore. First, the country

[16] Press Statement, "Release of the Energy Efficiency in Singapore Report", at http://www.nccc.gov.sg/Newsroom/IACEE%20report.shtm [1 Jan. 2007].

[17] "Energy Conservation: Educate the Consumer", *Business Times*, 30 Oct. 2006.

[18] *Ibid.*

[19] *Ibid.*

constantly faces the competitive pressure of keeping energy costs low in order to attract foreign investment. Second, the issue of energy security has occupied the main policy agendas of many countries in Asia, especially China and India, but also in Southeast Asian states such as Malaysia, Indonesia and Vietnam. Third, unstable geopolitics in the Middle East, coupled with another peak in worldwide oil prices in 2006, have raised the stakes and renewed the country's resolve to provide an adequate response to an increasing sense of global energy insecurity. Fourth, energy demand management is now a paramount policy concern given the country's commitments to addressing climate change since the middle of 2006.

Singapore's governance of energy consumption and efficiency can be interpreted more clearly through the broader lens of the country's governance philosophy, especially for the environment, with differential policy shifts over time to address particular needs according to national interest or international trends.

According to the Civil Service College, a statutory board under the Prime Minister's Office which supports the public service through research and training, Singapore observes governance imperatives that emphasise a proactive leadership which is constantly reflexive in order to optimise the country's limited resources and stay relevant in the international community, whilst safeguarding values and identity intrinsic to the fundamentals of the polity.[20] Such principles underlie the Government's general level of policy pragmatism towards environmental governance, especially in energy resource governance:

> "Singapore lacks natural resources. To date, there are no renewable energy sources that Singapore could harness in an economically viable way to reduce its reliance on fossil fuel.... With today's technology, using gas turbines for power generation is still the most cost-effective solution on the Singapore market. The country has limited options as far as a fuel mix policy is concerned; it has no indigenous supply of fossil fuels, such as coal, oil or gas".[21]

[20] Singapore Civil Service College (CSC) Principles of Governance, at http://www.cscollege.gov.sg/page.asp?id=61 [1 Jan. 2007].

[21] Singapore's National Assessment Report for BPOA+10, 2004, pp. 22–23; Vivian Balakrishnan, "The Role of Singapore in Regional Gas Markets", *LNG Review, 2005*.

Thus economic imperatives also underlie the governance logic of long-term sustainability agendas and goals. Policy pragmatism however, is also filtered through a constant reflexivity and corresponding policy innovation in energy resource governance to ensure that the country stays "nimble and flexible" to changes in the global economy and national sustainable development.[22] Such policy pragmatism and reflexivity find their expression in the following avenues: the energy industry reform process, multi-sectoral engagement and the establishment of specialised multi-agency bodies to promote energy conservation and efficiency.

Reforming the Energy Industry

In 1995 the Government embarked on its first reform of the electricity industry — traditionally vertically-integrated and government-owned — by corporatising the electricity undertakings of the Public Utilities Board (PUB), which was formed in 1963 to provide water, electricity and piped gas to the general population.The aim was to allow "market forces rather than central planning to drive investment, production and pricing decisions".[23] Furthermore, the Government has deliberately "avoided the policy of fuel subsidies that can and do distort the market and lead to over-consumption of energy [since] artificially low energy costs [would discourage] businesses and consumers…to conserve energy".[24] In turn, the negative consequences would lead to rendering Singapore "less attractive to investors especially energy intensive industries like petrochemicals and semiconductors".[25]

In 1999, a comprehensive review of the electricity industry led to the further restructuring of the electricity and gas industry "to put in place a competitive market framework to complement the liberalisation of the electricity industry".[26] A new statutory body, the Energy

[22] CSC Principles of Governance.

[23] Energy Market Authority (EMA), "Introduction to the Singapore NEM", 2006, at http://www.ema.gov.sg/doc/introduction_to_the_singapore_NEM.pdf [1 Jan. 2007].

[24] Speech by Mr S. Iswaran.

[25] *Ibid.*

[26] EMA, "Introduction to the Singapore NEM".

Market Authority (EMA) was created in April 2001 under the Ministry of Trade and Industry to oversee the regulation of the electricity and gas industries and secure operation of the power system, while the PUB was restructured into a comprehensive water authority under the Ministry of the Environment and Water Resources (MEWR). Additionally, the Ministry also oversaw the establishment of the National Environment Agency (NEA) in July 2002 to implement environmental policies.[27]

In Singapore's National Assessment Report for the Barbados Plan of Action (BPOA) + 10, the Government noted its continued reliance on fossil fuels for primary activities such as power generation and transport, due to a lack of viable alternative energy options.[28] However, the present energy resource strategic policy is two-pronged: for the EMA to "continue to look into ways to diversify Singapore's sources of energy" and the NEA to "continue to promote the use of cleaner energy and energy efficient technologies".[29]

Multi-Sectoral Engagement

Finally, the Government's reflexivity has also led to an acknowledgement of the need for engagement beyond the public sector to include partnership with the private and people sectors. This policy innovation in environmental governance occupies the second key thrust of the Singapore Green Plan 2012 — the country's blueprint in setting out broad directions and strategic thrusts in the next ten years towards sustainable development — as '3P (People, Private, Public) Partnership for Environmental Sustainability'.[30]

---

[27] Economist Intelligence Unit, "Organising an Investment: Environmental Law", part 20, 2006.

[28] Singapore's National Assessment Report for BPOA + 10, 2004, pp. 28–30.

[29] Speech by Mr S. Iswaran.

[30] Speech by Mr Lim Swee Say, Acting Singapore Minister for the Environment and Minister of State for Communications and Information Technology at the launch of the Public Consultation on the Singapore Green Plan 2012 at http://app.mewr.gov.sg/press.asp?id=SAS726 [1 Jan. 2007].

The Government's 3P engagement philosophy achieves greatest resonance in the country's efforts towards water resource security, by engaging the 3P sectors to "generate greater awareness of the importance of conserving, valuing and enjoying water and develop a sense of shared ownership of our water resources".[31] Yet, the 3P approach also extends towards the governance of energy conservation and efficiency, and supports the establishment of 3P departments in the MEWR, PUB and NEA for more effective implementation.

Specialised Multi-Agency Government Bodies

To bridge the apparent gap in energy efficiency vis-à-vis developed nations, the Government's initial policy move was to establish the Inter-agency Committee on Energy Efficiency (IACEE) in 1998. Chaired by Associate Professor Koo Tsai Kee, then Senior Parliamentary Secretary (National Development), and Mr Low Puk Yeong, then Deputy Secretary (National Development), the 11-agency body comprising Government ministries, statutory boards and academic institutions was tasked to study and propose policy measures to improve energy efficiency in Singapore.

While the IACEE completed its report on Energy Efficiency in Singapore in 1999, the Committee noted that the timing was inappropriate for the report's release, since Singapore was "just recovering from the economic crisis in the region, and the main concern of the Government was to help Singapore tide over the crisis." Such sentiments leading to the report's release only a year later, demonstrate in part the economic imperative underlying the governance logic of long-term sustainability agendas and goals. The IACEE report relied upon the potential for energy efficiency to enhance 'cost competitiveness' as well as the need to address the energy consumption rate in terms of fulfilling 'international obligations' with respect to the country's carbon dioxide emissions.

[31] Singapore Ministry of the Environment and Water Resources, "The Singapore Green Plan 2012 Executive Summary", p. 10, at http://www.mewr.gov.sg/sgp2012/index.html [1 Jan. 2007].

To properly achieve energy efficiency, the Singaporean authorities also recognised the systemic need to create a sustainable "environment that is conducive to the identification and adoption of energy efficiency measures." The IACEE proposed the adoption of three strategic thrusts:

a. Strengthen the regulatory and institutional framework;
b. Improve the market environment;
c. Use the public sector as the leading edge.

A year after the IACEE report came out, the MEWR announced the restructuring of the IACEE, specifically the expansion of its scope as of 1 April 2001. The IACEE was also subsequently renamed the National Energy Efficiency Committee (NEEC). The NEEC is comprised of four sub-committees and a research and development workgroup.[32] Its key thrusts are to:

a. Promote energy conservation through the efficient use of energy in the industrial, building, transportation and consumer sectors;
b. Promote the use of cleaner energy sources and renewable energy;
c. Promote Singapore as a location for the pilot test-bedding of pioneering energy technologies and as the hub for the development and commercialisation of clean energy technologies.

The IACEE's expansion included the taking of a "differentiated approach" to enhance energy conservation across the "various verticals [of] households, buildings, industries and transportation sectors".[33] The focus on energy conservation allows for "the twin effect of curbing growth in demand and encouraging sustainability".[34] There is also a new emphasis on R&D for economic development in

[32] Singapore's National Assessment Report for BPOA + 10, pp. 28–30.
[33] Speech by Mr. S. Iswaran.
[34] "Singapore Inter-Ministry Group to Formulate Energy Policies", *Business Times*, 8 Sept. 2006.

green sectors and fulfilling both the EMA and the NEA's tasks of increasing the diversification of energy sources and promoting the use of cleaner energy and energy efficient technologies, even if current alternative energy options are still not viable on a large scale basis.

More recently, the Government's concern for energy conservation and efficiency has been pursued under the broader discourse of energy security which has occupied the policy agendas of numerous countries in Asia. In June 2006, the Government established an inter-ministry Energy Policy Group (EPG) to "keep the local cost of energy competitive to underpin an attractive and stable environment for Singapore's long-term economic growth".[35] The EPG adopts a holistic approach to cover energy issues in terms of Energy Security, Economic Competitiveness, Environmental Sustainability and Energy Industry Development.

Speaking at the Financing Energy Projects in Asia Conference on 18 October 2006 in Singapore, Minister of State for Trade and Industry, S. Iswaran pursued the 3P discourse by drawing linkages between the development of "keen partnerships" between the private and public sectors with the enhancement of energy security by ensuring "sufficient supplies of energy to support growth and development".[36] "The Government's role," he added, "is to create a conducive environment for energy investments [from the private sector]."

Apart from energy security, the Government has endeavoured to position the energy efficiency and conservation measures proposed by the IACEE and NEEC within the current global challenge of climate change, as a way to draw closer relevance to the needs of an increasingly integrated global community and energy footprint as a "responsible global citizen" and since "the country's primary greenhouse gas (GHG) emissions can be attributed to carbon dioxide generated from energy use".[37]

---

[35] *Ibid.*

[36] Speech by Mr S. Iswaran.

[37] See NCCC website at http://www.nccc.gov.sg/aboutnccc/about.shtm [1 Jan. 2007]; "Singapore to Stem the Flow; Agrees to Join Climate Treaty to Cut Emission of Greenhouse Gases", *TODAY*, 8 Mar. 2006; "Addressing Climate Change to be Priority in Years Ahead For Environment Ministry", *Channel NewsAsia,* 3 Nov. 2006.

Coinciding with Singapore's plan to accede to the Kyoto Protocol as an non-Annex I country — in effect, agreeing to pursue the (non-binding) target of reducing carbon intensity by 25 per cent compared to 1990 levels by 2012 — in April 2006, the NEEC was renamed the National Climate Change Committee (NCCC) to reflect a scope extension a second time round.[38] Retaining the four-sector focus on households, buildings, industries and transportation, as well as adding three more workgroups (electronics, pharmaceuticals, chemicals) along with the R&D workgroup, the NCCC now includes both government and private sector participation.

The NCCC sets out to address climate change by:

a. "Promoting greater energy efficiency and less carbon-intensive energy in key sectors;
b. Raising awareness amongst the people, private and public sectors on the impacts and opportunities arising from climate change, and the actions they can take;
c. Building competency in Singapore to better respond to climate change such as through promoting research and development of low-carbon technologies;
d. Understanding Singapore's vulnerability to climate change and facilitating the adaptation actions needed".[39]

In other words, the NCCC aims to meet its GHG reduction targets through energy efficiency and conservation measures.

Whilst Singapore is a latecomer to the Kyoto scene, with other Southeast Asian states (with the exception of Brunei) having already ratified the Protocol over the past four years, the country has witnessed a flurry of initiatives that may provide the necessary stimulus for catching up in a region that remains ill-informed and ill-prepared to deal with climate change. When Singapore acceded to the Kyoto

---

[38] Inter-agency Committee on Energy Efficiency, "Energy Efficiency Policies of Singapore", 2006, at http://www.nccc.gov.sg/aboutneec/report.shtm [1 Jan. 2007].

[39] See NCCC website at http://www.nccc.gov.sg/aboutnccc/about.shtm [1 Jan. 2007].

Protocol on 12 April 2006, the NEA partnered with the Singapore Environment Council — a local environmental NGO — to launch a public awareness campaign to draw linkages between climate change and energy conservation.[40]

A month later, Singapore launched the National Climate Change Strategy on 30 May 2006. The move was followed up on 21 August with a two-year research study on the impact of climate change on Singapore. Back in April 2005, the NEA had already announced several energy-efficiency initiatives to eliminate up to 190,000 tons of carbon-dioxide emissions by 2012 and to help achieve the target of 25 per cent improvement in carbon intensity for 1990–2012.

Another agenda undergirding the Government's rapid policy shift is the desire to make Singapore a regional hub for carbon trading, allowing top carbon-emitting countries to attain their Kyoto Protocol targets by purchasing other countries' carbon emission quotas, often by sponsoring the UN's Clean Development Mechanism (CDM) projects. The Sustainable Energy Association of Singapore (SEAS) was launched on 12 July 2006, to "promote energy efficiency to industries and disseminate information on renewable energy technologies and services," not simply within Singapore but reaching throughout East Asia.[41]

The liberalisation of the electricity and gas industries — leading to the establishment of the EMA, along with the IACEE's scope expansion to NEEC and 3P multi-sectoral engagement, laid the cornerstone of new policy shifts. The recent policy moves of the EPG and NCCC mark a consistent chapter in the overall energy governance account, albeit with an additional thrust of linking the environmental sustainability and economic development imperatives so as to benefit from carbon trading and alternative energy options.

KEY INITIATIVES FOR POLICY IMPLEMENTATION

The initiatives revolve around the Government's four-sector pronged approach for households, buildings, industries and transportation,

---

[40] "Singapore to Sign Global 'Green' Pact This Year", *Straits Times*, 8 Mar. 2006.

[41] "Move to Turn S'pore into Carbon Emissions Trade Centre", *Straits Times*, 13 Jul. 2006.

as well as targeting power generation (as the primary energy consumption source) and R&D and development of new energy technologies.

To remain economically competitive, energy-efficient and environmentally sustainable at the same time, the Government realises the growing importance of seeking partnerships with the private and people sectors under the 3P approach, as well as the need to provide up-front investment funding for a change in the general mindset (both corporate and consumer) towards green practices, and for removal of cost as a prohibitive factor.

Such needs are being expressed increasingly forcefully and matched with compatible responses from the relevant industry players and grassroots action. Much more development needs to be made in the various sectors however, given the country's relatively late start especially with respect to alternative energy (with the appropriate infrastructural support) and the amount of time required to effect a mindset change.

Power Generation

The EMA launched the New Electricity Market (NEM) on 1 January 2003 to promote a competitive energy market and cope with increasing power demand so as to sustain an efficient energy industry. According to the EMA, customers have benefited from this competition, with average electricity prices falling by 10.5 per cent since 2002 (assuming a constant fuel price). Companies have also been exploring more efficient options for generating electricity, such as the use of natural gas for power generation and combined-cycle gas turbines (CCGTs), over the less efficient oil-fired steam plants.

This direction also befits the Government's present focus on developing a USD500 million LNG terminal by 2010 to augment its current imports of around 1.2 million standard cubic feet of gas daily via pipeline from Indonesia and Malaysia (including imports by Keppel Energy and Island Power), which supplies about 80 per cent of the country's electricity (for power generation and as a feedstock for petrochemical production).[42] The estimated operational capacity of the new terminal is three million tons.[43]

In future, the Government also plans to develop a regional hub for LNG imports to "broaden Singapore's fuel options and encourage a more resilient and robust market" as feasibility studies have shown that natural gas has a "good geopolitical distribution" to enhance energy security. It also emits lower carbon emissions than other fossil fuels and is estimated to be the "fastest growing energy source", with gas consumption expected to double by 2020. Furthermore, the "buffer stock of gas from an LNG terminal could offer arbitrage opportunities among LNG buyers and even outprice fluctuations".[44]

Further reforms are also underway to restructure the gas industry. The EMA introduced a Gas Network Code in October 2005, to provide "common terms and conditions for players to access the gas pipeline network in a fair and open manner".[45]

Consumers and Household Sector

The NEA launched a household appliance energy labelling initiative, the Singapore Energy Labelling Scheme, in conjunction with a local environmental non-governmental organisation called the Singapore Environment Council (SEC) in April 2002. The initiative covers refrigerators and air-conditioners, the two main energy-intensive household appliances. Whilst voluntary at the outset, the NEA will render the initiative mandatory and possibly include other appliances such as cloths dryers, dishwashers and water-heaters by the middle of 2007, as the original scheme was "limited in its usefulness to consumers" with

---

[42] The remaining 20 per cent is generated from fuel oil, orimulsion and waste incineration. Refer to Vivian Balakrishnan, "The Role of Singapore in Regional Gas Markets"; "Singapore Stresses Importance of Energy Security", *Xinhua News Agency*, 18 Oct. 2006.

[43] Speech by Mr S. Iswaran.

[44] Vivian Balakrishnan, "The Role of Singapore in Regional Gas Markets".

[45] EMA, "The Energy Connection", Annual Report 2005/6, p. 3, at http://www.ema.gov.sg/attachments/download/UPLOAD_20061121160819.pdf [1 Jan. 2007].

only 121 air conditioner and 77 fridge models labelled by January 2006.[46]

To promote public awareness for energy conservation and efficiency, SP Services Ltd., an electricity licensee, established the Electricity Efficiency Centre, providing various activities to target students and households, including a permanent energy education exhibit. Elsewhere, the NEA, EMA and SEC also seek to create partnerships with the private and people sectors to organise activities at the community level to spread the message of energy conservation.

At the grassroots level, the Northwest Community Development Council teamed up with Singapore Power, the largest electricity and gas utility company in Singapore, on 5 November 2006 to train 200 young volunteers to provide energy-saving tips to residents in April 2007.[47]

Industry Sector

Since July 2002, the NEEC (now NCCC) has developed an Energy Audit scheme to improve the energy efficiency of major industrial consumers (typically exceeding 10,000 terajoules annually), such as oil refineries and petrochemical plants. Companies have also started using their in-house staff or external audit specialists to conduct energy audits once every three to five years.

The Government also introduced an Energy Efficiency Improvement Assistance Scheme, a co-funding scheme (capped at S$200,000 per facility over five years) for the manufacturing and building sector, where 50 per cent of the funds will go towards conducting investment-grade energy appraisals, recommending specific measures for the participating companies to implement. As of 27 September 2006, 30 applications have been approved, with estimated annual savings (based on a preliminary audit) of around S$10.3 million or 202 GWh.

---

[46] "Singapore To Sign Global 'Green' Pact This Year", *Straits Times*, 8 Mar. 2006.

[47] "High Utility Bills? 'Green Envoys' Can Give You Tips", *Straits Times*, 6 Nov. 2006.

Another industry initiative is the Accelerated Depreciation Allowance for Energy Efficient Equipment, which encourages Singapore-registered companies to replace inefficient pieces of equipment with more efficient pieces and to invest in energy-saving devices. Since 1996, the Government has received 35 applications.

Finally, as a demonstration of Singapore's resolve to encourage industry efforts to become more 'green,' the country hosted the inaugural United Nations Global Business Summit for the Environment (B4E) in April 2007, and energy diversification was a central theme. Elsewhere, the Singapore Institute of International Affairs (SIIA) teamed up with Shell Singapore in the same month to organise a Forum on Corporate Social Responsibility and the Environment.

Building Sector

The Building and Construction Authority (BCA) introduced the Energy Efficient Building Award (EEBA) in October 2001 to recognise building owners, architects and engineers who have integrated energy efficiency into the design of buildings. The initiative was later replaced by the 'Green Mark for Buildings' programme in January 2005, to expand the assessment criteria beyond energy efficiency to include "building management, water conservation, indoor environmental quality and environmental protection in a building".[48] The overall aim is to promote sustainable development in the construction industry. Elsewhere, the BCA also conducts an annual banding exercise of some 444 public sector buildings, which are evaluated on energy consumption and energy efficiency and ranked into three bands. The Housing Development Board, Singapore's public housing authority, has incorporated energy efficiency into the design of public housing.

In April 2002, a university-led initiative by the Energy Sustainability Unit (ESU) of the National University of Singapore (NUS) developed an online performance-based building energy

[48] See BCA website at http://www.bca.gov.sg/Awards/EnergyEfficient/energy_efficient_awards.html [1 Jan. 2007].

benchmarking tool called the Energy Smart Building Scheme, to recognise buildings whose energy performances are among the nation's top 25 per cent and maintain a healthy and productive indoor environment.

According to Professor Lee Siew Eang, head of the ESU, the office sector in Singapore could annually save up to S$108 million and reduce carbon dioxide emissions by 360,000 tons, if all office buildings joined the scheme.[49] In December 2005, the Scheme was extended to office buildings, called the Energy Smart Office which is modelled after the United States' successful Energy Star Programme.[50] Smart Office aims to encourage building management to seek voluntary labelling by providing performance toolkits for buildings' performance levels and energy saving potential to be assessed.[51] In late 2006, the Scheme was extended to cover hotels.

In 2007, backed by a S$70 million Government incentive fund, the BCA announced its Green Building Masterplan, mandating all new public buildings and those undergoing major retrofitting to earn the 'Green Mark' as proof of energy- and water-efficiency, with good indoor environments.[52]

Of the S$70 million Government funding, developers can draw cash incentives of up to S$3 million per project from the S$20 million Green Building Incentive Scheme, depending on the building's Green Mark rating (gold for 70 to 79 points; gold plus for 80 to 84 points; platinum for 85 to 100 points).The rest of the S$50 million will sponsor R&D green technology projects for the building industry.[53] According to BCA chief executive John Keung, this initiative was a

---

[49] "Energy Bills Could Be Slashed: NUS", *Straits Times*, 25 Oct. 2006.

[50] "Energy Star, Singapore Style Launched", *TODAY*, 17 Dec. 2005.

[51] Speech by Associate Professor Koo Tsai Kee, Senior Parliamentary Secretary, Ministry of the Environment and Water Resources and Singapore Ministry of Defence at the Launch of the Energy Smart Buildings Programme, 16 Dec. 2005, at http://app.mewr.gov.sg/press.asp?id=CDS3436 [1 Jan. 2007].

[52] "New Public Buildings to Go Green from 2007", *Straits Times*, 15 Dec. 2006; "Govt Encourages Firms, Developers to Be More Environment-Friendly", *Channel News Asia*, 14 Dec. 2006.

[53] "$50m R&D Fund to Boost Energy Efficiency", *Straits Times*, 15 Dec. 2006.

critical milestone for Singapore as "the incentives would encourage more green buildings, when increased demand will create more competition and prices for materials and services will fall".[54] At present, 34 buildings have Green Mark certification, with 200 more being targeted in the next three years. Apart from funding, the BCA will also embark on public education campaigns beginning in 2007.

Transportation Sector

The Government's policy engagement to reduce energy use in the transportation sector has been to reduce the usage of cars by promoting public transport and setting road usage quotas and charges. The NEEC (now NCCC) also introduced a Fuel Economy Labelling Scheme for passenger cars in June 2003 to raise public awareness of fuel economy in cars and to encourage car traders to promote more fuel efficient cars.[55] Currently, 80 car models have participated.

To encourage the use of more fuel efficient green vehicles such as electric and hybrid cars, the Land Transport Authority (LTA) and the NEA established the Green Vehicle Rebate (GVR) in January 2001. The GVR is equivalent to 40 per cent of the car's Open Market Value (OMV) to narrow the cost differential between green vehicles and conventional vehicles. In December 2005, the Government extended the GVR to December 2007, and as of June 2006, there were 129 hybrid cars and 238 green vehicles on the road, albeit many of them were CNG taxis.[56]

R&D and Development of New Energy Technologies

Whilst Singapore is a relative latecomer on the green industry and technology scene, the Government has in recent years channelled R&D efforts towards the development of viable alternative energy options in order to address the urgent need to diversify the nation's energy

[54] *Ibid.*

[55] Singapore's National Assessment Report for BPOA + 10.

[56] Speech by Mr S. Iswaran.

sources. Furthermore, the Economic Development Board (EDB), the organisation responsible for the continued economic success of Singapore, has identified alternative energy as a "potential growth sector ... from an industry development standpoint".[57] To this end, the EDB announced in March 2007, a S$350 million 'clean energy' fund for research and development, test-bedding and pilot projects in Clean Energy, and promoting Singapore as a 'Global Clean Energy Hub'.

Some hotels and major food catering facilities switched to solar thermal energy for hot-water applications, and the roof of the German European School has solar cells built by the German firm Sunset Energietechnik GmbH and its Singaporean subsidiary, Sunseap Enterprises. But the use of solar energy in the country is still limited. Research has been conducted on photovoltaic (PV) cells even though current technology levels render the costs of implementation overly prohibitive at "three times the average Singapore electricity pool price".[58] The first silicon cell solar manufacturing plant in the country was established by homegrown Solar Energy Power, which began production of PV cells in January 2006, for export to Germany, Taiwan and India.[59]

As for biodiesel, wind, hydrogen fuel cell and trigeneration, the Government has successfully facilitated the location of multinational energy corporations' regional headquarters in Singapore. Germany's Peter Cremer has announced plans to invest up to S$34 million in the Jurong Island plant, which will have a capacity of 200,000 metric tons per annum.[60] Australia's Natural Fuel will also build the world's largest USD130 million biofuels plant in Singapore at 1.8 million tons by 2012 and work with an oil major on producing environmentally friendly, ultra-low sulphur diesel.[61]

---

[57] *Ibid*.

[58] Vivian Balakrishnan, "The Role of Singapore In Regional Gas Markets".

[59] See MEWR website at http://www.mewr.gov.sg/nccs/competency-building.htm [1 Jan. 2007].

[60] "Singapore to Host Two Biodiesel Plants, Investments Total over S$80m", *Channel News Asia,* 26 Oct. 2005.

[61] "Natural Fuel Eyes Tie-Up with Oil Major; Partner May Take Stake in its S'pore, Other Biofuel Plants", *Business Times*, 10 Nov. 2006.

British Petroleum has also established two hydrogen refuelling systems which DaimlerCrysler is presently using to test its fuel cell vehicles. In 2005, Rolls-Royce signed an agreement with a Singaporean consortium of companies to jointly invest USD100 million in fuel cell research, to render the technology commercially viable and environmentally friendly. On 13 October 2006, Singapore's Agency for Science, Technology and Research (AStar) announced plans to collaborate with Netherlands-based Shell Hydrogen BV to develop improved hydrogen-powered cars. Finally, Vestas, world leader in wind technology, also opened its Asia Pacific headquarters in Singapore with an investment of S$500 million for an R&D centre.[62]

Trigeneration (stream is generated from waste heat in the conversion of natural gas to electricity) is used to contribute to manufacturing processes and render conventional generating plants more energy efficient. TPGS Green Energy, a joint green venture by Tuas Power and gas importer Gas Supply Pte Ltd, will build an S$8 million tri-generation plant for pharmaceutical company Pfizer Asia Pacific to cut estimated annual utility costs by around eight per cent.[63]

On the domestic front, the NEA formed an Innovation for Environmental Sustainability Fund (IES) in late 2001 to encourage and assist Singapore-registered companies undertake innovative environmental projects to fulfil long-term sustainability goals. Between 2003 and 2005, the number of IES projects increased from 18 to 33 (S$4.4 million to S$10.6 million). In August 2003, the Ministry of the Environment (now Ministry of the Environment and Water Resources) and the EDB also jointly launched the Environmental Test-bedding Initiative (ETI) that saw five projects increasing to 18 in 2005.[64]

---

[62] Speech by Mr S. Iswaran.

[63] "Tuas Power in Tie-Up To Sell Tri-Gen Plants", *Business Times*, 8 Nov. 2006; "Pfizer to Cut Energy Bill in Singapore with Trigeneration", *Dow Jones Energy Service*, 7 Nov. 2006.

[64] See MEWR website at http://app.mewr.gov.sg/press.asp?id=SAS107 [1 Jan. 2007]; and MEWR, *Key Environmental Statistics 2006* at www.mewr.gov.sg/soe/kes2006.pdf [1 Jan. 2007].

On the international front, the Government has also made headway with joining the Renewable Energy and Energy Efficiency Partnership, a 160-strong international alliance of governments, non-government organisations and businesses dedicated to accelerating and expanding the global market for renewable energy and energy-efficient technologies. According to the NEA, the partnership focuses on the development and support of legislative and regulatory frameworks that accelerate the marketplace for renewable energy and energy efficiency, and has more than 58 active projects worldwide "targeting the development of policy or financial models that can be replicated by governments and project developers worldwide".[65]

Whilst much headway has been made to create a new industrial economy in alternative energy, the Government is taking a precautionary stance towards maintaining energy security because the country is still so dependent on fossil fuels. Construction of an underground oil storage facility and the planning of a S$500 million LNG terminal in 2006 to be operational by 2010, will coincide with the establishment of a new multi-agency body of the EPG. As Minister of State for Trade and Industry, S. Iswaran explained, "LNG is part of Singapore's diversification effort, with implications on energy security and competitiveness, while the rock caverns [underground oil storage] are part of the infrastructure development effort to attract more investors".[66]

Recently, the Jurong Town Corporation (JTC), the national developer for industrial infrastructure, has been assigned to study the environmental impact and feasibility of underground hydrocarbon storage on Jurong Island.[67] Elsewhere, the Government has embarked on its latest venture to gain a foothold in the LNG market and broaden the country's fuel options, to exploit the prediction of a "doubling of worldwide LNG consumption over the next ten years... shift in balance from a sellers' to a buyers' market for LNG in Asia

[65] "Singapore Joins Energy Partnership", *Power in Asia*, 6 Jul. 2006; "Singapore Chosen as Base for Renewable Energy Projects", *Straits Times,* 23 Jun. 2006.

[66] Speech by Mr S. Iswaran.

[67] "Singapore Inter-Ministry Group to Formulate Energy Policies", *Business Times*, 8 Sept. 2006.

[and] new growth markets for Asian LNG exporters," such as China and India.[68]

Currently, EMA is evaluating feedback on the planned LNG project from industry players, as well as pre-qualifying consultants to assist in the preparation of a Request for Proposals. An EMA spokeswoman is reported to have said that the feedback received will help EMA "fine-tune the policies and regulatory framework" for the LNG terminal.[69]

ENERGY RESOURCE GOVERNANCE POLICY: WHITHER PRAGMATISM AND REFLEXIVITY?

The industrial and household sectors have responded to government initiatives, though challenges remain to induce a more cohesive mindset change and integrate both bottom-line priorities with green goals. For instance, the global oil price hike in mid-2006 prompted companies to turn to energy audits — taking advantage of the Government's Energy Audit funding scheme — to lower their utility bills. From the first half of 2006, 17 companies have signed, as compared to three in the previous six months.[70] However, in the first instance, the current motivation is still led more by economic logic instead of energy efficiency standards.

The Government's efforts in the building sector hold promise, with active initiatives such as the Energy Smart Building Scheme to target office buildings as the principal benchmark for other building types to follow, such as hotels and shopping malls. The recent Green Building Masterplan, backed by a S$20 million Green Building Incentive Scheme and S$50 million R&D fund from BCA in 2007 will provide additional incentive for organisations to go green. Currently, the only regulations governing electricity use in Singapore pertain to building design, which falls under the purview of the BCA.

---

[68] Vivian Balakrishnan, "The Role of Singapore in Regional Gas Markets".

[69] "Feedback on LNG Terminal Under Study", *Business Times*, 11 Dec. 2006.

[70] "More Firms Now Keen on Energy Audits", *Straits Times*, 14 Jun. 2006.

The Government's reflexivity goes beyond policy-making to receiving feedback from the public to ensure that governmental plans stay relevant to emerging environmental challenges. Engagement platforms range from the Singapore Green Plan 2012 (initiated in April 2005) and internet surveys, to public consultation for the National Climate Change Strategy. Such platforms have been useful in obtaining ideas that were heeded by the Government, such as mandatory labelling for high energy-intensive household appliances.[71]

Yet, the sombre reality is that NEA survey results indicate that "of the 1,860 Primary 4 to JC2 students polled, around 90 per cent knew about environmental issues such as global warming, energy conservation and anti-littering. But only six in 10 actually put their knowledge to practical use...and nearly 42 per cent of the total believed that maintaining the environment is the Government's job".[72] Moreover, another nationwide poll on littering conducted before the survey also revealed similar results, as "about a third of the respondents...felt it was up to the management of coffee shops, hawker centres and bus interchanges to keep their areas spotless".[73]

In response, the NEA is engaging schools to launch the Environmental Champions programme, where appointed student ambassadors will help promote green projects and encourage anti-littering habits in their schools. New environmental NGOs are also an emerging player in Singapore's civil society with the likes of the Nature Society Singapore, WaterWays Watch Society, ECO Singapore and Climate Change Organisation, promoting environmental conservation thought to evoke social change.

Apart from the above shortfalls and promising developments, there are two ways to view the challenges that the Government is facing despite ongoing and projected initiatives to promote energy conservation and efficiency in Singapore. First, the prevailing perceptions held by Singaporean society of what constitutes an 'affluent lifestyle'

---

[71] "Can 'Robin Hood' Taxes Help S'poreans Go Green?", *Straits Times*, 28 Jul. 2005; "S'poreans Voice Environmental Concerns in Survey", *Business Times,* 26 Oct. 2005.

[72] "NEA Blues: Students Don't Go Green", *Straits Times*, 15 Nov. 2006.

[73] *Ibid.*

may present a persistent stumbling block to the country's energy sustainability efforts. For example, most Singaporeans (in fact most Asians) associate car ownership with an affluent lifestyle and regard it as a symbol of prestige when in fact the reverse trend is occurring in the wealthiest parts of the world. A Singaporean blogger's comment quoted by *Reuters* sums up the situation well: "If you love cycling in Singapore, you have to accept the status of a secondary citizen, many places are 'restricted zones' and you are simply not welcome".[74] Apart from setting road usage quotas, promoting public transportation use, and introducing fuel economy schemes and the Green Vehicle Rebate, and in light of the country's climate change drive, the Government can explore the use of alternative means of transportation, such as more attractive car-pooling schemes especially for the city districts, and the 'fold-up bike' concept to conserve physical space and circumvent the need to restructure the road/transport infrastructure system. Encouraging a consumer/citizenry mindset change in the transport sector will complement and contribute positively towards the Government's green promotion efforts in other sectors.

Second, the Government's present aggressive drive to promote industry development (e.g., the LNG terminal project) and R&D in alternative energy (e.g., biofuels) may be interpreted as fulfilling the country's economic imperative of rendering green initiatives economically viable. But caution should be taken to avoid the pursuit of such initiatives at the expense of other options that can provide greater systemic change in energy conservation and efficiency across both private and people sectors in the long run. Examples are a roundtable organised by the SIIA and sponsored by Shell, on 29 September 2006. Participants occupying leadership positions within the corporate, NGO and government sectors proposed several measures for the provision of subsidies or 'carrots' to the private sector to fulfil Singapore's energy conservation and efficiency needs:

- "The Government can reduce the price of climate friendly products to provide a pull factor, in conjunction with push factor

[74] "Singapore Fold-Up Bike Goes Against Asian Tide", *Reuters,* 20 Dec. 2006.

efforts in the supply chain to allow climate-friendly products to be produced at a lower cost. The private strata can take the lead in such an agenda."

- "Environmental aims do not necessarily incur economic costs. Pro-environment measures may receive government assistance at the beginning, but at some point the private sector takes over. As industry finds a technology-driven, cost-efficient way to meet mandated standards, people are awakened to the benefits, and it eventually becomes an accepted practice. The tough challenge is how to be a 'leading edge' and not a 'bleeding edge' company in the process."
- "The best way forward for solar energy might be for the Government to set targets in terms of X% of electricity generated using renewables by year Y, and give industry performance-based incentives to reach it, such as double tax reliefs, tariffs for feeding the renewable energy into the grid, or capital subsidies, to be eliminated gradually over time."

CONCLUSION

Singapore stands out from other countries in Southeast Asia, in that there is strong political will — filtered through policy pragmatism — to ensure the country achieves the goal of matching energy sustainability with economic viability. Furthermore, Singapore's political will is translated into action via investments for R&D and implementation of green schemes, public education, and most importantly, policy innovation. However, challenges remain in developing initiatives that will change consumer/citizenry mindsets, especially towards the use of automobiles. At the time of writing this was the weakest link in the Government's four-sector policy approach. Initiatives that promote industry development can also be tempered with other options that can provide greater systemic improvement in energy conservation and efficiency over the long run. With the Government's reflexive stance and flexible governance which seeks participatory voices from the 3P, the future prospects are promising.

Chapter

12

Energy Conservation Policy Development in Thailand

Surapong Chirarattananon

INTRODUCTION

The success story of energy conservation in Thailand has been publicised internationally.[1] Following this has been a large number of activities undertaken over a couple of decades as a result of the impetus of the oil crises and the action of contemporary governments.

Economic and energy policies have been formulated under the guidance of successive five-year economic plans called National Economic and Social Development Plans (NESD Plans) administered by a secretariat office and governed by a board chaired by the Prime Minister.

When the First NESD Plan was launched in 1962, Thailand was importing all of its commercial energy needs. Traditional energy

[1] Keyuraphan and Thongyai Na Ayudhya Ungbhakorn, published in proceedings from a forum organised by the International Institute of Energy Conservation, 1997.

resources such as fuelwood and charcoal were mainly used in households and small-scale rural industries. In all, more traditional energy was consumed than imported commercial energy (in the form of refined oil products), measured in terms of energy units (megajoules or tons of oil equivalent). Establishment of modern industries was encouraged by the successive NESD Plans, leading to increasing consumption of imported energy. Electricity generation relied heavily on imported oil, with a small quantity coming from domestic hydro generation.

The country began to face the problem of dependency on imported oil when the first oil crisis occurred in 1973, at the time of the Third NESD plan. When oil prices were rising and oil became scarce during the 1970s and early 1980s, successive Thai governments called for the exploration and development of domestic oil resources. In addition, they attempted to secure long-term oil supply agreements with oil exporting countries through diplomatic means. Supply security was the issue of the period. Short-term measures to reduce oil consumption were also implemented. When natural gas was discovered in the Gulf of Thailand and was piped on shore by 1990, the country faced another oil crisis when the Gulf War erupted in the Middle East. However, by this time, the government paid more serious attention to energy conservation. Plans were drawn to implement demand-side management (DSM) programmes and to pass a law to promote energy conservation. In 1991, the Government approved a DSM plan proposed by the Electricity Generation Authority of Thailand (EGAT) and the two electricity distribution utilities. In 1992, the Energy Conservation Promotion Act (ENCON Act) was passed by parliament. It called for mandatory and voluntary means to achieve energy conservation and develop renewable energy resources. The ENCON Act also created an Energy Conservation Promotion Fund (ENCON Fund) to be used in implementing activities sanctioned by the Act. This fund is sourced almost exclusively from a levy on consumption of all types of oil, except jet fuel, at the level of up to 0.1 baht per litre. Intensive activities commenced after the passage of the ENCON Act.

THE ECONOMY AND ENERGY CONSUMPTION

Consumption of Energy and Electricity

In 2004, GDP stood at 3,685,944 million baht (MB) in 1988 constant prices or 91,586 million US dollars, also in 1988 constant prices. Table 1 gives a breakdown of GDP into four major sectors for 2000 to 2004. It also provides sectoral GDP, final energy consumption, ratios of sectoral final energy consumption to sectoral GDP in US dollars and the ratio of overall final energy consumption to total GDP. Here, final energy consumption includes energy from oil, natural gas, coal, electricity and other sources all converted to tons or kilograms of oil equivalent (ktoe and kgoe).

The values of 0.66 to 0.74 for 2000 and 2004 for the ratio of overall final energy consumption to total GDP are similar to those of Indonesia, Malaysia, the Philippines and the East Asian region as a whole.[2]

Some 80 per cent of primary commercial energy supply is imported. This imported energy is mainly crude oil, natural gas (piped from Myanmar), coal, petroleum products (mainly diesel for motor vehicles) and electricity (from Laos). Major domestic energy sources are natural gas, lignite, crude oil, condensate from natural gas production and hydropower. Table 2 shows the value of each in tons of oil equivalent and percentages. It also gives the breakdowns of final energy consumption. Transportation is the largest consuming sector, followed by industry.

Petroleum products account for 64 per cent of the final energy consumed. These are mainly consumed by the transport, industry and agriculture sectors. Electricity accounts for 19 per cent of final energy consumption and is used mainly in the industrial, commercial and residential sectors.

The transportation sector has accounted for about 66 per cent of the total petroleum product consumption since the late 1970s and early 1980s. In the Fifth NESD Plan, it is mentioned that the

---

[2] Asian Development Bank Energy and Industry Department, *Energy Indicators of Developing Member Countries*. Manila: ADB, 1992.

Table 1: Final Energy Consumption and GDP (1988 prices) by Sector

Year	Industry			Transportation			Agriculture			Residential and Commercial			Overall
	Energy (Ktoe)	GDP	Energy/ GDP (Ktoe/ USD)	Energy (Ktoe)	GDP	Energy/ GDP (Ktoe/ USD)	Energy (Ktoe)	GDP	Energy/ GDP (Ktoe/ USD)	Energy (Ktoe)	GDP	Energy/ GDP (Ktoe/ USD)	Energy/ GDP (Ktoe/ USD)
2000	16,442	30833.4	0.533	18,022	7239.8	2.489	2,791	7727.4	0.361	10,551	26770.6	0.394	0.659
2001	17,143	28190.3	0.608	18,632	6978.3	2.670	2,847	7202.4	0.395	10,920	24481.1	0.446	0.741
2002	18,934	31195.4	0.607	19,636	7711.1	2.546	3,032	7526.8	0.403	11,377	26365.0	0.432	0.728
2003	20,255	35461.8	0.571	20,927	8284.4	2.526	3,308	8473.7	0.390	11,799	28411.3	0.415	0.698
2004	21,961	39623.8	0.554	22,812	9199.3	2.480	3,520	8350.6	0.422	12,667	30974.8	0.409	0.692

Source: Thailand Department for Alternative Energy Development and Energy Efficiency, 2006.

Note: GDP is in million US dollars (USD).

Table 2: Primary Supply and Final Consumption of Energy in 2004

Primary	Ktoe	Per cent	Final	Ktoe	Per cent
Domestic	**14362**	**19.7**	Industry	17394	34.3
Lignite	5610	7.7	Transportation	22812	45.0
Crude oil	4297	5.9	Agriculture	3520	6.9
Condensate	3115	4.3	Commercial, Residential and Others	7020	13.8
Hydro	1340	1.8	Total	50746	100.0
Imports	**58721**	**80.3**	**Type**		
Stock	1029	1.4	Petroleum products	32684	64.4
Crude oil	43536	59.6	Electricity	9803	19.3
Electricity	289	0.4	Coal and its products	5918	11.7
Petroleum products	1511	2.1	Natural gas	2341	4.6
Coal	4749	6.5	Total	50746	100.0
Natural gas	7607	10.4			
Total primary	73083	100.0			

Source: Thailand Department for Alternative Energy Development and Energy Efficiency, 2006.
Note: Unit is kilotons of oil equivalent (ktoe).

transportation sector consumed 42 per cent of primary supply of petroleum products.[3]

Almost half of the electricity generated is used by the industrial sector (46 per cent), followed by the commercial sector (32 per cent) and residential sector (22 per cent). The agricultural and transport sectors use less than 1 per cent each. A small amount of diesel, fuel oil and renewable resources contribute to power generation. The main sources are natural gas, lignite and hydro. Natural gas is used extensively by EGAT's generation and by the independent power producers, as well as by the small power producers (SPPs). A number of SPPs use rice husks and bagasse in cogeneration of heat and power. By the end of 2006, only one independent power producer used imported coal for power generation. So far, EGAT has used domestic lignite for power generation, and has no power plant using imported coal.

[3] National Economic and Social Development Board (NESDB), 1982, at http://www.nesdb.go.th/econSocial/macro/NAD/2_ni/ni_1998-2005/NI2005. [Jan. 2007].

Some 70 per cent of the power generated is from natural gas, 17 per cent from lignite, 9 per cent from hydro and 3 per cent from oil. Renewable sources, including biomass and geothermal, contribute around 1 per cent.

Peak power requirements in 2005 reached 20,538 megawatts (MW) and electricity generated reached 134,827 gigawatt-hours (GWh). The load factor improved to around 75 per cent over the year, due mainly to innovative tariffs. Except for 1998 and 1999, energy and power requirements increased by around 5,000 to 9,000 GWh and by 500 to 1200 MW every year for over almost two decades.

Energy Consumption and GDP Growth

Energy consumption and intensity of economic activity as measured in terms of GDP are clearly correlated in Thailand. Figure 1 portrays per capita final energy consumption and the ratio of final energy consumption to GDP. It also provides the corresponding information for electricity consumption. Final energy consumption per capita increased slowly but steadily from 1983 to 1994. There was a marked increase from 1995 to 1997 coinciding with the phenomenal increase in motor vehicle sales. Final energy consumption dropped in 1998 after the Asian Financial Crisis erupted in 1997. After 1998, consumption rose slowly in 1999 and 2000, but steadied thereafter. Electricity consumption per capita increased more smoothly from 1983 to 1997. In 1998, consumption dropped, but from 1998 to 2000 the pattern was similar to that of final energy consumption. After 2000, the pattern again became steady.

The graphs of the ratio of final energy consumption to GDP and of the ratio of electricity consumption to GDP markedly showed the effect of the change in GDP that occurred from 1998 to 2000. The ratios exhibit sharp increases during the period. The ratio of final energy consumption to GDP increased from 0.2 in 1995 to 0.6 in 1998. Similarly, for electricity consumption, the ratio increased from 0.6 in 1995 to 1.2 in 1998.

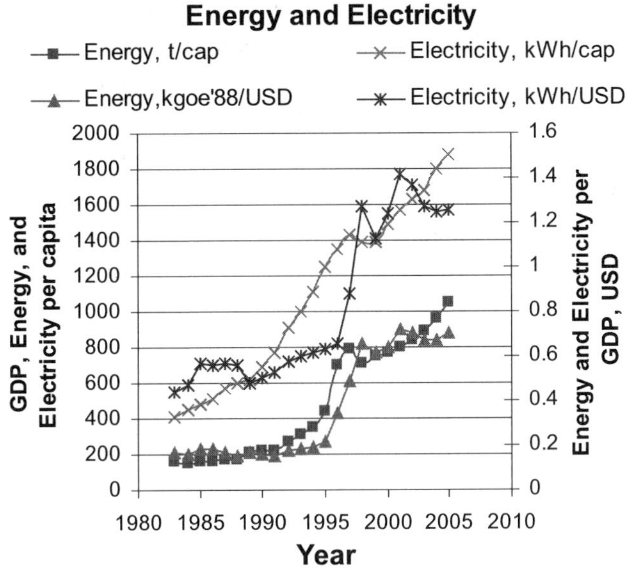

Figure 1: Development of Energy and Electricity Consumption, 1983–2005 (in 1988 USD)

Source: Plotted using data from Asian Development Bank, Energy and Industry Department, *Energy Indicators of Developing Member Countries.* ADB: Manila, 1992; Energy Policy and Planning Office (EPPO) website at http://www.eppo.go.th/info/index.html; and NESDB (2005) at http://www.nesdb.go.th/econSocial/macro/NAD/2_ni/ni_1998-2005/NI2005.

Energy Conservation Activities in the Pre-ENCON Act Period

The development of energy conservation activities can be discussed in terms of three periods. The first is identified as the period before the passage of the ENCON Act. Two years after the beginning of the First NESD Plan, Bangchak Oil Refinery started to refine imported crude oil with a capacity of 5,000 barrels per day.[4] In the same year Thai Oil Plc. also began its oil refining operations with a capacity of 35,000 barrels per day.[5] These events signify the beginning of

[4] See the Bangchak Petroleum Company website at http://www.bangchak.co.th [Jul. 2007].

[5] See the Thai Oil Company website at http://www.thaioil.co.th [Jul. 2007].

Thailand's industrialisation and shift to a high energy economy. The Electricity Generating Authority of Thailand (EGAT) was created in 1968 as the sole state enterprise for power generation, while the Metropolitan Electricity Authority (MEA) and the Provincial Electricity Authority (PEA) were also created for power distribution in the same year. Manufacturing grew at about 10–12 per cent annually during the 1960s and 1970s, fuelling a corresponding growth in energy consumption.[6]

1973-1980 Period[7]

During this period there were continual challenges from uncontrollable events related to oil price hikes and supply disruption. Successive governments reacted to find alternative supply channels for oil and to curtail consumption.

The Organisation of Petroleum Exporting Countries (OPEC) was formed in 1973 and was able to control the world oil price. At that time, the oil price rose to over USD5 per barrel from much lower levels. There was a shortage of crude oil. The government of the day passed a royal enactment (a law royally endorsed for application under emergency, but must be approved by parliament when it convenes) to prevent and act against oil shortages.

From 1973 to 1980, there were three governments. When a new government was formed, new laws and directives on matters relating to oil were issued. One government forged cooperation with the People's Republic of China and was able to purchase a certain quantity of crude oil at a 'friendship' price.

Shortages of oil remained a perennial problem in addition to price fluctuations. In order to reduce the effects of the fluctuation, an oil fund was created. When the price fell, an extra charge was added to the retail sale of oil and was used to contribute to the fund. When

---

[6] See http://www.nationsencyclopedia.com [Jul. 2007].

[7] This section is sourced mainly from Thienchai Chongpipien (ed.), *History of Energy Conservation*. Manila: Energy Policy and Planning Office, Ministry of Energy, 2004.

the price rose, the fund was drawn to subsidise the final consumers. This mechanism was used beyond this period.

In the latter part of this period, attention was turned towards exploring and developing indigenous energy resources, and controlling the use of oil and electricity by government offices while other measures were also enforced. All government offices were required to reduce both their electricity and oil consumption by 10 per cent. Another directive required government offices to turn off air conditioners. Towards the end of this period, more specific measures were issued and attention was turned to developing alternative energy resources and energy conservation. These included an initiative to develop technology for production of ethanol from sugar for blending with gasoline and to utilise solar energy in low temperature applications. In case it was necessary to use air conditioning, be it in government offices or in private households, the temperature was to be set above 27°C. A speed limit of 100 kilometres per hour was imposed on all roads and highways. Apart from government officials, people were encouraged to use public transport and reduce use of electric lighting in homes.

1981-1986 Period

This was a period of transition and preparation of energy conservation plans. At the start of this period, the international price of oil reached USD34 and peaked at almost USD40. By the end of the period, the oil shortages and tight oil prices eased off. Thailand was able to utilise its natural gas from the Gulf of Thailand through the first pipeline.

The Fifth NESD Plan (1982-1986) specifically addressed energy production and utilisation. In reviewing the situation, the Fifth NESD Plan (1982) referred to the rapid growth in energy consumption due to the changes in the economy. At the inception of the First NESD Plan, commercial energy consumption and demand for electric power stood at 2,760 million litres (Ml) and 268 MW, respectively. By 1981, before the launch of the Fifth plan, the corresponding quantities reached 17,960 Ml and 3,736 MW, respectively. Imported commercial energy resources contributed to 75 per cent of total energy consumption. The high and increasing energy consumption was deemed in part the result of past attempts by successive governments to delay the impact of energy price rises, when the price of crude oil rose to 12 times the 1973 level. The value of oil imports reached 68,000 MB, equal to 42 per cent of the value of foreign currency earnings from exports.

To address the problem of heavy energy consumption in the transport, industry, and commerce sectors, the Plan addresses the three sectors separately. The common policy is to reduce energy consumption per unit, be it per unit of product, per unit distance travelled or per unit area of building. The Fifth NESD Plan devotes particular attention to energy supply, demand and development. It outlines active support for exploration and development of indigenous petroleum resources, lignite, natural gas, geothermal energy and hydro power.

This Plan also discusses support for research & development (R&D), production and utilisation of non-conventional energy resources. Another dimension is the call for development of education curricula for manpower development to serve and manage the energy industries.

The final part of the Plan calls for amalgamating departments and units that deal with energy into a common chain of command, and expresses support for the development of an energy master plan under preparation by the National Energy Administration (NEA), the energy planning and implementing unit that was changed to the Department of Energy Development and Promotion (DEDP), and Department for Alternative Energy Development and Energy Efficiency (DEDE).

The United States Agency for International Development (USAID) assisted the NEA in a study on rural energy resources and energy uses. Field surveys were undertaken across all regions of Thailand, in order to develop a database on rural energy.[8] This database was to be

---

[8] Surapong Chirarattananon "Analysis of Rural Energy Development in Thailand", in *Rural Energy to Meet Development Needs,* ed. by M. Nurul Islam, Richard Morse and M. Hadi Soesastrob. Boulder CO: Westview Press, 1984.

used in sectoral energy planning models.[9] The USAID also provided technical assistance to the NEA to assess solar radiation and wind energy. This culminated in publication of maps of solar radiation and wind speeds for all locations in the country.[10]

On the industrial front, a private organisation called the Technological Promotion Association (Thai-Japanese) supported by Japanese industrial agencies, contributed to training Thai educational institutes and industries, by providing experts for training and disseminating knowledge on energy efficiencies in industry. The Association also published a number of translated Japanese technical manuals.[11]

During this same period, NEA received support from the Japan International Cooperation Agency (JICA) to send in a number of Japanese experts to conduct energy audits and report on techniques for energy conservation and energy efficiency improvement in Thai industries.[12] The objective was to transfer knowledge about energy auditing and analysis of industry to NEA personnel. During this period, the NEA and the Association of Thai Industries jointly created an industrial service unit called the Energy Conservation Centre of Thailand (ECCT) that was modelled after a similar unit in Japan. The ECCT was established as a state-private sector funded unit to assist Thai industry through conducting energy audits and provide advice to industry.

---

[9] Kriengkorn Bejraputra, "Sectoral Energy Planning Models: Supporting Studies of Thailand's Energy Master Plan", submitted to the NEA, Ministry of Science Technology and Energy under the Renewable Nonconventional Energy Project of the Royal Thai Government with support from the USAID, 1984.

[10] Bunterng Suwantragul, Warunee Watabutr, Kwan Sitathani, Visa Tia and Pichai Namprakai, "Solar and Wind Energy Potential of Thailand", Renewable Nonconventional Energy Project USAID Project No. 493-0304 (Solar/Wind Resource Assessment Component) submitted to Meteorological Department and King Mongkuts' Institute of Technology Thonburi, Thailand, June 1984.

[11] B. Rojarayanon, S. Pumwuthisan and J. Maekampornpong (eds), "Electrical Energy Conservation Techniques" (in Thai), translated from a corresponding manual written by Motoki Matsuo, Technological Promotion Association (Thai-Japan), 1982.

[12] Keisuke Arita (President, JICA), "Report of the Study on Energy Conservation Project in the Kingdom of Thailand", Japan International Cooperation Agency, 1984.

The NEA also received other support from the Asian Development Bank (ADB) to assess the potential for improving energy efficiency in Thai industries and the investment required to achieve such improvement in each industrial subsector. The study identified measures for improvement at three levels: house-keeping where little or no investment is required (such as boiler tuning), replacement of low efficiency equipment such as motors, and changes in the process of production. The three levels of measures require successively higher levels of investment.

Within the Association of Southeast Asian Nations (ASEAN), a Working Group on Non-conventional Energy Research was formed under the Committee on Science and Technology. ASEAN forged cooperation with a number of 'dialog partners' on various issues. One of the issues that was well-supported by Australia and the US, the two important dialogue partners during the 1980s, was energy conservation in commercial buildings. An ASEAN-US cooperative project on energy conservation in commercial buildings included a component on the development of an energy code for buildings. A team from the Lawrence Berkeley Laboratory of the University of California was engaged by USAID to work with the ASEAN Working Group. The output of the project included a draft or model building energy code for Indonesia, Malaysia, the Philippines and Thailand.[13] Singapore had already adopted a working code and the project produced a draft revised code for Singapore.

1987-1992 Period

Oil prices collapsed in 1986. This led to a decade long economic boom in Thailand at the end of which the economy had tripled in size. The measures carried out during the Fifth NESD Plan strengthened the economy amid a global economic downturn. The Gulf War erupted when Iraq invaded Kuwait in 1990 causing another round of global oil shortages. All industries grew substantially during the

[13] M. D. Levine, J. F. Busch and J. M. Loewen (eds), "ASEAN-USAID Buildings Energy Conservation Project Final Report", vol. I-III, Lawrence Berkeley Laboratory, 1992.

decade. Automobile production grew 43 per cent annually between 1986 and 1990 with output increasing from 76,162 to more than 310,000 units. Amidst such economic growth, energy consumption of course also increased. Moreover, the caution and frugalness that prevailed and accompanied the austerity measures in the Fifth NESD Plan began to erode. However, the impetus of efforts undertaken earlier led to adoption of a DSM plan by the electric utilities and passage of the ENCON Act by the end of this period.

The Sixth NESD Plan (1987–1991) largely addressed acceleration of the development of domestic energy resources to substitute for imported oil and further shift to the use of domestic resources for electricity generation. Deregulation of oil product prices and abandonment of the oil fund were specifically addressed. It was noted that energy use in transportation was a major cost and it recommended that the fuel efficiency of road vehicles be improved, and other modes of transport be introduced in Bangkok. The plan mentions that households were switching from charcoal to liquefied petroleum gas (LPG). The Plan took the recommendation of the Energy Management Centre study to note that industries could adopt three levels of measures and that energy intensity in industry could decline by 14 per cent. It also recommended that commercial and government buildings use more efficient air-conditioning and lighting equipment.

As an outcome of the outputs from the JICA and ADB projects, as well as its own initiatives, the NEA produced an Electricity Savings Plan for Industry, Commercial Buildings and Households. The Plan also drew upon information and experiences the NEA had accumulated from energy audits that commissioned consultants had carried out during the period. Table 3 lists some major items of improvement that the Plan proposed for industry, commercial buildings and households. It estimates energy saving and power demand savings against the cost of replacement of each specific item of improvement. In most cases, the Plan compares existing equipment with a more energy efficient alternative and estimates the monetary savings based on electricity savings at the existing tariff rate and the assumed usage pattern, and produces a figure for the payback period. Only those items with payback periods of less than five years are included in the

Table 3: Energy Efficient Measures in the Electricity Savings Plan

System	Energy Efficiency Measures		
	Industry	Commercial Buildings	Households
Lighting	1) Replace 40W fluorescent lamps by 36W lamps and 20W fluorescent lamps with 18W lamps 2) Remove unnecessary lamps 3) Use daylight	1) Same as for industry 2) Replace 40W incandescent lamps with 7W compact fluorescent lamps 3) Replace existing fixtures with reflective fixtures 4) Use room key cards in hotels	1) Same as for industry
Cooling	1) Improve cooling system by reducing air leakage, replacing chillers with more efficient ones, and using heat exchangers 2) Improve cooling systems in ice-making factories by insulating chilled water pipes, and cleaning condensers, cooling towers, and ice containers	1) Same as for industry 2) Install venetian blinds or curtains at windows, or use reflective film on windows 3) Use room key cards in hotels	1) Use electronic thermostats 2) Same as for commercial buildings 3) Use energy efficient air-conditioners 4) Use refrigerators with no auto defrosting

(Continued)

Table 3: (*Continued*)

System	Energy Efficiency Measures		
	Industry	Commercial Buildings	Households
Heating	1) Insulate steam pipes 2) Improve furnaces 3) Improve electric heaters 4) Replace furnaces	1) Recover waste heat from condenser to produce hot water	
Others	1) Improve air compressors 2) Improve transformers 3) Improve power factors 4) Improve conveyer belts 5) Replace motors 6) Improve other equipment		

Plan. Table 4 combines the electricity savings and investment required for the items categorised under lighting, cooling, heating and others for the industrial, commercial and household sector. The Plan dwells neither on how the improvement could be implemented, nor how much penetration it could expect in each case and category.

During the same period, the International Institute for Energy Conservation (IIEC), a non-government organisation based in the US, had taken an active role to promote energy efficiency in Thailand (it later expanded its activities to cover a major part of Asia). It convened a conference on Energy Efficiency in Pattaya on 3–4 March 1988, to highlight the issue.[14]

One of the outputs from the ASEAN-US Buildings Energy Conservation Project, was a draft Building Energy Code for Thailand that was submitted to the NEA in 1987.[15] An estimate was made of potential energy savings if it became mandatory for new large commercial buildings.

In 1991, EGAT with financial assistance from the Global Environmental Facility (GEF) of the World Bank, engaged the IIEC to develop a Demand Side Management Plan that it adopted with support from the Government. Table 5 gives a summary list of the programmes, their costs and savings.

In 1992, after repeated calls in the Fifth and Sixth NESD Plan for reorganising all energy agencies under one command, the Government decided to create a central command unit called the National Energy Policy Office that acted as the secretariat for the National Energy Policy Committee chaired by the Prime Minister. Its role is to coordinate all activities related to energy. Its creation also removed energy policy from the NEA and changed the NEA into the

---

[14] Deborah Lynn Bleviss and Vanessa Lide (eds), "Energy Efficiency Strategies for Thailand: The Needs and Benefits", report of a conference held 4–6 March 1988 in Pattaya, Thailand, organised by the International Institute for Energy Conservation, 1989.

[15] Levine, Busch, and Loewen, (eds), "ASEAN-USAID Buildings"; Surapong Chirarattananon and Limmechokchai, "A New Building Energy Efficiency Law in Thailand: Impact on New Buildings", *Energy-The International Journal*, 19 (1994): 269–278.

Table 4: Electricity Savings and Investment, 1991

System	Industry			Commercial Buildings			Households		
	Savings		Investment	Savings		Investment	Savings		Investment
	GWh/Y	MW	MB	GWh/Y	MW	MB	GWh/Y	MW	MB
Lighting	33.9	7.7	86.6	885.5	334.1	1608.5	140.9	80.4	86.8
Cooling	127.5	29.1	82.2	612.2	204.3	2249.2	89.0	30.9	789.5
Heating	323.7	73.9	1453.4						
Others	279.7	63.9	316.7	1.8	0.5	299.7			
Total	765.0	174.7	1939.0	1499.5	538.5	4157.4	229.9	111.3	876.3

Table 5: Programme List, the Costs and Savings of the DSM Plan

Programme	Cost of Programme (MB)	Annual Energy Savings (GWh/Y)	Avoided Peak Power (MW)	Cost of Saved Energy (B/kWh)	Cost of Avoided Peak (B/kW)
Commercial					
New buildings	158.29	55.2	14.2	0.25	11,100
Lighting — Large buildings	131.18	23.1	3.3	0.75	39,500
Lighting — Small buildings	8.11	2.2	0.3	0.48	25,300
Industrial motors	198.04	40.8	5.4	0.57	36,500
Residential					
Insulation and air-conditioning	44.56	3.4	2.6	1.17	17,500
High efficiency air-conditioning	62.19	8.6	3.3	0.85	19,100
Efficient refrigerators	14.92	4.0	0.5	0.58	29,700
32W lighting	4.37	0.8	0.3	0.75	14,600
Total for all DSM rebate programmes	621.66	138.0	30.0	0.53	20,600
Total for all DSM activities	771.38	138.0	30.0	0.65	25,500

Department of Energy Development and Promotion (DEDP) under MOSTE. At the same time, the Government in 1992 passed the Energy Conservation Promotion Act. The ENCON Act aims to promote energy conservation and development and use of renewable energy, as well as conservation of the environment that is related to the use of energy, through mandatory and voluntary means. It specifically addresses energy conservation for industry and buildings. Some of the clauses reflect specific statements in the Policy on Energy in the Fifth and Sixth NESD Plans. The Act also creates an Energy Conservation Promotion Fund (ENCON Fund) which can collect a levy on the retail sale of gasoline, diesel, fuel oil and kerosene at a rate to be determined by an ENCON Fund Committee. The Committee is chaired by a Deputy Prime Minister. The ENCON Fund can be used as a grant to promote R&D and to support all works sanctioned by the ENCON Act for government agencies, state enterprises and non-profit organisations. The ENCON Fund Committee is entrusted to recommend allocation of the use of the ENCON Fund to the National Energy Policy Committee. The ENCON Act leaves specific details pertaining to definition of energy performance requirements and what to do to achieve energy conservation to by-laws issued as ministerial regulations and announcements.

ORGANISATION OF ENERGY CONSERVATION ACTIVITIES FROM 1992 TO 2004

The sections of the Act that created NEPO, the Act that transformed the NEA into DEDP, and the ENCON Act changed the organisation of energy administration to the structure shown in Figure 2. The scope of the NEPO's responsibilities includes oil, gas, coal and other commercial energy sources such as LPG as well as energy conservation and renewable energy sources, but those of DEDP are limited to energy conservation and renewable energy resources including small hydro. Licensing of power generation by an entity for its own use at capacity beyond 200 kVA also comes under the responsibility of the DEDP.

The ENCON Fund Committee advises the National Energy Policy Committee on the use of the ENCON Fund, the levy on consumption

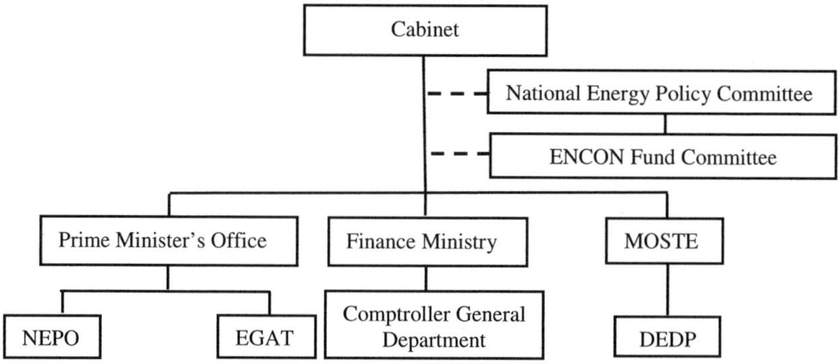

Figure 2: Organization of the Energy Agencies

of recommended types of petroleum products and endorses requests and recommendations from NEPO, DEDP and other agencies. It sets regulations on fund disbursement, and approves requests on the use of funds by NEPO, DEDP and other agencies that fall under approved policy and guidelines.

The National Energy Policy Committee makes recommendations to the cabinet regarding energy development and conservation, and passage of laws and regulations on energy matters. It is empowered to set levies on petroleum products and to set guidelines on the use of the ENCON Fund. At the time the ENCON Act was enacted, EGAT was under the Prime Minister's Office while MEA and PEA were and still are under the Ministry of the Interior. The Comptroller General Department is responsible for keeping accounts of the fund collection, its disbursements, as well as auditing its uses.

After the ENCON Act was approved by the Parliament and gazetted, it took three years to complete the planning for use of the funds, drafting of some royal decrees and ministerial regulations before activities sanctioned by the Fund could begin. The ministerial regulations requiring designated buildings and factories to report energy consumption and set targets and plan for improvement in energy efficiency reflect the statements on energy conservation in the Fifth and Sixth Plans and the Electricity Savings Plan. This in turn illustrates the underlying belief that such actions must be mandatory and that these will lead to substantial energy savings.

No energy conservation activity by the DEDP, NEPO and most other agencies was conducted while the plans and regulations were being drafted. However, the electric utilities, particularly EGAT, became active and commenced their DSM programmes from 1992. In carrying out the DSM Plan, there was no need to rely on funding from the ENCON Fund as budgetary support from the utilities themselves as well as financial support from GEF and other sources were already available.

ENERGY CONSERVATION ACTIVITIES SUPPORTED BY THE ENCON FUND

The ENCON Fund was created to support energy conservation activities, renewable energy development and conservation, and protection of the environment resulting from the use of energy. It was decided by the ENCON Fund Committee to classify programmes into three plans: Mandatory, Voluntary and Complementary. A subcommittee was appointed to supervise the administration of projects and programmes under each plan.

The ENCON Fund Committee approved the energy conservation plan in two phases: 1995–1999 and 2000–2004, each phase covering a period of five years. The DEDP was given the responsibility of executing programmes supported by the ENCON Fund under the Mandatory Plan. As a government unit, DEDP was already allocated a normal budget for employment of officials and staff to carry out its functions. Additional funds from the ENCON fund were used to facilitate its work and enable it to extend and enhance its capacity. Similar arrangements also applied to NEPO, where it was given the responsibility to undertake both the Voluntary and Complementary Plans.

In October 2002, the Government merged departments and agencies responsible for matters relating to energy into an Energy Ministry, relegated the role of NEPO on national policy to the Ministry, moving it from the Prime Minister's Office to the Ministry, and changed its name to the Energy Policy and Planning Office with a status equivalent to a department. The Government also moved the authority to supervise EGAT and the Petroleum Authority of Thailand

(PTT) from the Prime Minister's Office to the Energy Ministry, while that for MEA and PEA was scheduled to be moved from the Interior Ministry to the Energy Ministry in due course. Later on, PTT was corporatised and privatised as a publicly listed company under supervision of the Ministry.

The Government also merged the three subcommittees under the ENCON Fund into one, and assigned the permanent secretary for the Energy Ministry to head it. In 2003, a new energy conservation plan was drafted while the original plan was being executed with minor changes.

Mandatory Plan

The DEDP was assigned to execute all programmes in the Mandatory Plan under the supervision of the Subcommittee for Mandatory Plan. The Subcommittee was tasked to direct activities, approve programmes and projects and monitor the execution of each project. Table 6 lists the four programmes, the target beneficiaries and the expenditures incurred from various activities.

Programme for Existing Designated Factories and Buildings

This programme supports designated factories and buildings in complying with the Ministerial requirements, through supporting the cost of energy audits and the preparation of targets and plans for energy efficiency improvement. The ENCON Fund supports the cost of the preliminary energy audit for each factory or building up to 100,000 baht per factory or building. For preparation of targets and energy efficiency improvement plans, a designated factory or designated building is expected to conduct a detailed energy audit, from which items of improvement are identified. The ENCON Fund would support one half the cost of this exercise, but the amount of support for each case is limited to 500,000 baht.

In practice, all designated buildings and designated factories claimed support for a preliminary audit to the ceiling amount of

Table 6: Number of Facilities and Actual Expenditures

Description	1995–1999		2000–2004	
	Number	Expenditure MB	Number	Expenditure MB
Programme for Existing Designated Factories and Designated Buildings				
Designated Buildings	878	306	686	191
Preliminary energy audit	94	41	872	614
Formation of target and plan			479	2,471
Support to cost of improvement				125
Administrative cost				
Designated Factories	318	32	1,645	164
Preliminary energy audit			385	75
Formation of target and plan			2	0
Support to cost of improvement				1,085
Administrative cost				
Programme for Designated Factories and Buildings under Construction			7	1
Improvement to building plan			8	10
Investment according to improved building plan				
Programme for Government Buildings				
Energy audit and retrofitting	573	1,740	794	1,619
Total		2,119		4,725

100,000 baht, no matter how small or how large the given factory or building. Much of the work involved in the conduct of a detailed audit is generally a repetition of the preliminary audit. It would be difficult for a consultant to claim payment of fees for a detailed audit for up to the ceiling amount. Furthermore, energy efficiency improvements allowed were limited to the items consistent with those in the Electricity Savings Plan shown in Table 5, otherwise the report would not be approved and the ENCON Fund could not be used to support such retrofitting.

During this phase, the number of designated factories as defined by the royal decree was few: 1,583 buildings in 1996, 285 factories with demand exceeding 10 MW in 1997, 877 factories with demand exceeding 3 MW in 1998, 1,285 factories with demand exceeding 2 MW in 1999 and 2,455 factories with demand exceeding 1 MW in 2000. There was no estimate of energy savings to be achieved by the Programme. However, each detailed energy audit report identified items of improvement, their costs and savings.

Programme for Designated Factories and Designated Buildings under Construction

This Programme aimed to provide grant support, at the design stage, to the developers of buildings or factories in order to encourage the adoption of designs that are more efficient than required by the building energy code. The attention of the DEDP in this initial phase was tuned to existing buildings. Conditions for financial support to new buildings not yet constructed were not announced until the second phase and only eight projects were supported.

Programme for Government Buildings

This programme aimed to improve energy efficiency in government buildings not designated and not required by law to comply with the requirements. An outcome of the programme was for the participating government buildings to demonstrate to the public how much energy efficiency in buildings could be achieved. Each of the retrofitting measures chosen was selected from those that offered an internal rate of return of over 9 per cent. The measures chosen were consistent with those in the Electricity Savings Plan in Table 3.

Apart from the programmes that were scheduled as described above, the DEDP also requested that funds allocated for administrative support be used to operate three projects that were not scheduled in the Plan. The subsidy project subsidised participating facilities up to 30 per cent of the cost of new equipment from an approved list to replace existing equipment, while for the ESCO project, an allocation from the ENCON fund was used as a revolving fund for loans at 4 per cent interest to an ESCO that was disbursed through financial institutions. The joint participation project required each participating facility to set up an energy conservation committee to identify energy conservation items that required little or no investment and jointly solved them together with an external expert.

Voluntary Plan

The NEPO was assigned to execute all programmes under the supervision of the Subcommittee for Voluntary Plans. Table 7 lists the programmes and expenditures for the 1995–1999 and 2000–2004 periods, respectively.

As the name of the Plan implies, all programmes were voluntary to the participants. However, while the Renewable Energy and Rural Industry Programme promoted development and use of renewable energy, and the Research and Development Programme promoted R&D of renewable energy and energy conservation technologies and practices, only the Industrial Liaison Programme dealt with promotion of energy conservation practices. Even then, most of the projects under the Industrial Liaison Programme were not aimed at achieving active and direct participation of industrial, commercial or household sectors. Yet estimates of energy savings are given from all programmes. Table 8 lists the number of projects.

Table 7: Voluntary Plan Programmes and Expenditures

Programme	1995–1999		2000–2004	
	Number of Projects	Expenditure MB	Number of Projects	Expenditure MB
Renewable energy and rural industry	9	603	21	1,345
Renewable energy-based SPP	NA	NA	NA	2,060
Industrial liaison	23	225	43	2,832
Research and development	58	436	127	784
Non-designated facilities	NA	NA	3	18
Total	90	1,264	195	7,038

Renewable Energy and Rural Industry Programme

The objective of this programme was to promote use of renewable energy resources that would have little or no adverse impact on the environment and at the same time promote energy conservation for both the agricultural and industrial sectors. Most of the projects resulted in the use of renewable energy resources to substitute for commercial energy sources.

The programme invited government agencies, state enterprises, private organisations and private enterprises in rural areas to develop and use renewable energy in rural agricultural or industrial activity. The conditions in each participating enterprise are: that its electric power demand not exceed 300 kW and that it bears its own investment costs. In return, the ENCON Fund would bare the interest on loans for the required investment. The ENCON Fund could be drawn to pay the costs of development of the project, costs of project management (not exceeding 20 MB per project per year), and a revolving fund for investment required in carrying out the project.

Table 8: Number of Projects Classified into Sectors or Types of Renewable Energy

Programme	Sector							
	Transport	Industry	Households	Solar, Wind, Hydro	Biogas	Biomass	Others	Total
Renewable energy and rural industry	3	3	—	11	12	1	—	30
Renewable energy-based SPP	—	—	—	—	—	20	—	20
Industrial liaison	8	10	3	15	—	2	28	66
Research and development	28	14	70	22	9	11	31	185
Non-designated facilities	—	3	—	—	—	—	—	3
Total	39	30	73	8	21	34	59	304

Renewable Energy-Based (Very) Small Power Producers

This programme was initiated in the second phase. Its objective was to promote use of renewable energy in power generation through purchasing electrical energy generated at subsidised levels using the ENCON fund to pay the subsidies.

Industrial Liaison

This programme aimed to promote energy efficient technologies and equipment, and utilisation of materials that could substitute for commercial energy sources. It would support those who manufacture or market energy efficient equipment and products through demonstration of the use of the equipment and products. A government agency, state enterprise, educational institute or non-profit organisation could undertake a project. The ENCON Fund could be drawn to provide full support for project development, project management or project demonstration. A total of 23 projects were supported under this programme.

Research and Development Programmes

This programme provided grants to government agencies, state enterprises and non-profit organisations to conduct studies, R&D on energy policies, energy conservation and renewable technologies. Emphasis was placed on application research and application of proven technologies. A total of 59 projects received support from this programme.

Even though some projects were not directly involved, individual participants in energy savings, notably R&D projects, for projects under Renewable Energy, Rural Industry and Industrial Liaison, there was an attempt to estimate the expected energy and monetary savings, with the results shown in Table 9.

Complementary Plan

This plan was comprised of programmes deemed to support energy conservation efforts. Its success was not measured in terms of energy saved (see Table 10).

Table 9: Expected Benefits from the Voluntary Plan

Programme	Savings					
	1995-1999		2000-2004		Total	
	Ktoe/Y	MB/Y	Ktoe/Y	MB/Y	Ktoe/Y	MB/Y
Renewable energy and rural industry	35.4	158	48.7	414	84.1	572
Renewable energy-based SPP	2.4	69	512.6	14,327	515.0	14,396
Industrial liaison	0.2	3	1.0	22	1.2	25
Research and development	—	—	140.0	4,078	140.0	4,078
Non-designated facilities	—	—	1.2	11	1.2	11
Total	37.9	230	702.5	18,852	740.4	19,082

Table 10: Programmes and Expenditures under Complementary Plan (MB)

Programme	1995-1999	2000-2004	Total
Human resource development	595	1,459	2,054
Information dissemination in support of Mandatory Plan	194	63	257
Public information and campaign	591	818	1,409
Administrative support	3,024	2,115	5,139
Total	4,404	4,455	8,859

Human Resource Development Programme

This programme was comprised of five projects (see Table 11). The project on Development of Curriculum and Teaching Media included development of a curriculum on energy at the school level to supplement the normal curriculum used in country-wide schools of the Ministry of Education. It also involved training of teachers in 600 schools. The project includes support to engineering and architectural schools in developing new courses on energy conservation and

Table 11: Projects for Human Resource Development and Their Expenditures (MB)

Projects	1995-1999	2000-2004	Total
Development of curriculum and teaching media	422.6	600.5	1,023.1
In-country short-term training	55.6	451.4	503.9
Overseas short-term training	16.9	9.0	25.9
Postgraduate scholarships	84.7	263.6	348.3
Support to thesis research	10.5	48.1	58.5
Others	7.7	86.4	94.1
Total	594.9	1,458.9	2,053.8

renewable energy application and strengthening of existing courses. The fund was also used to support various short-term training projects. The most sustaining project has been one that supports student theses at undergraduate, master and doctoral levels, with the bulk of the funds going to support master theses. The fund also provided scholarships for students to participate in international programmes in energy, both in-country and overseas.

Information Dissemination in Support of the Mandatory Plan

The ENCON Fund was used by the DEDP to conduct information dissemination to designated factories and buildings on the requirements of the law, regulations and mechanisms on energy reporting, energy audits, and formation of targets and plans for energy efficiency improvement. It was also used for information dissemination to support all other activities of the DEDP.

Public Information and Campaign

The NEPO has extensively utilised ENCON Funds to conduct energy conservation campaigns among the public, through all media such as television, radio, newspapers, booklet handouts, etc. This part was aimed at the public at large. Therefore, all campaigns have been to provide general information and education to the public. These include occasional campaigns to turn some roads temporarily into 'walking streets'.

Administrative Support

The ENCON Fund has been used to hire consultants to help with the administration of various projects. These include the use of consultants by DEDP to handle audits reports and reports targeting and planning energy conservation by designated factories and buildings. In-house consultants are used by NEPO to handle project proposals, project reports, fund accounts and coordinating with project principals who receive funding support.

REVIEWS OF THE MASTER PLAN FOR ENERGY CONSERVATION

The ENCON Fund Committee appointed a subcommittee to evaluate projects, programmes and plans carried out in each phase. One subcommittee was appointed in 1998 to evaluate phase one of the Energy Conservation Plan (ENCON Plan), while another subcommittee was appointed in 2002 to evaluate the Plan in phase two. Each subcommittee was assisted by consulting companies or agencies.

Review of Phase One

A review of the projects, programmes and plans was conducted by the Energy Research Institute of Chulalongkorn University as the lead consultant. The conclusions are summarised below.

Mandatory Plan

The programmes under this plan were undertaken in full in 1997, just before the Asian Financial Crisis erupted. Time was too short when it came for review. There were initially few trained personnel.

The consultants employed to conduct energy audits and the DEDP personnel were just learning to carry out their functions.

Voluntary Plan

The programmes in this plan emphasised development of renewable energy resources and demonstration of energy technologies. For industry, there was no standard of energy efficiency. Nor were there sufficient numbers of personnel with appropriate expertise. There was a need to create more projects to support industry directly.

Complementary Plan

This plan did not include programmes for training of personnel to support implementation of the Mandatory Plan. Specific short training courses should be conducted for consultants and for personnel from designated facilities in order to identify and analyse energy efficiency improvement items. Public information and campaigns had been adequate.

The review took note of another study headed by the Energy Research Institute, engaged by NEPO, which predicted future energy scenarios for Thailand to 2025 based on the situation in 1995. The study identified the potential for energy efficiency improvement in the transportation sector as highest, at 56 per cent, followed by the industrial sector at 30 per cent, and the commercial and residential sectors at 7 and 6 per cent, respectively (see Table 12). The agricultural sector had the lowest potential.

The recommendations were categorised into industry, commerce, household and renewable energy.

Review of Phase Two

The reviewer of the second phase of the Energy Conservation Master Plan began by stating that the implementation of the Plan was efficient and the impact was positive when viewed from the objectives and targets of each programme. However, when viewed

Table 12: Energy and Electricity Consumption in 1995 and Potential Savings in 2025

	Share of Energy and Electricity Consumption in 1995		Potential Savings in 2025			
			Energy		Electricity	
Sector	Energy (per cent)	Electricity (per cent)	Ktoe	(per cent)	Ktoe	(per cent)
Transportation	39	0	12,773	56	0.4	0
Industry	32	42	6,955	30	2,828	54
Commercial	4	25	1,638	7	1,535	30
Residential	21	28	1,381	6	878	17
Agriculture	3	2	252	1	16	0
Total	100	100	22,999	100	5,289.4	100

Source: Energy Research Institute of Chulalongkorn University, 2000.

in terms of energy savings and the scale of renewable energy used to replace commercial fuels, the implementation was ineffective as the amounts of energy saved and the size of renewable energy used were small.

Mandatory Plan

The ministerial regulation required that all designated buildings and factories improve their energy performance to achieve compliance with the required energy performance levels in the regulation. However, all that was accomplished thus far were the detailed energy audits and the setting of targets and plans. This was still short of actual implementation. For the government building programme, metered savings from sampled retrofitted buildings fell short of estimated savings. It was recommended that a programme to promote ESCO be initiated to involve the private sector in providing a service that could lead to improvement in energy efficiency with minimal subsidisation from the state. The state through DEDP should provide training to increase manpower in this sector at various levels.

Voluntary Plan

The results of most projects under this plan were deemed to meet the objectives and targets positively, while the execution of some projects exceeded the set timeframe. Improvements in project conduct were visible. However, the energy efficiency improvement targets for projects from different sub-sectors were unclear. Most conservation projects were confined to specific groups and geographical locations, while the number of renewable energy projects was still few and too small to exert any impact at the national level. It was recommended that there be a more proactive approach in setting energy conservation targets and planning for provision of support to participants to achieve these targets. Priority should be placed on the sectors, subsectors or types of renewable energy that offered higher potential success, or higher potential for conservation and development.

Complementary Plan

The results of projects and programmes under this plan were deemed to be meeting the objectives and targets well. Problems, however, remained in the management of some projects to meet the set time schedules. The project to support pedagogical development for education in energy conservation and renewable energy in architecture might not lead to effective use of resources allocated. The processes involved in the use of consultants to assist DEDP in the execution of the mandatory programme might not lead to expected results. It was recommended that projects for human resource development be prioritised towards projects that would lead to significant energy conservation and be coordinated to meet the requirements of various plans and programmes. Priority support should be given to R&D projects that offered common benefits to the whole segment of industry rather than a specific factory.

DEMAND-SIDE MANAGEMENT

When the cabinet approved the DSM Plan in 1991, a Demand-Side Management Office (DSMO) was established as a unit within EGAT in

1992. A DSM Subcommittee was appointed to supervise the DSMO. This was comprised of representatives from the three electric utilities, the DEDP, NEPO and Office of Financial Economics, Ministry of Finance. With the official establishment of the DSMO, the following funds have been used to operate the DSM Plan:

- A USD15.5 million grant from the Global Environmental Facility and a USD9.5 million grant from the Government of Australia to be used between 1993 to 2000;
- A USD25 million loan from the Japan Bank for International Cooperation between 1994 and 2002;
- Charge to Fuel Adjustment Tariff (FT), a component in the electric tariff that accounts for change in fuel costs, costs of interests and other related costs, to a cumulative amount of 210 MB between 1993 to 2000;
- Charge to general operating fund of EGAT, when the cabinet approved a new tariff structure on 3 October 2000, the DSM cost was removed from FT and charged to the central operating fund of EGAT.

The programmes that were eventually implemented by the DSMO differed in content and method of implementation from those in the approved plan but were in consonance with the concept of DSM. All DSM programmes were voluntary. The DSMO spent almost one year developing its first programme for Phase One.

PHASE ONE OF THE DSM PLAN, 1993 TO 2001

Residential Programme

The underlying principle in this programme was market transformation, mainly through application of energy labelling to create a market for energy efficient products.

- The first energy efficient fluorescent lighting programme was launched by the DSMO by securing agreement from the manufacturers of fluorescent lamps in Thailand to cease production of the 40W long (1.2m) 'fat' (1.125 inch diameter) and the 20W short (0.6m) 'fat' lamps in favour of the thin (1.0 inch diameter) 36W long and 18W short lamps, with each corresponding type producing the same amount of light output. In return, the DSMO launched heavy public relations campaigns to inform the public that the 'thin' lamps produced the same light output as the corresponding 'fat' lamps, but saved 10 per cent of the energy;

- The DSMO also introduced energy labels on other electrical appliances. It educated the public on energy saving labels through public relation campaigns. In this period, energy labelling was applied to refrigerators, small unitary air-conditioners, ballasts for fluorescent lamps, unpolished rice, compact fluorescent lamps and electric fans.

Commercial Programme

The DSMO initiated activities in commercial buildings by including energy audits as part of its involvement with each building. It included load management through the use of demand control. Since each building differed in its use, a 'customer-oriented' strategy was used. The 'Green Building' project was initiated for new and old buildings that included energy conservation and environmental conservation as part of the label. Up to 450 buildings have participated in the project. The DSMO also initiated a cool storage air-conditioning project but this was discontinued due to financial problems at the time of Asian Financial Crisis in 1997. Success in the commercial programme has been limited because most building managers regard energy costs as only a small portion of the total cost of operating a building.

Industrial Programme

This programme was initiated after the DSMO had accumulated experience in operating other programmes. The DSMO offered energy auditing services and load management similar to what it offered for commercial buildings. It initiated a high efficiency motor programme that was not continued due to the Asian Financial Crisis. It conducted a pilot 'Energy Service Company' or ESCO project in four industries through funding support from the ENCON Fund. The success of this programme was also limited.

Value Creation Programme

This programme, launched in 1994, aimed to create a sense of participatory responsibility in using electricity efficiently. One method used was frequent exposure in the media public of the 'five stars' label (indicating the highest energy efficiency level) on appliances. The DSMO ran an educational project called 'green classroom' which went to over 230 schools from kindergarten to high school level. Another 120 schools were planned to be included.

RESULTS

The DSMO continually monitored and estimated the savings performance from all the projects and programmes. At the end of Phase One, the results of the DSM Plan were evaluated by an Independent Monitoring and Evaluation Agency engaged as a condition of GEF support. The output and outcome were rated 'highly satisfactory' as the savings were over two times the targets at a total cost of less than half the original budget (see Table 13).

It was rated to have a 'high international development impact, likely to sustain', and to be 'satisfactory to borrowers'. An internal evaluation report of the DSMO based on the labelling and public street lighting projects gives savings results to the year 2001 (see Table 14).

Table 13: Overall Savings of the DSM Plan, 1993–2000

Indicators	Target Savings	Estimated Savings
Electricity saved (GWh)	1,427	3,140
Power saved (MW)	238	566
Carbon dioxide reduction (Mton)	1.16	2.32
Expenditures (million US$)	189	59.3

Table 14: DSM Savings Results by June 2001

Year	Appliances and Projects												Carbon Dioxide Reduction (Mkg)
	Thin lamps		CFL		Refrigerators		Air-conditioners		Public lights		Total		
	MW	GWh	MW	GWh	MW	GWh	MW	GWh	MW	GWh	MW	GWh	
1994	7	36									7	36	26.4
1995	60	293			7	71					67	364	269.3
1996	102	495	1	8	18	186	9	35			130	724	535.4
1997	86	417	6	34	21	210	13	50	0	4	125	715	528.2
1998	64	313	2	12	14	140	22	85	0	13	102	562	415.0
1999	43	208	1	7	14	141	21	80	0	0	79	136	322.0
2000	40	195	1	3	18	184	24	91	0	0	83	473	349.8
2001	0	0	0	0	17	175	27	104	0	0	45	279	206.4
Total	401	1,957	11	64	109	1,106	117	445	0	17	638	3,589	2,652.5

BEYOND PHASE ONE

At the time of presenting its report, the DSMO on 2 August 2001 proposed continuing with a successive plan to be conducted from 2001 to 2005. But it was pending a cabinet resolution to restructure the whole electricity supply industry. As a consequence, the DSMO continued but there was no substantial plan and corresponding budget allocation to execute the Plan on the scale that had been done in Phase One. Table 15 gives the estimated savings to June 2004.

From 2004 to 2006, the Government required EGAT to corporatise and prepare to change itself into a publicly listed company. The future of the DSMO was uncertain during this period. However, the administrative court in 2006 ruled that the process of corporatising and privatising EGAT was illegal thus EGAT reverted back into a state enterprise. With a change in government in October 2006 and a new Energy Minister, the DSMO was re-vitalised.

THE THIRD ENERGY CONSERVATION PLAN

The cabinet approved a Third Energy Conservation Plan in its meeting on 2 September 2003. It was adopted as a part of the energy strategies agreed during a workshop organised by the Energy

Table 15: Estimated Savings from 1993 to June 2004

Projects	Savings		CO_2 Reduction
	MW	GWh	Mton
Thin lamps	401.5	1,957.5	1.45
CFL	10.0	57.2	0.04
Refrigerators	195.9	1,992.1	1.47
Air-conditioners	382.5	1454.5	1.07
Public lighting	—	17.2	0.01
Ballasts	7.0	43.4	0.03
High efficiency motors	0.2	1.2	—
Green buildings	2.63	10.3	0.01
Total	999.8	5,533.4	4.08

Ministry on 28 August 2003 and chaired by Prime Minister Thaksin Shinawatra.

The rationale was that Thailand consumed commercial fuels worth 800 billion baht (BB) in 2002, or equivalent to 14 per cent of GDP in that year. Of this expenditure, the value of imported oil accounted for 400 BB (and increased to 700 BB in 2006). Such expenditure and a possible oil supply disruption posed an energy security threat and erosion of competitiveness to the economy. The workshop adopted four energy strategies, two of which are relevant to energy conservation and renewable energy development. These are described below.

Strategy to Improve Efficiency of Energy Use

Economic development in the past had resulted in heavy consumption of commercial energy. The elasticity of energy to GDP had grown to 1.4:1. This was grossly undesirable when viewed against a ratio of 0.8:1 in the US and 0.9:1 in Japan. If the energy elasticity could be reduced to 1:1 by 2007, there would be savings of 3.1 trillion baht. Since transportation and industry accounted for 37 and 36 per cent, respectively, of final energy consumption, the strategy focuses on these two sectors.

Transportation

Up to 80 per cent of energy used in transport is for land transport, of which 78.6 per cent is for road transport and 4.6 per cent for rail. The following measures were adopted:

- Restructuring of passenger transport and the transferral of goods from small vehicles to rail by:
  - — increasing investment into rail systems and other mass transit systems in Bangkok and the adjacent provinces,
  - — accelerated construction of parallel rail tracks throughout the rest of the country.

- Development of multimodal transport systems and
  - (a) support use of energy efficient vehicles,
  - (b) use of planned city zoning to define routes for transport of goods,
  - (c) use of tax incentives for increasing the energy efficiency of transport

Industry

Industry in Thailand is energy intensive compared to that in more developed economies. The following measures were adopted:

- The Industry and Energy Ministries and the NESDB were to jointly restructure industry to increase its competitiveness, and to include the dimensions of energy and economic value in investment promotion;
- The Finance Ministry was to develop tax measures to support energy conservation in industry and transport through tax waivers;
- The Industry Ministry in collaboration with the Energy Ministry were to set minimum energy efficiency standards for electrical appliances and automobiles, create a system of energy conservation certification for industry and support efficient utilisation of energy such as employment of cogeneration and district heating.

Strategy to Develop Renewable Energy Resources

The development and utilisation of renewable energy resources would reduce the use of imported fossil fuel, increase economic benefits to communities, and reduce local and global environmental impacts. The target was to increase the contribution of renewable energy from 0.5 per cent in 2002 to 8 per cent in 2011 through the following measures:

- For all new power generation capacity to be connected to the grid, a renewable (solar, wind, or biomass) portfolio standard of 4 per cent was required;

- Incentives were to be provided in the purchase of power from renewable generation through tax credits, special rights and/or through subsidies to be provided from the energy conservation promotion fund.

FRAMEWORK FOR PLANNING AND ALLOCATION OF ENCON FUND

The Third Energy Conservation Master Plan was conceived as a rolling plan where a detailed plan and funding would be scheduled each year in anticipation of developmental results, changes in the socio-economic situation and shifts in government policy. The Master Plan is comprised of three standing plans: Plan for Increasing Energy Efficiency, Plan for Development of Renewable Energy Resources and Plan for Strategic Management. The weights or priorities for the plans are shown in Table 16.

The Third Energy Conservation Plan had been formulated as a part of the Thai energy strategy to reduce the ratio of elasticity of

Table 16: Weights or Priorities of Plans and Programmes

Plan	Weight (per cent)	Programme	Weight (per cent)
Increasing Energy Efficiency	35	R&D	30
		Development and demonstration	50
		Human resource development and public information	20
Development of Renewable Energy Resources	50	R&D	70
		Development and demonstration	20
		Human resource development and public information	10
Strategic Management	15	Policy study	33
		Management and administration	33
		Others	34

energy to GDP from 1.4:1 to 1:1 from 2005 to 2011. This implied a reduction of final energy consumption of 10,354 ktoe or 11.7 per cent in year 2011 from an estimated value of 91,877 ktoe to 81,523 ktoe. The plan also aims to increase the contribution of renewable energy in final consumption by 8 per cent in that year. This means a contribution of 7,530 ktoe, or 9.2 per cent of the total, from renewable resources in that year.

PLAN FOR INCREASING ENERGY EFFICIENCY

This plan has R&D, implementation measures, human resource development and public information components. It is expected to contribute to final energy savings of 10,354 ktoe. The breakdown of savings appears in Table 17. This plan does not have a component for commercial buildings.

PLAN FOR DEVELOPMENT OF RENEWABLE ENERGY RESOURCES

This plan aims to develop and use renewable energy resources to replace conventional commercial energy resources by 7,530 ktoe. The programmes and measures are given in Table 18.

STRATEGIC MANAGEMENT

This programme has structured a broad framework for studies that could be used to assist in policy decisions on various issues such as identification of weights or priorities for development of different types of renewable resources, priority between energy conservation and renewable energy development and the use of different instruments for voluntary energy conservation schemes.

CONCLUSION

Thailand as an energy importing country has faced serious economic and security problems from the fluctuation of energy prices and

Table 17: Measures to Reduce Final Energy Consumption

Sector	Saving		Measures
	Ktoe	Per cent	
Transportation	6,269	21	Use of mass transport, traffic management, goods transportation on rail and water ways, development of logistic depot, efficient goods transportation network, increase energy efficiency of vehicles. R&D on traffic management, transit linkages, logistics, energy efficient engines and innovative engines.
Industry	3,411	9	Use of energy efficient machines and equipment, good house practices, minimum efficient standards for appliances and industry restructuring. R&D on energy efficient equipment and on energy efficiency requirements.
Households	673	4	Promotion and use of energy efficient appliances, cooking stoves, and management practice. R&D efficient appliances and cooking stoves.
HRD and Public Information			Promotion of graduate, postgraduate, and pre-university education, and short training. Providing specific and general information.

Note: Per cent refers to percentage of consumption of the given sector.

supply since the time it embarked on an accelerated economic expansion. The Governments' response was initially mostly reactive, though at times on the alert as if at war. One government was voted out by parliament because of the increase in oil prices. The attempt by planning agencies to create a more systematic and structured means to deal with energy issues resulted in the passage of a law to promote energy conservation. The country became known for its success in energy conservation efforts, less from the ENCON Act and subsequent activities, but more from the DSM activities undertaken by the DSMO of EGAT, after it presented its laudable successes from its energy labelling programme.

Table 18: Measures for Development and Use of Renewable Energy Sources

	Saving		Measures
	MW	Ktoe	
Solar Energy	250	28	Remote solar home, remote schools and clinics, roof top solar, and renewable portfolio standard (RPS). R&D on PV technology and solar thermal.
Wind	115	19	RPS, low speed turbine, and water pumping. R&D wind energy assessment, development of low speed turbine.
Small Hydro	350	102	Remote hydro, low head turbine, and village turbine. R&D on community water management.
Biomass	955	3,441	Very small power producers, RPS, power generation from municipal waste and waste from agricultural produce. R&D on power generation from biomass and heat from biomass.
Biogas	51	1,625	Biogas from animal waste, from community waste, and from agricultural waste. R&D on biogas generation and utilization.
Biofuels		2,078	Ethanol to replace MTBE in gasoline blending and biodiesel from palm oil. R&D on cost reduction for ethanol and diesohol for diesel replacement.
HRD and Public Information			Scholarships for graduate studies on renewable energy, support for thesis research and curriculum development at all degree levels.

The promulgation of the ENCON Act and creation of the ENCON Fund illustrated the remarkable foresight and efforts of the planning authorities of the time. However, the implementation of the successive energy conservation plans thereafter cannot be rated as being totally successful. In the first phase, the agencies were experimenting and learning with sizeable funds at hand. However, the improvement in the execution of the Plan in the second phase was not impressive. The Mandatory Plan was carried out by a government agency used to its own bureaucratic means. It would approve an energy audit report that recommended only an approved list of energy improvement items, while the costs of actual improvement would be born by the building management. The so-called Voluntary Plan was comprised of mostly R&D projects. Projects that dealt with participants directly were given to various agencies that did not have the experience to carry out such activities. When the new government took the helm and attempted to steer the Third Energy Conservation Plan in another direction, it appeared as if the approach of using measures in the past was revived again.

Chapter

13

Energy Conservation Policy Development in Vietnam

Felix Gooneratne and Sumit Pokhrel

INTRODUCTION

Over the last decade, Vietnam's energy consumption has been increasing rapidly to fuel the unprecedented economic growth. Between 1990 and 2004, energy consumption increased at a rate of 11.2 per cent per annum, about 1.5 times higher than the GDP growth for the same period.[1] From 2000 to 2005, Vietnam experienced economic growth averaging 8.2 per cent and it is expected to continue to grow rapidly at 7.3 per cent annually through 2010, but will slow down thereafter to about 6.0 per cent annually until 2030. With expected higher economic growth through 2010 and relatively

[1] Asia Pacific Energy Research Center, *APEC Energy Demand and Supply Outlook 2006*. APERC, 2006; United States Agency for International Development ASIA, "Clean Energy Priorities of Asia: A Regional Imperative for Clean Development, Climate Change and Energy Security", Vietnam Country Report, USAID, Jan. 2007; International Institute for Energy Conservation, *Review of Cleaner Production, Energy Efficiency and Clean Energy in Vietnam, Cambodia, and Lao PDR*. Bangkok, IIEC, Nov. 2006.

moderate growth beyond, Vietnam will require substantial expansion of its energy sector. Primary energy demand is projected to increase more than three-fold from 42 million tons coal equivalent (mtoe) in 2002 to about 142 mtoe in 2030, increasing annually at 3.9 per cent.[2]

Vietnam has been a net energy exporter since 1990. Energy has become an important source of foreign currency revenue to the economy, accounting USD7.4 billion or 25 per cent of the total export revenue in 2005. It is expected to become a net energy importing economy beyond 2020, with energy import dependency projected to reach 15 per cent in 2030. The total investment requirement to 2030 in the energy sector is estimated at USD136–172 billion, of which four-fifths will be needed for electricity generation and transmission. To meet the projected electricity demand, 54 gigawatts (GW) of additional generation capacity is to be built which will increase installed capacity seven-fold over the 2002 level.[2]

Increased use of energy has raised environmental concerns. Though Vietnam is currently one of the lowest per capita emitters of CO_2, at 0.7 tons of CO_2 in 2002, emissions are expected to increase rapidly with increasing energy consumption. CO_2 emissions from the energy sector are projected to grow by 6.2 per cent per year, reaching about 301 million tons of CO_2 in 2030.[3]

Huge increases in energy demand and investment, plus the potential environmental consequences have caused the Government to regard energy efficiency and conservation as imperative for sustainable economic development. From 1990 to 2005, Vietnam saw its energy intensity grow at an average annual rate of 2.4 per cent. This might partly be the result of it being a transitional economy with changes in energy usage, and partly due to low energy efficiency due to poor management. By 2030, the energy intensity is expected to fall, decreasing by 1.8 per cent per year.[4]

---

[2] APERC, *APEC Energy Demand and Supply Outlook 2006.*

[3] USAID ASIA, "Clean Energy Priorities of Asia".

[4] APERC, *APEC Energy Demand and Supply Outlook 2006*; and Energy Information Administration, International Energy Annual, EIA, 2006.

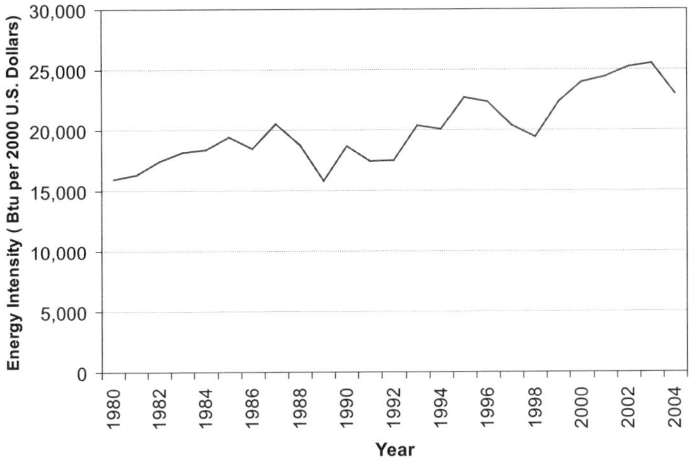

Figure 1: Energy Intensity

Source: EIA, *International Energy Annual*, 2006.

Generally, overall energy utilisation in Vietnam has been characterised by: (1) high losses due to obsolete technologies and equipment; (2) lack of demand side management practices; (3) lack of awareness on the part of the public regarding the social benefits of energy conservation; and (4) lack of a comprehensive national programme on energy efficiency and conservation (EEC) with clearly defined components and time-bound targets. Initial studies have indicated the potential for energy conservation in industries is as high as 20 per cent of current consumption.[5]

RECENT AND PROJECTED ENERGY CONSUMPTION PATTERNS IN VIETNAM

Though per capita energy consumption of Vietnam is one of the lowest in the APEC region at 0.36 toe per person in 2005, the energy demand growth is one of the fastest in the region.[6] Between

[5] Vietnam Ministry of Industry, *Vietnam Targeted Programme on Energy Efficiency and Conservation in Period 2006–2015*. MOI, 2006.

[6] APERC, *APEC Energy Demand and Supply Outlook 2006.*

Table 1: Energy Reserves, 2005

Energy Reserves	Total	Proven	Production	R/P ratio
Coal (million tons)	150	150	28	5
Oil (million barrels)	3119	3119	140	22
Natural Gas (billion cubic metres)	235	235	6	43

Source: USAID ASIA, "Clean Energy Priorities of Asia".

2000 and 2005, total primary energy consumption, excluding biomass, grew at an annual rate of 10.6 per cent, compared with 9.9 per cent per year during the previous decade. Oil production increased from 16.9 mtoe in 2000 to 19.1 mtoe in 2005, and coal production more than doubled from 6.5 mtoe in 2000 to 17.0 mtoe in 2005. Natural gas production surged six-fold from 1.1 mtoe in 2000 to 6.7 mtoe in 2005.[2] See Table 1 to view the present energy reserve situation.

The residential sector's share is projected to remain the largest, but will shrink substantially from 67 per cent in 2002 to 35 per cent in 2003. The industrial sector is expected to continue as the second largest share at 35 per cent, followed by transport at 24 per cent and the commercial sector at 6 per cent (see Figure 2).

ENERGY MIX

Total final consumption as given by International Energy Agency for 2004 was 44,627 thousand toe on a net calorific value basis. As shown in Figure 3, biomass-based energy (combustibles, renewable and waste) remain dominant followed by petroleum products.

Biomass, mainly wood fuels that now account approximately for 50 per cent of the primary mix, will decline as commercial energy becomes available, accessible and affordable with urbanisation and increases in income levels. Demand for commercial energy sources will increase rapidly at 6.7 per cent per annum, surpassing traditional

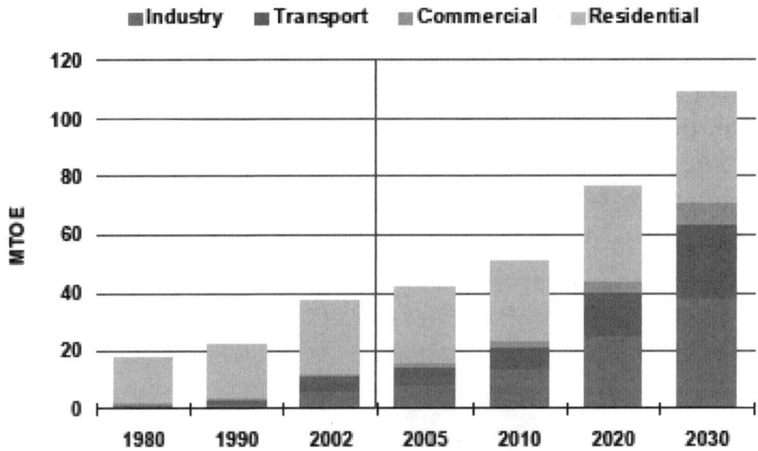

Figure 2: Final Energy Demand by Sector

Source: APERC Analysis, 2006.

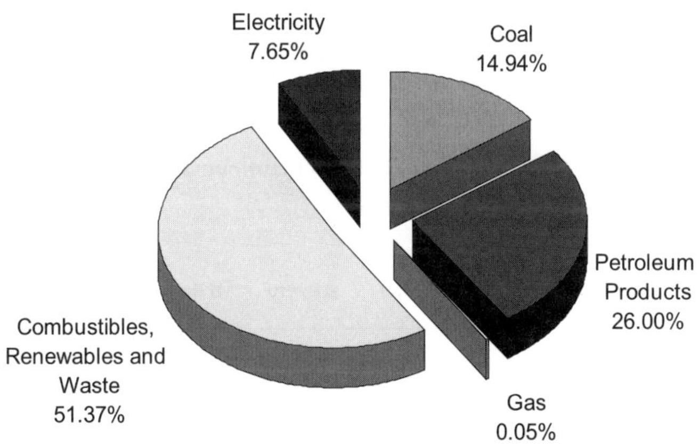

Figure 3: Total Final Consumption, 2004

Source: International Energy Agency, *Energy Balance for Vietnam*. Paris: IEA/OECD, 2006.

energy sources (mainly biomass) in 2005. Consequently the share of biomass will decrease substantially from 55 per cent in 2002 to 18 per cent in 2030.[7]

[7] APERC, *APEC Energy Demand and Supply Outlook 2006*; USAID ASIA, "Clean Energy Priorities of Asia"; National Statistics of Vietnam.

Consumption of refined petroleum products in 2003 amounted to 215 thousand barrels per day. The breakdown of refined petroleum products consumption is given in Figure 4. Among the fossil fuels, oil will continue to hold the largest share. Annual growth in oil consumption, driven by transportation and industrial sector demand, is expected to be 5.7 per cent between 2002 and 2030 (see Figure 5).

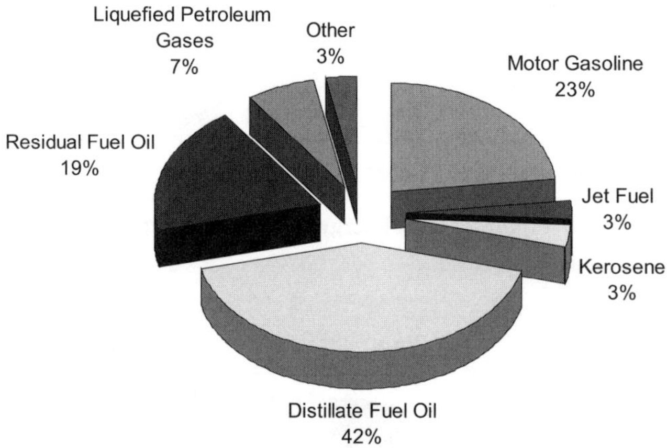

Figure 4: Consumption of Refined Petroleum Products, 2003

Source: EIA, *International Energy Annual*, 2006.

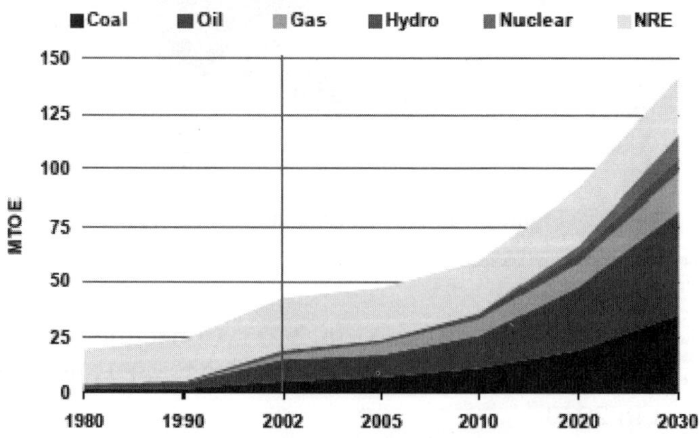

Figure 5: Primary Energy Demand

Source: APERC Analysis, 2006.

Even though Vietnam is an exporter of crude oil, in the absence of oil refineries, the country has remained a net-importer of oil products. The first oil refinery is expected to be commissioned in 2009, reducing the import dependence on oil products. However, with increasing economic growth and limited oil reserves, Vietnam's oil import dependence is expected to increase to 57 per cent in 2030 (see Figure 5).

Annual consumption of coal and natural gas, driven by the rapid development of industry and the electricity sector, will grow by 6.9 per cent and 7.3 per cent, respectively, from 2002 and 2030, accounting for 25 per cent and 12 per cent, respectively, of the primary energy mix in 2030 (see Figure 5).

Electricity demand is projected to grow by 7.8 per cent per year over the outlook period. The gross theoretical hydropower reserves amount to approximately 300 billion kilowatt-hours per year (kWh/yr). Of this total, 80,000 GWh/yr are estimated to be economically exploitable. This is equivalent to 18,000–20,000 MW capacity.[8]

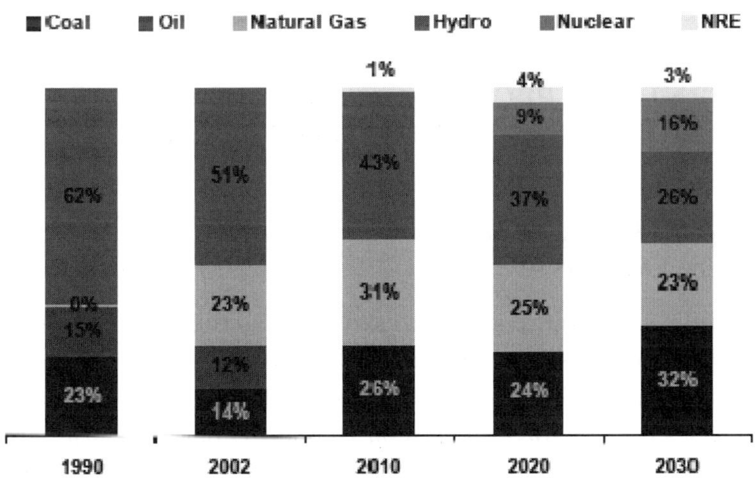

Figure 6: Electricity Generation Mix

Source: APERC Analysis, 2006.

[8] Electricity of Vietnam, *Power Development Master Plan*. EVN, 2006.

At present, the total capacity of existing hydropower plants and plants under construction is approximately 4,000 MW, capable of producing 18,000 GWh of electricity per year. This represents about 22.5 per cent of the total technically exploitable potential. By 2010, the total installed hydro capacity of Vietnam will reach 8,000 MW with total electricity production of over 30,000 GWh.[1] At this rate, as most possible locations for hydro are fully developed, the share of hydro will decrease considerably. The contribution of renewable energy, including large hydro and geothermal plants, in the total primary energy supply will decrease from 58 per cent in 2002 to 22 per cent in 2030 (see Figure 5). Hydropower production is projected to grow by 4.9 per cent per annum in 2002–2030.[9]

Excluding large-scale hydro, other types of new and renewable energy (NRE) such as mini-hydro, wind, solar PV, geothermal and MSW-landfill gas will continue to be promoted, raising the share of renewable energy to 18 per cent in 2030 (see Figure 5).

It is also likely that nuclear power will have a significant share by 2030. Nuclear power — to be introduced in 2020 — is expected to account for 8 per cent of total primary energy demand in 2030 (see Figure 5).

GOVERNMENT PROGRAMMES TO REDUCE ENERGY WASTAGE AND IMPROVE ENERGY CONSUMPTION EFFICIENCY

Overview of Energy Efficiency and Conservation (EEC) in Vietnam

Population and rapid economic growth over the next two decades will result in increased energy demand in Vietnam. Given the moderate supply potential of indigenous energy resources, future dependence on imports is inevitably threatening Vietnam's energy security. Among measures to strengthen national energy security, the Government continues to focus attention on diversification of supply sources and enhancement of energy efficiency on the demand side.

[9] APERC, *APEC Energy Demand and Supply Outlook 2006*.

After the approval of the Vietnam Environment Law by the National Assembly, the Ministry of Science, Technology and Environment (MOSTE) (now Ministry of Science and Technology (MOST)) in collaboration with the Ministry of Industry (MOI), began Vietnam's Energy Conservation and Efficiency Programme (VECP) in 1995. As a result of a number of information gathering studies, an assessment of energy conservation potentials was initiated. The most significant savings potentials were identified in the cement industry, ceramics industry, coal-fired power stations and on the demand side through lighting improvements.

From 1996, a number of energy conservation studies, energy audit trainings and energy efficiency public awareness and information programmes were conducted under the government-coordinated Vietnam Energy Conservation Programme (VECP). More importantly, the VECP established Energy Conservation Centres under the Department of Science and Technology (DOST) in five major cities: Hanoi, Ho Chi Minh, Hai Phong, Da Nang and Can Tho.

The VECP initiated the development of the Master Plan for EEC which was subsequently completed in 1998. Energy standards and labelling is another important initiative of VECP aiming to apply standards as a benchmark for production and harmonisation of national and international standards.

VECP has laid down a foundation for EEC in Vietnam that will lead to subsequent larger scale implementation of EEC projects and programmes and the promulgation of an EEC related regulatory framework. To date, the VECP programme has led to these follow-up projects:

Vietnam Demand-Side Management and Energy Efficiency (DSM&EE)

The DSM&EE project aims to promote more efficient use of electricity, reduce peak load and enhance energy access for the poor. A DSM assessment was completed in 1997 and identified potential savings of 770 MW generation capacity and 3,550 GWh/year electricity consumption.[10] Following the assessment, the DSM Programme Phase I was implemented by the national electric utility company, Electricity of Vietnam (EVN) in 2000 with the objectives of: (1) establishing DSM cells and developing an EE pilot programme; (2) developing capability on load research, load management and energy auditing; (3) establishing standards for equipment such as lighting and motors, and; (4) developing a building code for energy efficient commercial buildings. Following completion of Phase I, the DSM Programme Phase II, commenced in 2003 with these objectives:[11]

- develop and expand DSM business programmes and test new market transformation efforts within EVN;
- develop sustainable business models and mechanisms to support EE retrofit investments in commercial and industrial facilities.

The DSM Phase II project consists of two components. The first is management by EVN with the main focus of implementing EVN's DSM business plan developed in Phase I by (1) expanded time-of-use (TOU) metering by procuring and installing TOU meters; (2) pilot Direct Load Control (DLC) Programme introduced by using ripple control systems; (3) compact fluorescent lamp (CFL) promotion; (4) fluorescent tube lamp (FTL) market transformation; and (5) supporting programme and technical assistance for DSM efforts. The second component, a pilot commercial EE programme, is under MOI and seeks to test appropriate business models and mechanisms to catalyse a service market to support EE investments by supporting commercial service providers or 'project agents' in all phases of EE project identification, development and implementation.

[10] World Bank, *Vietnam-Demand-Side Management and Energy Efficiency Project Proposal Appraisal Document*. 30 May 2003.

[11] Electricity of Vietnam, *DSM Action Plan for Vietnam Phase 2 DSM Programme: 2004–2007*. EVN, Nov. 2004.

Vietnam Energy Efficient Public Lighting Project (VEEPL)

VEEPL is managed by the Ministry of Science and Technology (MOST). The project aims to a) improve lighting levels and lighting quality of public lighting, and b) reduce public lighting demand growth through improvements of light source, controllers and fixtures. Key project activities include policy development, technical support, financing and demonstration programmes.

Vietnam Energy Efficient Commercial Building Code (EECBC)

EECBC was launched in 2001 under the supervision of the Ministry of Construction (MOC). The objective of this project is to develop an energy efficient building code, mentioning the minimum mandatory energy efficiency requirements that commercial space, office buildings and hotels must obey. The output of this project is the Energy Efficient Commercial Building Code No. 40/2005/QD-BXD which was promulgated in 2005.

Energy Standard and Labelling (ES&L)

ES&L is an important initiative of the VECP. Consumer product labelling is at its initial stage in Vietnam. The VECP initiated a process for developing energy efficiency standards and labelling, overseen by a technical committee in industries. Test standards and minimum efficiency performance standards (MEPS) for incandescent bulbs, linear fluorescent lamps, ballasts and electric motors have been developed in draft form.

Recently, the Energy Efficiency and Conservation Office of the MOI has taken over the responsibilities of the ES&L and initiated a pilot programme. The goal of the pilot programme is the removal of barriers that have persistently hindered widespread development and application of the ES&L programmes. This project is expected to help in implementation of standards in three target products (thin-tube fluorescent lamps, compact fluorescent lamp (CFL) and air conditioners).

ENERGY CONSERVATION POLICIES, PROGRAMMES, AND REGULATORY FRAMEWORK

In addition to the Master Plan for Energy Efficiency and Conservation published in 1998, the Government has reinforced its commitment towards the EEC by issuing various legislation and decrees related to EEC implementation. Up to now, the Government and concerned ministries have issued various documents for the promotion and enforcement of EC&E to the country. The core regulatory framework is the Energy Efficiency Decree No. 102/2003/ND-CP published on 3 September 2003. Under this Decree, the MOI has responsibilities to conduct the energy conservation and efficiency programme. With the Prime Minister Decision No. 79/QD-TTG dated 14 April 2006 on Approval of the National Strategic Programme on Energy Savings and Effective Use proposed by MOI, the Government of Vietnam ratified the target savings of 3–5 per cent for total energy consumption during 2006–2010, and 5–8 per cent during 2011–2015.

A new Law on Energy Conservation and Efficiency is being drafted. It will help clarify the roles and responsibilities of agencies and ministries in carrying out the several decrees and policies now in place. MOI's Department of Science and Technology/Office of Energy Efficiency and Conservation is developing the draft law and will soon circulate it for comments and discussion. The draft will then be submitted to the National Assembly in December 2007. It is expected that the law will take effect in 2009.

Besides the above-mentioned regulations that focus on demand-side energy efficiency, the Power Sector Development Master Plan postulates supply-side efficiency improvement. This document calls for a shift towards higher-efficiency coal-fired base load units. The Institute of Energy is supporting EVN in studying the technologies required to meet the move from 300 MW subcritical coal steam electric power plants to 600 MW supercritical coal steam electric power plants.

The official decrees, circulars and decisions promulgated by the Government that address energy efficiency and conservation are listed in Table 2.

Table 2: Energy Efficiency and Conservation Decree, Decisions and Circulars

Legislation (Type)	Promulgator	Issuance Date	Effectivity
Energy Saving and Efficient Use of Energy (Decree)	National Government	3 Sept. 2003	Sept. 2003
Guiding the Saving and Efficient Use of Energy by Production Establishments (Circular)	MOI	2 July 2004	July 2004
Approving the 2006–2010 Electricity Saving Programme (Decision)	Prime Minister	14 Apr. 2006	May 2006
Approving the National Target Programme on Saving and Efficient Use of Energy (Decision)	Prime Minister	14 Apr. 2006	May 2006
Establishing the Energy Efficiency and Conservation Office at the MOI (Decision)	MOI	7 Apr. 2006	7 Apr. 2006
Establishing the Steering Committee of the National Target Programme on Saving and Efficient Use of Energy (Decision)	MOI	18 May 2006	18 May 2006
Promulgating the regulations on Selection of the Organisations, Individuals for Presiding over the Implementation of the Projects of the National Target Programme on Saving and Efficient Use of Energy (Decision)	The Steering Committee of the National Target Programme	20 Sept. 2006	20 Sept. 2006
Guiding the Order and Procedures for Energy Efficiency Labelling for Energy-Consuming Products (Decision)	MOI	16 Nov. 2006	Dec. 2006

Source: Electricity of Vietnam, *Power Development Master Plan*. EVN, 2006.

NATIONAL ENERGY POLICY

Vietnam developed and published a Master Plan for Energy Efficiency and Conservation in 1998 which indicated the importance of creating national and provincial Energy Conservation Centres (ECCs). It was partially addressed by VECP, and proposed the structure and roles of ECCs in various activities. The proposed National Energy Policy also specifically includes EEC measures aimed at reducing the energy elasticity from 1.46 in 2004 to 1.0 in 2010, 0.9 in 2020 and 0.8 after 2020. In particular, the policy gives priority to the development of low energy intensity industries and initiates the following measures:

- Energy management;
- Replacement of low energy efficiency equipment;
- Introduction of new technology;
- Manufacturing of high EEC equipment and devices.

For buildings and households, the policy recommends the promotion of DSM and the application of building codes for new building projects and the enforcement of MEPS for equipment and facilities.

GOVERNMENT DECREE ON ENERGY SAVING AND EFFICIENT USE OF ENERGY

In September 2003, the Government issued Decree No. 102/2003/ND-CP on Energy Saving and Efficient Use of Energy to encourage end-use energy efficiency. It defines an administrative structure and responsibilities for a coordinated set of directions, requirements, regulations, and enforcement for the application of energy efficiency and energy conservation measures in all segments of the economy. There are also sections that specifically address energy use in factories and buildings, and imports and exports. An important part of the decree will be the formulation of minimum energy efficiency requirements, testing, labelling and marking for compliance. Specifically, the decree:

- details EEC mandates for specified government agencies at the local and central levels;
- sets up the legal EEC requirements for large factories and enterprises (those with over 1,000 toe/year or 500 kW average electricity demand or 3 GWH/year electricity consumption) and for selected buildings with installed capacity over 750 kVA or annual energy use including electric power and thermal energy in aggregate of 10 million MJ or electric use over 2.8 GWH;
- provides incentives for the promotion of EEC;
- establishes mandatory energy consumption norms and limits, and voluntary labelling for designated equipment and facilities;
- provides for a national EEC programme.

The direct impact of this Decree will come from energy efficiency requirements for large factories, enterprises and buildings and the mandatory reporting and control of EEC by designated large end-users. This Decree on Energy Conservation and Efficiency also mentions that designated energy consumers must report to the concerned DOIs and MOI about their energy consuming situation as well as provide an action plan relating to energy conservation and efficiency. This is considered the legal basis for the EC&E activities.

Under Decree No.102/2003/ND-CP, the MOI has issued a Circular letter No. 01/2004/TT/BCN on 2 July 2004, in order to enforce the Decree from the Government. The Circular serves as a guideline to implement energy conservation and promote efficient use of energy. The guideline focuses on energy consumption in enterprises (industrial facilities) and specifies the definition of a designated enterprise as the facility that has total annual fuel and heat equal to, or more than one thousand tons in TOE, or have electric capacity equal to or more than five hundred kW, or electricity consumption equal to or more than 3.0 million kWh.

In August and September 2005, respectively, the MOI also issued circulars on financial incentive mechanisms to promote energy efficiency activities and on procedures and sequences for labelling energy efficiency equipment.

CODE FOR ENERGY EFFICIENT COMMERCIAL BUILDINGS

The Energy Efficient Commercial Building Code No. 40/2005/QD-BXD was promulgated on 17 November 2005. This code pertains to various commercial buildings having a total floor area of 300 square metres and above, which have an HVAC system and other energy consuming facilities. This code is promulgated to reduce energy loss in buildings and improve conditions for working/living.

TARGETED PROGRAMME ON ENERGY EFFICIENCY AND CONSERVATION, 2006–2015

In 2005, the MOI prepared a National Targeted Programme on Energy Efficiency and Conservation for the period 2006-2015. It was approved and enforced on 14 April 2006 by the Prime Minister and calls for concerted efforts to improve energy efficiency and cut down energy losses and also stronger energy conservation measures with the following programme objectives:

- Formulate and implement a framework for effective management of EEC at three levels — governmental level, enterprise level and public service and society as a whole;
- Establish a model for EEC management and apply the model to 40 per cent of the designated energy consumers/enterprises during the period 2006-2010 and 100 per cent of the designated consumers during 2011-2015;
- Formulate and enforce the energy efficient building code for all buildings to be constructed after 2006;
- Achieve 3-5 per cent reduction from 2006-2010 and 5-8 per cent reduction from 2011-2015 in total energy consumption in the

country. This is equivalent to 5 million toe and 13 million toe in the two periods respectively, based on the business as usual scenario;

- Reduce energy consumption in industry by 5 and 8 per cent for the two periods, 2006–2010 and 2011–2015, respectively (equivalent to 2.6 and 5 million toe); and achieve energy conservation equivalent to a 20 per cent reduction in energy consumption of new buildings;
- Effectively use road, waterway, railway, airway and seaway networks to optimally exploit transport capacity, minimise fuel consumption in transportation, experiment with substitute fuels in some big cities and provinces and restrict exhaust fumes from transport for environment protection.

The programme includes six content groups:

1: Enhance government control on economical and effective use of energy; organise management systems for energy efficiency and conservation with one project;
2: Enhance education, information dissemination, mobilise communities, improve perceptions and strengthen economical and effective use of energy, environment protection with three projects;
3: Develop, popularise high-productivity facilities for energy saving, gradually eliminate low-productive facilities with three projects;
4: Energy efficiency and conservation activities in industrial enterprises with two projects;
5: Energy efficiency and conservation activities in building with two projects;
6: Energy efficiency and conservation activities in transportation with one project.

The six content groups recommend eleven concrete projects to be carried out between 2006 and 2015:

- Complete the legal framework on EEC in all activities;
- Propagandise to raise the community's awareness of energy saving and effective use;

- Integrate education about energy saving and effective use into the national education system;
- Campaign on the establishment of a model named 'Energy saving in each household';
- Develop standards and afix energy-saving labels on some targeted products;
- Provide technical assistance to domestic manufacturers who meet the standards on energy efficiency;
- Establish energy management systems in enterprises;
- Assist factories to upgrade, improve and rationalise production lines for energy saving and effective utilisation;
- Improve capability and integrating EEC in designing and managing buildings;
- Construct pilot buildings and put the EEC activities in buildings into discipline;
- Optimise transport capacity, minimising fuel consumption and reducing emissions.

INSTITUTIONAL SET UP

Various government agencies contributed to the implementation of EEC projects and programmes in the early stages of EEC implementation in Vietnam. The EEC decree in 2003 detailed EEC mandates for specified government agencies at the local and central levels, and roles and responsibilities of government agencies related to EEC. According to the Decree, the MOI is the focal coordinator on EEC activities responsible for policy development and the energy sector.

The Decree also indicates specific duties for the MOI: cooperating with other government agencies to set up national targets; listing designated energy-using products and equipment; coordinating state-management on EEC with ministries, the private sector and provincial governments; issuing regulations on manufacturers' responsibilities to ensure the required specifications; stipulating procedures for the labelling of energy efficient products, and; stipulating competence consulting bodies working in the fields of energy conservation and efficiency. Under the EEC decree, the local people's committees and provincial governments are responsible for disseminating and implementing EEC activities in their own responsible areas with assistance from the provincial departments of industry.

The decision to establish the Steering Committee of the National Target Programme on Saving and Efficient Use of Energy was issued on 18 May 2006 headed by the MOI. The Steering Committee members are representatives from the Ministries of Construction, Transport, Education and Training, Culture and Information, Science and Technology, Planning and Investment, Justice, Finance, and the Union of Vietnam Association of Science and Technology.

Similarly, the EEC Office within the MOI was established on 4 April 2006 by the MOI Minister. The roles of the EEC office are to: (1) propose the mechanism, policies and measures for promoting EEC activities to submit to the Prime Minister for approval and enforcement; (2) set up the short-, medium and long-term EEC programmes to submit to the Prime Minister for approval and enforcement; (3) work as a focal point to organise and develop EEC management systems from Ministries to the localities for effective implementation of the EEC programme at all local levels; (4) cooperate with other governmental organisations to monitor and treat the problems arising during implementation of the programmes; (5) organise the EEC Awareness and Propagation Programme from the enterprises and community; (6) establish an energy consumption database for management; and (7) cooperate with international organisations to raise cooperation in EEC. The involvement of other institutions in EEC in Vietnam are summarised in Table 3.

EVALUATION OF THE SUCCESS OF ENERGY CONSERVATION EFFORTS: WHAT YET NEEDS TO BE DONE

Over the last decade, Vietnam has made considerable progress in instituting EEC activities. The Master Plan for Energy Efficiency and Conservation published in 1998, has been instrumental in gearing up policies, regulations and programmes to achieve energy conservation goals. Significant work has been carried out in priority areas set by

Table 3: Involvement of Other Institutions in EEC

Agency	Responsibility
Ministry of Science and Technology	Issuing of national standards and related regulations pertaining to energy-using equipment
Ministry of Construction	Issuing of norms for EEC in the construction of high-rise buildings, regulation for building and construction materials, and guidance for consumers
Ministry of Transport	Cooperating with the MOI to instruct EEC regulations in the transportation sector
Ministry of Finance	Guiding implementation of financial tools for encouraging EEC
Electricity of Vietnam	Overseeing the operations of the entire power sector in Vietnam

the Plan, especially in establishing a regulatory framework, setting of standards and implementation of programmes on DSM and energy efficiencies.

Perhaps the most important achievement is the Decree on Energy Saving and Efficient Use of Energy (102/2003/ND-CP). It sets forth the roles and responsibilities for all actors in government and society with respect to energy efficiency and conservation. Circular No. 01/2004/TT-BCN is the follow-up legal document for the implementation of the above decree. With detailed guidelines for various entrepreneurs to follow, the Circular took effect in 2004. Since then, significant numbers of designated factories have annually reported their energy consumption and energy intensities.

The promulgation of the Energy Efficient Commercial Building Code No. 40/2005/QD-BXD, setting minimum mandatory requirements for energy efficiency to which commercial and office buildings, and hotels must comply is an important achievement in EC&E in the construction sector.

Recently, the National Targeted Programme on Energy Efficiency and Conservation has launched a comprehensive plan for the establishment of the necessary institutional mechanisms for sustained operation of EEC. As a result, the State Steering Committee and Office of Energy

Efficiency and Conservation within the MOI have been established. These have prepared the initial work, such as the action plans and detailed programmes in cooperation with other governmental organisations to kick off and successfully implement the National Programme.[12]

Many Ministries are involved in the enforcement of the EC Decree. Moreover, small and medium enterprises have come under local control (prefecture level). Energy Conservation Centres to train and carry out energy audits are being set up around the country.

Despite this progress, the EEC objectives are far from being met. The main problems are lack of follow-up guidelines, inadequate financial support, inadequate inter-agency coordination, shortage of human resources, lack of capacity among officials to apply the policies and lack of knowledge amongst institutions.

There is an Energy Conservation Centre in Ho Chi Minh City and others under the Department of Science and Technology. Five centres have been established, but only one performs properly. The offices are supposed to provide the link from the central government to the provinces through five provincial offices for running energy efficiency and conservation training programmes and promotion and awareness campaigns. These implementation organisations have problems with their roles and budget, since the current EC Decree does not define the roles of organisation for energy conservation. Moreover, the existing centres have a problem with local human resources and their capacities. Currently, the activities of these offices are limited to energy auditing and service providing, while EC&E activities are considered a secondary function. Success in achieving EEC goals in Vietnam depends on:

Market Demand Drivers

Electricity Prices

Raising electricity prices in Vietnam will encourage industries to adopt higher energy efficiency in their production processes.

---

[12] Vietnam Ministry of Industry, *National Strategic Programme about Economical and Effective Use of Energy in the Period 2006–2015*. MOI, 2006.

Fiscal Concessions

Several fiscal concessions in energy-efficiency are being provided by the Government which could induce energy-intensive industries to take advantage of cost-effective opportunities for investment in energy-efficient equipment:

- Full tax exemption on profits for the first 2 years and 50 per cent exemption for the next two years (for new projects);
- Preferential income tax rates of 15, 20 or 25 per cent in lieu of the normal income tax rate of 30 per cent (for new projects);
- Power plants and electrical networks implementing energy-efficient technology and solar energy may be fully exempted from income tax in the 1st year of operation and receive a 50 per cent exemption during the next 2 years on income gained;
- Import tariff exemption for energy-efficient equipment and materials for construction of high technology zones or high-technology projects.

Sector-Specific Demand

Industry

Large amounts of electricity (40 per cent of total consumption) are consumed by the industrial sector. Generally, low or medium levels of technology and equipment are employed with low energy efficiency. The efficiency of boilers is especially low. It is recognised that industrial motors consume about 60–75 per cent of the total industrial consumption of electricity, with air compression systems and boilers another 15–20 per cent and lighting, 5 per cent.[13]

It is estimated that 15 per cent of the present electricity consumption could be saved by the industrial sector if it implements EEC. These savings represent 6 per cent of the total electricity consumption in the national power system. The potential to reduce

[13] IIEC, *Review of Cleaner Production, Energy Efficiency and Clean Energy in Vietnam*.

energy related expenses in production processes is the driving force for the industrial sector to make EEC investments.

Examples of energy efficiency improvements in specific industry sub-sectors include:

- **Cement** — a big energy user, consuming 40 per cent of the total 12 per cent of electricity consumed by all industry. An example of EEC improvement in the cement sub-sector is the adoption of air quenching coolers (these can increase efficiency by 7 per cent) and waste heat recovery to use for local power generation;
- **Ceramics** — incurs an energy cost of 33.54 per cent of the total intermediate production costs. High energy consumption is associated with the use of traditional kilns. The replacement of inefficient coal-fired kilns with new gas-fired kilns that are insulated with ceramic fibre could reduce energy use by 50 per cent;
- **Steel** — consumes 56 per cent of the total electricity used by industry. Technology used in this industry is very backward and leads to a very high energy consumption. A specific example of EEC improvement in the steel industry is the replacement of electric arc furnaces with fuel burners, oxygen injection, and preheating turn dishes. It is estimated that EEC measures in steel making could reduce energy use by 11 per cent;
- **Pulp and paper** — potential energy savings up to 10 per cent.

Power

Transmission and distribution losses in the Vietnamese power system are rampant. Coupled with increasing load demand this causes many restrictions in the power system. The situation will force the EVN to implement load management and investment in transmission and distribution.

Commercial and Services Sectors

The commercial and services sector, i.e., commercial buildings, hotels, offices, hospitals and schools have very low energy efficiency due to their equipment being old and their inefficient building designs. Increased demand for air conditioning and ventilation in new offices, hotels, and apartment buildings will increase the need for EEC given the escalating electricity prices.

Transportation

The increasing demand for transport in Vietnam warrants the adoption of EEC in the form of encouraging the establishment of more public transport in large cities. Motorcycles and other forms of private transport are already causing congestion in large cities while raising oil import requirements. Public transport in Vietnam meets only five per cent of travel demand.[14] The Government will try to improve the situation by promoting bus enterprises with fixed routes and frequent and dependable service. Specific measures to promote bus transport include:

- Import tax and excise tax exemption on buses for public transportation;
- Tax exemption on capital investment;
- Exemption from parking fees, tolls, registration fees and license fees;
- Provision of free land for bus stations and bus stops.

Financing for these incentives will come from higher tariffs levied on imported cars and motorbikes. Additionally, excessive use of private transport is discouraged through increased taxes on gasoline and diesel fuel, compulsory insurance for motorcycles, and parking fees on public roads and walkways.

Barriers

Scaling-up energy efficiency and conservation in Vietnam requires overcoming several categories of barriers — market (limited information,

[14] *Ibid.*

perceived risk or unfamiliarity, awareness), price (subsides or price distortion), financial (transactions cost, access to financing), institutional and technical. These barriers require active intervention (regulation and reforms).

Market (Information, Knowledge and Skills Barriers)

Insufficient information on potential EE improvements, costs and benefits of EE equipment, potential low-cost measures and new technologies/practices are some market barriers. Similarly, lack of technical expertise by end-users, manufacturers/suppliers and potential service providers of modern efficient technologies and practices, efficiency potentials, energy audits and inspections, actual performance of EE measures as well as limited understanding of third party EE services (e.g., ESCOs) are also limiting effective EE improvements.

FINANCIAL, INSTITUTIONAL AND REGULATIONS AND REFORMS

High capital investment costs, due to the higher costs of EE equipment as well as limited local manufacturing capability, which currently discourage end-users from selecting high efficiency equipment despite their overall lower life-cycle costs are discouraging the adoption of EE technologies. Also, high project development costs, due to audits and technical studies required to properly determine investment requirements and ensure appropriate project design; perceived risks of projects developed by auditors with limited track record; perceived risks on technologies/equipment with limited tested performance under Vietnamese conditions; lack of affordable financing, due to lack of commercial lending culture in Vietnam, the weak banking sector and very restrictive lending terms, relatively small investment sizes for EE, and limited credit available to residential sector; and poor customer credit worthiness, due to the poor financial status of many of the state owned enterprises is hindering EE implementations.

Limited interest of end-users, due in part to a production or core business priority bias, exacerbated by the limited financial significance of the operating cost reductions from energy savings is limiting EE implementation. At the same time limited local EE equipment, given the current manufacturing capability within Vietnam and low domestic demand for high-efficiency products is also a hindrance in EE improvement.

Though significant achievements have been made in the form of the Decree on Energy Saving and Efficient Use of Energy (102/2003/ND-CP) and the National Targeted Programme, Vietnam still lacks follow-up guidelines and institutional capacity to enforce these regulations.

SUMMARY

Over the last decade, Vietnam has experienced rapid population growth and economic transition. Continuation of this unprecedented economic growth over the next two decades is likely to cause energy consumption to increase sharply, and the country will likely change its status as an energy net exporter to energy net importer. In recent years, energy has been a source of foreign currency revenues, but the Vietnamese Government must begin to take strong measures to ensure future energy security. Moreover, increased energy consumption will continue to raise various environmental concerns.

Tremendous increases in energy demand and investment, plus the concern for energy security and environmental consequences have forced Vietnam to regard energy efficiency and conservation as imperative for sustainable economic development. As a result, several energy efficiency programmes have been initiated. After the approval of the Vietnam Environment Law by the National Assembly, the MOSTE (now MOST) in collaboration with the MOI, began VECP in 1995. VECP with its Master Plan for Energy Efficiency and Conservation published in 1998 along with other subsequent programmes (e.g., DSM&EE, VEEPL, EECBC, and ES&L) have laid a foundation for EEC in Vietnam leading to larger scale programmes and projects. Significant work has been carried out in priority areas set by the Plan, especially in establishing a regulatory framework, setting of standards and implementation of programmes on DSM and energy efficiencies.

In addition, the Government reinforced its commitment to EEC by issuing various pieces of legislation as well as decrees related to EEC implementation. Perhaps the most important achievement is the Decree on Energy Saving and Efficient Use of Energy (102/2003/ND-CP), which sets forth the roles and responsibilities of all actors in government and society with respect to energy efficiency and conservation. The subsequent promulgation of other follow- up circulars and codes (e.g. Circular No. 01/2004/TT-BCN — with detailed guidelines for various entrepreneurs; Energy Efficient Commercial Building Code No. 40/2005/QD-BXD, setting minimum mandatory requirements for energy efficiency to which commercial, office buildings, hotels must comply; and recently, the National Targeted Programme on Energy Efficiency and Conservation) are important achievements.

Despite this progress, the EEC objectives are far from being met. The main problems are lack of follow-up guidelines, inadequate financial support, inadequate inter-agency coordination, shortage of human resources, lack of capacity among officials to apply the policies and lack of knowledge amongst institutions. Success in achieving the EEC goals in Vietnam depends on addressing issues such as:

- appropriate electricity pricing;
- financial concession for EEC (tax exemption, preferential income tax, import tariff exemption);
- targeting priority sectors and sub-sectors based on energy efficiency demand, e.g., for industry (cement, ceramic, steel, paper and pulp), power (transmission and distribution losses), commercial and service sector (commercial buildings, hotels, offices, hospitals and schools) and transportation.

Scaling-up EEC in Vietnam requires overcoming several categories of barriers: market (limited information, perceived risk or unfamiliarity, awareness), price (subsides or price distortion), financial (transactions cost, access to financing), institutional and technical. These barriers require active intervention in the form of regulations and reforms.

List of Editors and Contributors

Youngho Chang is Assistant Professor in the Division of Economics at the Nanyang Technological University (NTU), Singapore and a Senior Fellow at the Energy Studies Institute at the National University of Singapore. He is also an Adjunct Senior Fellow at the S. Rajaratnam School of International Studies at NTU.

Surapong Chirarattananon is Associate Professor in the School of Environment, Resources and Development's Energy Technology Program at the Asian Institute of Technology in Thailand.

Gavin Chua, based in Singapore, is a Senior Consultant at Control Risks, an independent, specialist risk consultancy with 31 offices on five continents.

Mui Pong Goh is a Research Fellow at the Royal Institute of International Affairs/Chatham House in London.

Felix Gooneratne is the Asia Director for the International Institute for Energy Conservation in Thailand.

Bob Sugeng Hadiwinata is Professor of International Relations at the Parahyangan Catholic University, Bandung, Indonesia, and Georg Forster Research Fellow of the Alexander von Humboldt Foundation, Germany.

Jae-Seung Lee is Associate Professor in the Division of International Studies at Korea University. He is also Managing Director of the Center for Energy, Resources and Environment in the Institute for Sustainable Development (ISD) at Korea University.

Sumit Pokhrel is a Greater Mekong Sub-region Environment Operations Center climate change specialist with the Asian Development Bank.

Yasuo Tanabe is Deputy Director-General, Economic Affairs Bureau of the Japanese Ministry of Foreign Affairs.

Elspeth Thomson is a Senior Fellow at the Energy Studies Institute at the National University of Singapore.

JianJun Tu is a Vancouver-based senior energy and environmental consultant, and a Research Associate at the Canadian Industrial Energy End-Use Data and Analysis Centre.

Peter Lee U is Dean of the School of Economics at the University of Asia and the Pacific (UA&P) in the Philippines.

Zhou Fuqiu is Deputy Director of the Energy Efficiency Centre in the Energy Research Institute of the National Development and Reform Commission in China.

Index